Plant Virology
The Principles

The Microbe

The Microbe is so very small
You cannot make him out at all,
But many sanguine people hope
To see him through a microscope.
His jointed tongue that lies beneath
A hundred curious rows of teeth;
His seven tufted tails with lots
Of lovely pink and purple spots,
On each of which a pattern stands,
Composed of forty separate bands;
His eyebrows of a tender green;
All these have never yet been seen —
But Scientists, who ought to know,
Assure us that they must be so. . . .
Oh! let us never, never doubt
What nobody is sure about!

The Bad Child's book of Beasts
verses by H. Belloc

Reprinted by permission of A. D. Peters and Co.

Plant Virology
The Principles

Adrian Gibbs

Research School of Biological Sciences, Australian National University, Canberra

Bryan Harrison

Scottish Horticultural Research Institute, Invergowrie, Dundee

A HALSTED PRESS BOOK

JOHN WILEY & SONS
New York

© Adrian Gibbs and Bryan Harrison, 1976

First published 1976
by Edward Arnold (Publishers) Ltd.,
London

Published in the U.S.A.
by Halsted Press, a Division
of John Wiley & Sons, Inc.
New York

Library of Congress Cataloging in Publications Data

Gibbs, A. J.
    Plant virology.

    "A Halsted Press book."
    Includes bibliographical references and index.
    1. Plant viruses. 2. Virus diseases of plants.
I. Harrison, Byran D., joint author. II. Title.
[DNLM: 1. Plant viruses. SB736 G442p]
QR351.G5   1976   576′.6483   76-1924
ISBN 0-470-15040-8

Printed in Great Britain.

Preface

We were both introduced to the enjoyment of work with viruses by the late Sir Frederick Bawden, whose infectious enthusiasm, wide knowledge and clear thinking were influences that few young research workers can be so fortunate to experience.

When, later, we were asked to pass on our knowledge to others we found that there was no book covering the general framework of plant virology at the undergraduate or early post-graduate level, and we suspect that this lack has contributed to the relative lack of interest in plant viruses at universities in Great Britain and, possibly, elsewhere. Although other volumes have since appeared, these are either more specialized or more detailed than seems necessary for a grounding in the subject. Our book is an attempt to fill the gap, and is intended not only for those specializing in virus research, but also for those interested in general biology for we feel that experiments with plant viruses can introduce students in an interesting and practical way to many aspects of the handling, biochemistry, genetics and ecology of macromolecules and organisms; aspects that are often more difficult to examine using other materials.

We have aimed to describe and discuss the principal phenomena found, and methods used, and have not attempted to detail the whole range of knowledge about plant viruses, nor have we discussed the viruses of particular crops. The references quoted are listed at the end of the book; some of these are to the original report, others to a more recent summary, and, although our selection is therefore somewhat arbitrary, most important papers are included. A list of reviews, of books, and of additional information for those wanting to work with viruses is given in an Appendix. In general, the book complements both the C.M.I./A.A.B. Descriptions of Plant Viruses, which summarize the properties and detail the data on particular viruses, and also the set of practical exercises described by Noordam (1973).

Throughout the book we use the vernacular names of plant viruses collated by Martyn (1968, 1971), and we also use and list cryptograms to give a thumb-nail sketch of each virus. The viruses mentioned in the book are listed, together with their cryptograms, in the Virus Names Index. Throughout, tobacco mosaic virus is abbreviated to TMV, ribonucleic acid to RNA, and deoxyribonucleic acid to DNA.

We have many people to thank for their help in producing this book. Colleagues in our own and other laboratories have generously provided many of the illustrations, several of which have not been published before, and these are appropriately acknowledged in the legends. We are particularly indebted to Ian Roberts of the Scottish Horticultural Research Institute and Roy Woods of Rothamsted Experimental Station for most of the unattributed electron micrographs. Other illustrations that are not original have come from collections at Rothamsted Experimental Station, the Scottish Horticultural Research Institute, and the Australian National University, and we thank the photographic staff and scientific illustrators of these establishments for their painstaking help. We also thank the several people who have helped by typing, indexing and proofreading the text, but particularly Margaret Campbell at Invergowrie and Teresa Radford at Canberra. Finally, we thank Pat Gibbs and Elizabeth Harrison for their great forebearance during the book's lengthy gestation period.

Adrian Gibbs
Bryan Harrison

Canberra and Dundee, 1975.

Acknowledgements

In addition to the people we have acknowledged in the legends for their generosity in supplying illustrations used in this book, we wish to acknowledge our indebtedness to the following Journals and Publishers for their courtesy in allowing us to use illustrations or data published previously.

The relevant publications are cited in the legends. Unattributed illustrations are used by courtesy of the Australian National University, Rothamsted Experimental Station and the Scottish Horticultural Research Institute.

Advances in Virus Research (Academic Press)
American Chemical Society Monographs (Reinhold Book Corporation)
Annals of Applied Biology (Association of Applied Biologists)
Biochimica Biophysica Acta (Elsevier Publishing Co.)
Cambridge University Press
Ciba Foundation
Faber and Faber
Hilgardia (California Agricultural Experiment Station)
Journal of Bacteriology (American Society for Microbiology)
Journal of Biophysical and Biochemical Cytology (Rockefeller University Press)
Journal of General Microbiology (Cambridge University Press)
Journal of General Virology (Cambridge University Press)

Journal of Immunology (Williams and Wilkins Co.)
Journal of Molecular Biology (Academic Press)
National Agricultural Advisory Service Quarterly Review (H.M.S.O.)
Nature (Macmillan)
Naturwissenschaften (Springer-Verlag)
North-Holland Publishing Co.
Pest Articles and News Summaries (Ministry of Overseas Development, London)
Phytopathologische Zeitschrift (Paul Parey)
Phytopathology (American Phytopathological Society)
Plant Disease Reporter (U.S. Government Printing Office)
Proceedings of the National Academy of Sciences of the United States of America (U.S. National Academy of Sciences)
Rijksmuseum, Amsterdam
Virology (Academic Press)

Contents

Chapter 1

The history and scope of plant virology

1.1 Virology and viruses

The word 'virology' means 'the study of viruses'. Hence plant virology is the study of viruses that infect plants. The effects, or symptoms, produced by these viruses are of many types and are found in many plant species, including most that are economically important. For example, infected plants may be stunted in comparison to virus-free ones, their growth form abnormal and their pigmentation altered so that the leaves may have a

Fig. 1.1 Still-life painting by Ambrosius Bosschaert (1619). The tulip shows petal striping of the type induced by tulip breaking virus. (Courtesy Rijks museum, Amsterdam.)

mosaic of light and dark green areas. These, and many of the other symptoms of virus infection, are described in Chapter 3.

The origins of viruses are unknown and there is no reason to believe that those we find today have recently come into being. Some of the earliest writings refer to a human disease which seems to be smallpox, and other virus diseases also have been recorded for several centuries. The first reliable record of a plant virus disease is of flower-breaking in tulips. No virus disease has been more beautifully illustrated, for infected tulip blooms were commonly included in the still-life pictures painted by Dutch masters in the early seventeenth century (Fig. 1.1). At this time the striped flowers produced by infected plants were greatly prized by the Dutch, far more so than the uniformly coloured blooms of virus-free plants. Dubos (1958), who describes this fashion as 'tulipomania', tells how a single infected bulb was bartered for oxen, pigs, sheep, tons of grain, and 1000 lb of cheese; and another for a mill; and how one lucky girl had an infected bulb as her dowry.

Some Dutch tulip growers knew that striping of the flowers could be induced in a normal tulip by grafting its bulb to that from a broken-flowered plant, but it was not until 1886 that the first experiments on transmission of a plant virus were described. Adolf Mayer, a German working in the Netherlands, found that he could induce mosaic disease in tobacco plants (Fig. 1.2) by injecting their leaf veins with sap taken from affected plants. He found also that boiling the sap inactivated the infectious agent, but concluded from these and other experiments that tobacco mosaic was caused by a bacterium.

The next crucial experiments were those of Dmitrii Ivanowski, a Russian working at that time in the Crimea. He reported in 1892 that he had confirmed Mayer's transmission experiments, and had found in addition that the causal agent of tobacco mosaic disease could pass through a filter with pores small enough to retain bacteria. However, Ivanowski did not realize the importance of his results, and suggested that his Chamberland

Fig. 1.2 Tobacco leaf (*Nicotiana tabacum*) showing mosaic symptom induced by tobacco mosaic virus.

disease of cattle could pass through a bacteria-retaining filter, and other plant and animal diseases were soon found to be caused by similar agents. The nature of these ultrafilterable viruses, as they were initially called, remained obscure for more than thirty years, and even in the early 1930s the search for viruses in plant extracts was likened to trying to find 'a black cat in a dark cellar without being certain that it is there'.

1.2 The nature of viruses

Unfortunately, for almost fifty years the study of the virus diseases of plants, animals and bacteria developed independently, and some animal-virus experts hotly denied that animal viruses were at all like those of plants or bacteria (also called bacteriophages or phages). More recently the trend has been towards reunification, and many common features have been recognized among viruses affecting different kinds of host. Because it is relatively easy to obtain and purify large quantities of the particles of some plant viruses, much of the pioneering work on the structure of virus particles has been done with those of plant viruses and, in particular, of tobacco mosaic virus

filter-candle may have been faulty, even though he found that 'liquids, most favourable to the development of bacteria, remained entirely unchanged for several months after filtration through this candle'. His scepticism was understandable at a time when Pasteur's ideas on the role of bacteria in causing disease were pre-eminent. It was left to Martinus Beijerinck, a Dutchman, to conclude from his own and Ivanowski's results that the causal agent of tobacco mosaic was something novel. Beijerinck (1898) confirmed Mayer's observations and found, like Ivanowski, that the tobacco mosaic agent could pass through a porcelain filter; he also showed that it could diffuse into an agar gel. These and other experiments convinced him that the elusive pathogen was not a bacterium but a *contagium vivum fluidum*. So Beijerinck (Fig. 1.3) is considered to be the real Father of Virology.

In the same year Loeffler and Frosch (1898) showed that the causal agent of the foot and mouth

Fig. 1.3 Martinus Willem Beijerinck (1851–1931), the 'Father of Virology'. (Courtesy J. P. H. van der Want.)

(hereafter called TMV). By contrast, the pioneering work on the genetics of viruses and on the biochemistry of virus replication has been done mainly with viruses of bacteria and animals, because their host cells are more readily grown in culture and more readily manipulated than those of plants.

The particles of all plant viruses, and also those of most other viruses, are too small to be seen in a light microscope. However indirect evidence that the particles of TMV are long thin rods was obtained from studies of the optical birefringence (Takahashi and Rawlins, 1932), sedimentation behaviour and viscosity (Lauffer, 1938) and X-ray diffraction (Bernal and Fankuchen, 1937) of preparations of the virus. This was confirmed in 1939 by Kausche, Pfankuch and Ruska, who, using the newly-developed electron microscope, obtained pictures of TMV particles (Fig. 1.4), which were later shown by others to be rods about 300 nm long and 16 nm in diameter (Fig. 1.5). Other viruses have particles of other characteristic shapes and sizes. Many are approximately spherical with diameters of 17 to 75 nm, whereas others range from flexuous filaments of 2000 × 10 nm to

bacilliform particles of 250 × 70 nm. These structures and their underlying 'design principles' are discussed in Chapter 5.

Similarly, much of the pioneering work on the composition of virus particles has been done with those of plant viruses. Helen Purdy (1929) showed that serological methods could be used to detect TMV particles, suggesting that they contained protein and/or polysaccharide. Then Stanley (1935) working in the United States and using newly introduced methods for purifying proteins, particularly enzymes, succeeded in purifying TMV particles and showed that they contained protein. For this achievement he was later awarded a Nobel Prize. But Stanley overlooked the essential component of TMV. This was first detected in 1937 by Bawden and Pirie, working in Britain. They noted that preparations of TMV contained phosphorus and found that this came from a nucleic acid of the same type as that previously described from yeast, namely ribonucleic acid (RNA). TMV particles contain 5% RNA and 95% protein. With a few exceptions, the particles of all other plant viruses examined contain these two components, as also do those of many viruses

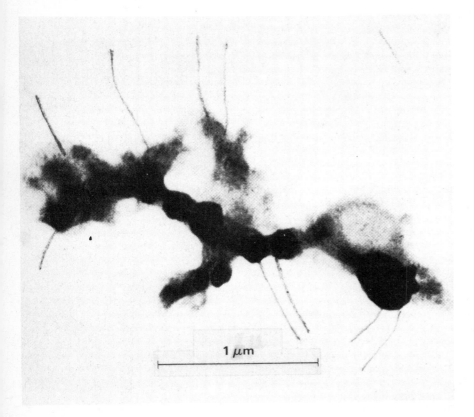

1 μm

Fig. 1.4 Early electron micrograph of tobacco mosaic virus particles. (After Kausche, Pfankuch and Ruska, 1939.)

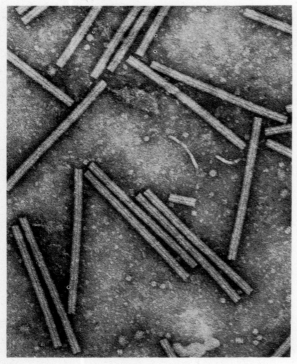

Fig. 1.5 Recent electron micrograph of tobacco mosaic virus particles, which are 300 nm long. Compare with Fig. 1.4. (Courtesy J. T. Finch.)

that infect animals or bacteria. The particles of cauliflower mosaic virus (Shepherd, Wakeman and Romanko, 1968), and many other animal and bacterial viruses, however, contain deoxyribonucleic acid (DNA), instead of RNA, a result foreshadowed by Schlesinger's work on bacteriophage, in preparations of which he detected DNA in 1934.

Some of the early purified preparations of TMV particles were paracrystalline, that is, they contained particles packed at regular spacings in two dimensions. Then, in 1938, Bawden and Pirie obtained true, three-dimensional crystals of the particles of another virus, tomato bushy stunt (Fig. 1.6). These findings stimulated much heated controversy about whether virus particles are living or not, and where they stand on the mystical border between the animate and the inanimate, but it was a sterile discussion and depended simply on the criteria of life used.

In 1949 Markham and Smith found that the spherical particles in preparations of turnip yellow mosaic virus were of two types, only one of which contained nucleic acid, and it was very significant that only the particles containing nucleic acid were infective. Another pointer to the importance of nucleic acid was the finding of Hershey and Chase (1952) that the DNA in the particles of bacteriophage T2 entered bacteria at infection whereas most of their protein did not. However, not until 1956 did Gierer and Schramm, and Fraenkel-Conrat (1956), show that a virus nucleic acid, the RNA of TMV, can be infective on its own, provided that precautions are taken to protect it from inactivation. When inoculated to plants, TMV-RNA induces the synthesis not only of more virus RNA but also of virus protein, and the two assemble into typical TMV particles. The main role of the protein in plant virus particles is apparently to provide a protective coat for the nucleic acid. TMV-RNA seems to be the sole bearer of genetic information of the virus; when the RNA is altered chemically the virus may be modified genetically. Thus Gierer and Mundry (1958) detected mutants of TMV after reacting inocula of TMV-RNA with nitrous acid, a treatment which changes the bases in the RNA (§12.2.2).

Turnip yellow mosaic is not the only virus to produce more than one type of particle, and others produce more than one type of nucleo-

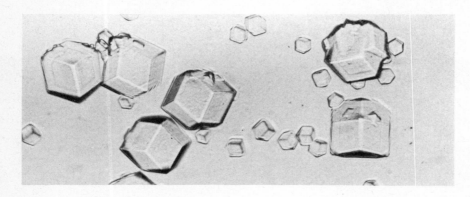

Fig. 1.6 Crystals of tomato bushy stunt virus particles. (Courtesy F. C. Bawden and N. W. Pirie.)

protein particle. For example, the tubular particles of the CAM strain of tobacco rattle virus are mostly either about 50 nm long or about 200 nm long. For some time it was not known whether the short particles have any function but more recent work has shown that the genome of the virus is in two parts, one part in the short particles, the other in the long. Many other plant viruses, including some with isometric or bacilliform particles, have also been shown to have multipartite genomes. This division of the genome is of particular interest because it increases the potential range of variability arising from a limited number of mutations.

It is clear that ideas on the nature of viruses, and hence the meaning of the word 'virus', have undergone many changes. To the ancient Romans *virus* meant poison. In the nineteenth century the word was used to mean 'the poisonous element by which infection is communicated' or simply 'a micropathogen'. However, since the turn of the century the word has been applied, with increasing specificity, to a heterogeneous group of small obligately parasitic pathogens. During this time viruses have been defined in various ways, but none of the published definitions (Bawden, 1950; Lwoff, 1957; Pirie, 1962) is ideal for they might individually exclude viruses with large particles such as the poxviruses, or include mycoplasma-like organisms, or the transferable episomes and plasmids of bacteria. The main features they emphasize are those of the virus particle such as its small size, potential pathogenicity, infectiousness, possession of nucleic acid and inability to multiply when apart from living organisms.

For the purpose of this book, however, we will define viruses as 'transmissible parasites whose nucleic acid genome is less than 3×10^8 daltons in weight and that need ribosomes and other components of their host cells for multiplication'. The figure of 3×10^8 is large enough to include the bacteriophages, poxviruses and iridoviruses (Chapter 16). Pathogenicity is not mentioned because not all viruses have deleterious effects on their hosts and there seems little point in separating the non-pathogenic ones from essentially similar pathogenic types. The necessity for living cells also is not specified because many of the steps of virus synthesis have now proved feasible in the test tube, with the aid of cell components instead of intact cells. Our definition covers the pathogens that have been called viroids (Diener, 1972), such as that causing potato spindle tuber disease, whose transmissible stage consists of RNA of 10^5 daltons or less, and which does not have a protein coat.

Although a single definition can be devised that encompasses all the diverse biological entities that virologists study, there is no reason to believe that all viruses are phylogenetically related to one another; the converse is more likely. Hence viruses should perhaps be considered not so much a family, more a way of life. However, it is increasingly obvious that many plant viruses fall into well-defined groups. Brandes and Wetter (1959) showed how the plant viruses with tubular or filamentous particles could be grouped according to the most common length of their particles, and that viruses grouped by this character shared many other characters, such as their mode of spread and stability. Many of the known plant viruses can now be grouped (Harrison *et al*, 1971), and although the generalizations made possible by such groupings can dangerously oversimplify, the advantages outweigh the disadvantages (Chapter 12). We describe several of these groups in Chapter 2.

Not all sub-microscopic plant pathogens that are transmissible by grafting and/or by other organisms are viruses. The causal agents of aster yellows and other similar diseases were long thought to be viruses and studied as such. However, Doi *et al* (1967) found that thin sections of infected plants contain bodies that resemble in the electron microscope those of *Mycoplasma* spp. and L-forms of bacteria. It is also interesting that the causal agent of primary atypical pneumonia of man, which for more than twenty years was thought to be a virus was eventually shown to be a mycoplasma. This illustrates the danger in making unjustified assumptions; indeed, of the 600 or more plant pathogens thought to be viruses and named (Martyn, 1968), more than two-thirds have not been rigorously proved to be viruses and could be agents of other kinds. In this book we shall distinguish pathogens that are proved to be viruses from those on which critical information is lacking by referring to the latter as 'virus-like agents' or some other similar phrase. The properties of some agents confused with viruses are described in Chapter 17.

Angiosperms are not the only plants known to be infected by viruses. Viruses have been detected in gymnosperms and pteridophytes, and there is increasing evidence of viruses, and virus-like particles, in fungi and even algae. These and viruses of other types of host will be discussed in Chapter 16.

1.3 Special characteristics of plant virus infections

It is now recognized that plant viruses have many properties in common with those which infect animals or bacteria. This recognition was greatly speeded by the discovery in 1952 by Black and Brakke that clover wound tumor virus could multiply not only in plants but also in the leafhopper (*Agallia constricta*), the insect responsible for carrying the virus from plant to plant. The host range of individual viruses can therefore overlap the boundary between plant and animal kingdoms; but although several such overlaps are known they are relatively uncommon and there are no confirmed experiments to show that any one virus can go through its complete life cycle in both plants and vertebrates, or plants and bacteria.

Virus infections of plants differ from those of animals and bacteria in various ways. One difference is in the way in which the host cells are infected. Plant viruses normally infect plants only through wounds. This is because the tough outer cell wall of plants is impermeable to virus particles, and must be broken for infection to occur; thus plants are not infected when a virus suspension is gently sprayed onto them. By contrast most animal viruses can infect unwounded animal cells, and pass into the cell by various routes without causing damage. Some bacteriophages infect in a similar way using bacterial sex pili, whereas others have a particle that is, in essence, an apparatus for injecting the nucleic acid into the host cell. Plant cells freed by enzyme treatment from their cell walls (and then called protoplasts) behave like animal cells in that they may be infected by adding virus to the medium in which they are suspended (Takebe and Otsuki, 1969).

In the laboratory, viruses can be transmitted from plant to plant by various methods. One of these is to graft virus-free plants with scions from infected plants of the same or a related species. Many viruses can be transmitted to plants by wiping fluids containing their particles over leaves because this action makes minute wounds in the leaf cells through which the particles infect the plant. It is much quicker than grafting but relatively inefficient because at least 10^5 virus particles have usually to be inoculated to infect each plant. By comparison, animal cells can be infected with only 10 to 100 virus particles, and 1 to 10 phage particles will infect a bacterium; the differences probably mainly reflect the relative inefficiency of producing wounds of the right type in the walls of plant cells. Experimental transmission of plant viruses is described in more detail in Chapter 4.

The greatest advance in the quantitative assay of plant viruses was that made by Holmes (1929) who discovered that TMV produced necrotic spots in inoculated leaves of *Nicotiana glutinosa* (Fig. 4.5a). The number of spots, which represent sites of infection and are called local lesions, depends on the concentration of virus particles in the inoculum. This local-lesion method, and several other assay methods, are described in Chapters 7, 8 and 9.

A characteristic feature of plant virus infections is that a plant, once infected, usually remains so for the rest of its life. In contrast to vertebrates, plants do not produce circulating antibodies which can inactivate virus within the organism. Hence many viruses survive indefinitely and cause much damage in vegetatively propagated plants, such as fruit trees, potatoes, strawberries and sugarcane. Typically, viruses invade all tissues of their plant hosts, with the possible exception of the meristematic regions of root and shoot, though some have other tissue specificities. Also, most plant viruses are not, fortunately, passed on to progeny seedlings through the true seed. The diverse and characteristic effects of virus infection on plants are discussed in Chapter 3.

1.4 How plant viruses spread in nature

Although individual infected plants often can act as virus sources for most of their lives they are static, and some mobile agent, called a 'vector', is needed to carry the viruses from plant to plant. The first evidence for such vectors came from Japan. Fukushi (1965) has told how Hashimoto, a rice grower in the Shiga Prefecture noticed dwarfing of his rice in 1883 and suspected that leafhoppers were involved because the trouble was common in fields where these insects were abundant. In 1893 he put young rice plants which were growing in a pot, into a cheesecloth cage together with leafhoppers. The rice plants became infected, and he confirmed this in the following year, but did not publish his results. This was left to Takata, who in 1895 named the leafhopper as 'inazuma-yokubai' (*Inazuma dorsalis* (Motsch.) Ishihara). However, in 1901 Takami summarized work showing that the rice dwarf pathogen, which he did not realize is a virus, was carried by 'tsumaguro-yokubai' (=*Nephotettix cincticeps* Uhl.), and *I. dorsalis* was discounted. Not until 1937 was the work on *I. dorsalis* confirmed, by Fukushi.

Early in the twentieth century, yellow fever, which was thought to be caused by a virus, was shown to be transmitted to man by mosquitoes and the idea of vectors was established. Subsequent work on plant viruses has implicated a whole range of plant sucking insects as virus vectors – aphids, coccids, leafhoppers, etc., and some biting insects such as leaf-eating beetles. In 1927, the first evidence was obtained that an eriophyid mite (*Phytoptus ribis*) is the vector of the virus-like agent that causes reversion of blackcurrants. Much later, vectors of quite other kinds were found, living in the soil instead of on the aerial parts of plants. Hewitt, Raski and Goheen (1958) showed that the nematode *Xiphinema index* transmits grapevine fanleaf virus and in 1960–1 workers in several laboratories found that the chytrid fungus *Olpidium* was involved in the spread of tobacco necrosis virus, and of the virus-like agents causing tobacco stunt and lettuce big-vein diseases. Several other examples of vectors in each of these groups are now known.

Despite the great range of organisms able to transmit plant viruses, it is remarkable that where a virus has more than one vector species, these are almost always closely related taxonomically. For example, viruses that have aphid vectors are not transmitted by leafhoppers, and vice versa. The different kinds of vectors, their relative importance and the different ways in which they carry viruses are discussed in Chapter 13.

The rapidity and pattern of virus spread in nature obviously depend on the relation between virus and vector, on the numbers and activity of the vectors and on many other factors. How these interact is discussed in Chapter 14. The knowledge of how viruses spread leads to attempts to prevent this happening, and control measures are often aimed at the vector. However, there are many viruses which are apparently maintained without the aid of vectors, by vegetative propagation of infected plants. With vegetatively multiplied crop plants it is often worth while to obtain a nucleus of virus-free plants and to propagate them in conditions in which infection does not occur. Chapter 15 deals with the difficult problem of controlling virus spread.

1.5 Importance of plant virology

Plant virology is important in two main ways. On the one hand, its study should lead to the prevention of crop losses from virus disease. This is one aspect of plant pathology. On the other hand it provides information that should shed light on the general properties of viruses and their hosts, and on the structure and function of macromolecules of biological importance, especially proteins and nucleic acids. This is molecular biology. A few examples will illustrate the scope and importance of these two aspects.

Most major crops are subject to damaging, and often prevalent, virus diseases. North American stocks of potatoes were at one time so generally infected with viruses that potato virus X was known as the 'healthy potato virus' because the healthiest potatoes contained only potato virus X, which causes a loss of yield of about 10–15%, whereas less healthy ones contained other viruses as well, and the yields were even lower.

Many other vegetatively propagated and perennial crops, such as fruit trees, also suffer from virus infections. In citrus, the aphid-transmitted tristeza virus is widespread, and in Sao Paulo State, Brazil, alone it killed 6 million trees (75% of the total) of sweet orange growing on sour orange root-stocks, in a period of twelve years. In Ghana, more than 100 million cacao trees have been cut down since 1945 in attempts to stop the spread of cacao swollen shoot virus, which has mealy bug vectors. Grapevine fanleaf virus, which has nematode vectors and is also maintained by vegetative propagation of grapevine, is found in many parts of the world. It decreases the yield of fruit of individual plants by 10 to 50%, and every plant of some grapevine varieties is infected.

Wheat, barley and oats are infected with barley yellow dwarf virus in many cool temperate countries, the incidence of infection varying greatly from year to year with differences in the numbers and activity of its aphid vectors. Wheat mosaic virus, which is fungus-transmitted, can cause large losses in the wheat crop in North America. In 1957, before disease-escaping varieties were widely grown, losses in Kansas State, U.S.A., alone were estimated to value 4 million dollars. In sugar beet in Britain, the aphid-transmitted beet mild yellowing and beet yellows viruses cause yearly losses of 1.5×10^4 to 1.5×10^5 tons of sugar.

We now know how to avoid some of these crop losses but many others present unsolved problems. With the continuing increase in world population, and the increasing strain on supplies of food and natural products, it is imperative that plant virus diseases should be intensively studied and ways found of minimizing the losses they cause.

Interest in what is now called molecular biology

has increased dramatically in recent years. Plant viruses have important advantages as subjects for some studies of this kind and several can be obtained in relatively large amounts, and sufficiently pure, to permit detailed chemical and structural analysis, and they can be handled without risk to the experimenter. The particles of TMV and turnip yellow mosaic virus are better understood structurally than those of any other viruses – the way in which their protein subunits are packed to form coats for the RNA, the location of the RNA, etc. Using TMV and cowpea chlorotic mottle virus, it has been possible to find out how helical and near-spherical virus particles can be taken to pieces and put together. The principles revealed by this kind of work can be applied more generally to the problem of how other, more or less complex, structures are built from macromolecules of one or a few types.

Chemical work has revealed the sequence of amino acids in the coat protein of TMV, and, by determining the amino-acid composition of TMV mutants, indirect confirmation has been obtained that plants and bacteria use the same genetic code – the way in which the sequence of nucleotides in a nucleic acid determines the sequence of amino acids in proteins (see Chapters 5 and 12). Thus plant viruses also provide a tool for studying the biology of plants and plant cells. But so far only the surface has been scratched and much remains to be explained – the biochemical steps involved in virus multiplication and disease development, the features that determine ability to infect one species but not another, the specificity between viruses and their vectors, and much else. The molecular biological approach to these and many other problems is perhaps the most promising one in prospect.

Chapter 2

Some plant viruses and their names

2.1 Nomenclature

Most plant viruses are first recognized by the diseases they cause in plants, and it is thus not surprising that they are usually named from these diseases. For example, the commonest virus that causes a mosaic of light and dark green areas in the leaves of tobacco plants is called tobacco mosaic virus. However, early work, particularly on the viruses infecting potatoes, showed that several viruses can infect the same plant species, and so in 1927 James Johnson suggested that plant viruses should simply be named after the host in which they were first found, and given a distinguishing number. Thus the first virus obtained from tobacco became tobacco virus 1, the second, tobacco virus 2, and so on. K. M. Smith in 1937 suggested that these names would be made more acceptable internationally by using the latinized generic name of the host (e.g. *Nicotiana* virus 1), and finally Holmes (1939) and McKinney (1944) proposed Linnaean-style latinized binomials (e.g. *Marmor tabaci*) tied to classifications based on the natural vectors of the viruses, and on the types of symptom the viruses caused. However for various reasons these and other schemes for devising names were not adopted widely, and nowadays most virologists still prefer vernacular names of the kind 'tobacco mosaic virus'.

The vernacular names of many plant viruses are long, so that in use they are often abbreviated to a few letters, such as TMV. However, these abbreviations have not yet been standardized and, to avoid confusion, have to be redefined whenever used; otherwise AMV, for instance, could equally well refer to alfalfa mosaic virus, arabis mosaic virus or even apricot mottle virus. In this book we will use the vernacular names of viruses collated by Martyn (1968, 1971).

We and others (Gibbs *et al*, 1966a; Gibbs and Harrison, 1968) have suggested that vernacular names can be made more definitive, when that is needed, by adding a cryptogram, which gives in code form some of the properties of the virus. Many viruses fall into definite groups, and the properties included in the cryptogram are those that seem most useful for distinguishing between all the different groups, which include viruses infecting animals, bacteria, fungi, etc. (Chapter 16) as well as those infecting seed plants. In this book the cryptograms used consist of four pairs of symbols, the meanings of which are given in Table 2.1. For example, the cryptogram for tobacco mosaic virus and some closely related viruses is R/1: 2/5: E/E: S/0. Cryptograms may also be used with the names of groups of viruses to indicate the characters of the group, but it is best to distinguish group cryptograms in some way because the properties encoded in them may have been determined for only one or a few of the viruses included; in this book they are given in parentheses.

Virus groups are usually named after the best-known virus of the group (e.g. the tobacco mosaic virus group or TMV-group). In this book we will use shortened group names derived from the vernacular name of the type member of the group so that, for example, the tobacco mosaic virus group will be called the tobamovirus group or tobamoviruses (Harrison *et al*, 1971). Group names can also be used when giving the full name of a virus that is known to belong to a particular group; for instance bottlegourd mosaic virus, */*: */*: E/E: S/*, *tobamovirus group*.

2.2 Some viruses and virus groups

Some plant viruses do not seem to be closely related to any other known virus, whereas many others fall into well-defined clusters, the largest of which contains more than fifty distinct viruses and includes potato virus Y. The ways of classifying and grouping viruses will be discussed in Chapter 12. Suffice to say here that just as it is best to use several characters to describe individual viruses, so also it is best to define each group of viruses by several properties, none of which is adequate on its own (Harrison *et. al*, 1971).

In this chapter some of the better studied or more important viruses or virus groups are briefly described, especially those mentioned frequently

Table 2.1 Meaning of the symbols used in the cryptograms

Each cryptogram consists of four pairs of symbols (for example, tobacco mosaic virus, R/1: 2/5: E/E: S/0) with the following meanings:

1st pair. *Type of nucleic acid/Strandedness of nucleic acid*
Symbols for type of nucleic acid
R = RNA: D = DNA
Symbols for strandedness
1 = single-stranded: 2 = double-stranded

2nd pair. *Molecular weight of nucleic acid (in millions)/Percentage of nucleic acid in infective particles*
This term gives the composition of infective particles. The genome of some viruses is divided. When different pieces of the genome occur together in one type of particle the symbol indicates the total molecular weight of the pieces in the particle (e.g. clover wound tumor virus, R/2: Σ16/22: S/S: S, I/Au), but when the pieces of the genome occur in different particles, the composition of each particle type is listed separately (e.g. tobacco rattle virus, R/1: 2.3/5+0.6 to 1.3/5: E/E: S/Ne).

3rd pair. *Outline of particle/Outline of nucleocapsid* (the nucleic acid plus the protein most closely in contact with it)
Symbols for both properties
S = essentially spherical
E = elongated with parallel sides, ends not rounded
U = elongated with parallel sides, end(s) rounded
X = complex or none of above

4th pair. *Kind of host infected/Kind of vector*
Symbols for kind of host
A	= alga	P = pteridophyte
B	= bacterium	S = seed plant
F	= fungus	V = vertebrate
I	= invertebrate	
M	= mycoplasma	

Symbols for kind of vector
Ac = mite and tick (Acarina, Arachnida)
Al = white-fly (Aleyrodidae, Hemiptera, Insecta)
Ap = aphid (Aphididae, Hemiptera, Insecta)
Au = leaf-, plant-, or tree-hopper (Auchenorrhyncha, Hemiptera)
Cc = mealy-bug (Coccidae, Hemiptera)
Cl = beetle (Coleoptera, Insecta)
Di = fly and mosquito (Diptera, Insecta)
Fu = fungus (Chytridiales and Plasmodiophorales, Fungi)
Gy = mirid, piesmid, or tingid bug (Gymnocerata, Hemiptera)
Ne = nematode (Nematoda)
Ps = psylla (Psyllidae, Hemiptera)
Si = flea (Siphonaptera, Insecta)
Th = thrips (Thysanoptera, Insecta)
Ve = vectors known but none of above
0 = spreads without a vector via a contaminated environment
Symbols for all pairs
* Property of the virus is not known
( ) Enclosed information is doubtful or unconfirmed
[ ] Enclosed cryptogram gives information about a virus group

in this book. Some properties of these viruses are given in Table 2.2 and further details can be found in the C.M.I./A.A.B. *Descriptions of Plant Viruses*, and in the review by Gibbs (1969). At the end of this book, we list the names and cryptograms of the plant viruses mentioned, together with the name of the group to which each belongs where this is known.

2.2.1 *Viruses with helically constructed tubular or filamentous particles*

There are many viruses of this type and most cause mosaic symptoms. Typically they are sap-transmissible, that is, they are transmitted when the leaves of a healthy plant are rubbed with the sap from an infected plant. Brandes and Wetter (1959) showed that many of these viruses are readily distinguished by the modal lengths of their particles; and that different viruses with particles of closely similar or indistinguishable lengths are usually also related in other ways. Here we have arranged them in order of increasing particle length.

TOBRAVIRUS GROUP
[R/1: 2.3/5+0.6 to 1.3/5: E/E: S/Ne]

The type member of this group is *tobacco rattle* virus, and both it and the closely related pea early-browning virus are transmitted by trichodorid nematodes. These viruses have straight tubular particles mainly of two lengths (Fig. 2.1d, Fig. 5.14c and d); for example, those of the CAM strain of tobacco rattle virus are 197 nm and 52 nm long. The short particles will not infect plants alone, but seem to contain the genetic information for making the coat protein, which is of one kind and is found in both long and short particles. Many isolates of these viruses are readily transmitted mechanically, and have a wide host range (i.e. they infect plants from many different families).

Table 2.2 Properties of some well-known plant viruses

Virus	Cryptogram	Approx. no. of similar viruses	Usual symptoms	Stability¹ when heated	Concentration² in sap	Particle size (nm)
Rod-shaped or filamentous						
Tobacco rattle	R/1: 2.3/5+0.6 to 1.3/5: E/E: S/Ne	1	Necrosis	Very stable	Moderate	190×22+45 to 110 ×22
Tobacco mosaic	R/1: 2/5: E/E: S/0	6	Mosaic	Very stable	Large	300×18
Potato virus X	R/1: 2.2/6: E/E: S/0, (Fu)	8	Mosaic	Stable	Large	520×13
Red clover vein mosaic	R/1: */6: E/E: S/Ap	7	Mosaic	Stable	Moderate	650×12
Potato virus Y	R/*: */*: E/E: S/Ap	70	Mosaic	Stable	Moderate	740×11
Beet yellows	R/1: 4.3/5: E/E: S/Ap	4	Yellows, phloem necrosis	Unstable	Small	1200×10
Isometric						
Brome mosaic	R/1: 1.1/23+1.0/22+0.75/22: S/S: S/*	2	Mosaic	Very stable	Large	25
Cucumber mosaic	R/1: 1.3/19+1.1/19+0.8/19: S/S: S/Ap	2	Mosaic	Stable	Large	30
Barley yellow dwarf	R/1: 2/*: S/S: S/Ap	6	Yellows	Stable	Small	25
Tobacco necrosis	R/1: 1.5/19: S/S: S/Fu	0	Necrosis	Very stable	Moderate	26
Satellite	R/1: 0.4/20: S/S: S/Fu	0	None	Very stable	Moderate	17
Tomato bushy stunt	R/1: 1.5/18: S/S: S/*	4	Mosaic necrosis	Very stable	Large	30
Turnip yellow mosaic	R/1: 2.0/34: S/S: S/Cl	12	Mosaic	Very stable	Large	28
Cowpea mosaic	R/1: 2.3/34+1.5/28: S/S: S/Cl	7	Mosaic, necrosis	Stable	Large	28
Tobacco ringspot	R/1: 2.4/42+1.4/29 (or Σ2.8/46): S/S.S/Ne	10	Ringspot, mottle	Stable	Moderate	28
Prunus necrotic ringspot	R/*: */16: S/S: S/0	3	Necrotic ringspot	Unstable	Moderate	25
Pea enation mosaic	R/1: 1.6/28+1.3/28: S/S: S/Ap	0	Enations, mottle	Stable	Moderate	30
Cauliflower mosaic	D/2: 4.5/16: S/S: S/Ap	4	Mosaic	Stable	Moderate	50
Clover wound tumor	R/2: Σ16/22: S/S: S,I/Au	6	Tumours	Stable	Moderate	70
Tomato spotted wilt	(R)/*: */*: S/*: S/Th	0	Mosaic, necrosis	Unstable	Moderate	80
Bacilliform						
Alfalfa mosaic	R/1: 1.1/16+0.8/16+0.7/16: U/U: S/Ap	0	Mosaic, necrosis	Stable	Large	58×18+52×18 +42×18
Cacao swollen shoot	*/*: */*: U/U: S/Cc	?3	Mosaic, tumours	Unstable	Small	130×28
Lettuce necrotic yellows	R/1: 4/*: U/E: S,I/Ap	⎱ 20	Yellows, necrosis	Unstable	Moderate	230×70
Potato yellow dwarf	R/1: 4.3/0.4: U/E: S,I/Au	⎰	Yellows, dwarfing	Unstable	Moderate	380×75

¹ Stability when heated in sap: very stable = infective after 10 min at 75°C, stable = infective after 10 min at 55°C but not 75°C, unstable = infectivity lost in 10 min at 55°C.
² Concentration in sap: large = >100 mg/l sap, moderate = 5–100 mg/l sap, small = <5 mg/l sap.

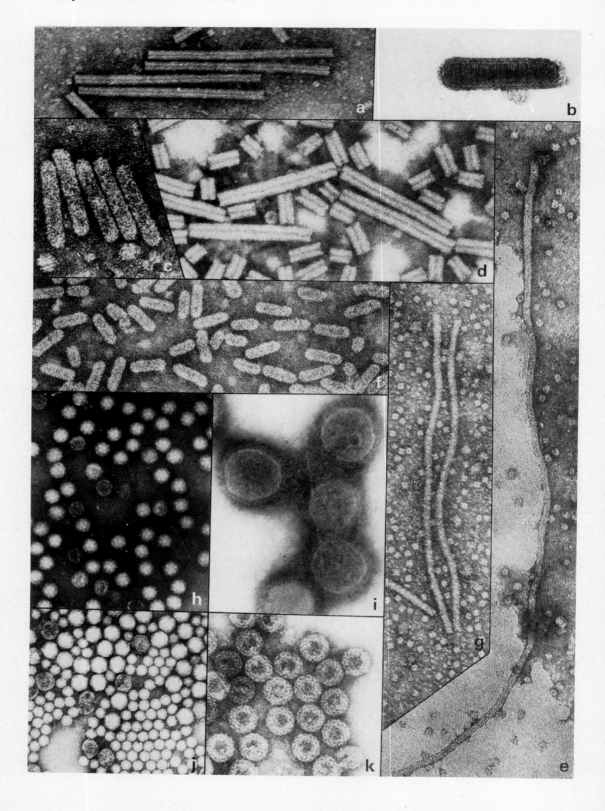

SOIL-BORNE WHEAT MOSAIC VIRUS
R/1: 2/(5): E/E: S/Fu and
POTATO MOP-TOP VIRUS (R/1): */*: E/E: S/Fu

Soil-borne wheat mosaic virus once caused much damage to wheat crops in North America but disease-escaping varieties of wheat have been selected and are now widely grown. Potato mop-top virus occurs in potatoes in Western Europe. These viruses have particles similar to those of tobacco mosaic virus but most are either 100 to 160 nm or about 300 nm long. They have narrow host ranges and are transmitted by zoospores of plasmodiophoromycete fungi; the wheat virus by *Polymyxa graminis*, and the potato virus by *Spongospora subterranea*.

TOBAMOVIRUS GROUP [R/1: 2/5: E/E: S/0]

This group of viruses includes *tobacco mosaic* and tomato mosaic viruses, together with various viruses of tropical legumes (e.g. sunnhemp and cowpea); also viruses of cucurbits, orchids, cacti, etc. Tobacco mosaic virus is the most fully studied of all plant viruses. It has straight tubular particles 300 nm long (Fig. 2.1a) and is readily transmitted mechanically to a wide range of plants in which it causes mosaic and necrotic symptoms.

POTEXVIRUS GROUP R/1: 2.2/6: E/E: S/0]

This group includes *potato virus X*, white clover mosaic virus and several others. They have flexuous filamentous particles 480 to 580 nm long (Fig. 2.1g). Like the tobamoviruses, they are often transmitted when an infected plant rubs against a healthy one; there is also a report that potato virus X is transmitted by the fungus *Synchytrium*. Potexviruses have narrow host ranges, and cause mosaic symptoms.

◀——————————————————

Fig. 2.1 A collection of electron micrographs of the particles of various plant viruses chosen to illustrate the variety of particle types. All micrographs are at almost the same magnification: the particles of tobacco mosaic virus are 300 nm long. The particles are of:
  a, tobacco mosaic virus; **b,** lettuce necrotic yellows virus; **c,** cacao swollen shoot virus; **d,** tobacco rattle virus CAM strain; **e,** beet yellows virus; **f,** alfalfa mosaic virus; **g,** potato virus X; **h,** turnip yellow mosaic virus; **i,** tomato spotted wilt virus; **j,** tobacco necrosis virus (larger particles) and satellite virus (smaller particles), and **k,** cauliflower mosaic virus.
  (Courtesy Rothamsted Experimental Station and Scottish Horticultural Research Institute.)

CARLAVIRUS GROUP [R/1: */6: E/E: S/Ap]

This group is alternatively known as the potato virus S group and includes *carnation latent* virus, potato viruses S and M (also called potato paracrinkle virus), red clover vein mosaic virus and at least eight other viruses. They have almost straight filamentous particles about 650 nm long and are readily transmitted mechanically. They have limited host ranges, and several are transmitted in the non-persistent manner by aphids (see Chapter 13).

POTYVIRUS GROUP [R/1: 3.5/5: E/E: S/Ap]

This group, named after *potato virus Y*, is the largest group of plant viruses known at present and, like the two preceding groups, includes many viruses that cause economically important diseases of crop plants, such as bean yellow mosaic (and the related bean common mosaic and pea mosaic), cowpea aphid-borne mosaic, lettuce mosaic, plum pox, sugarcane mosaic, turnip mosaic and numerous others. These viruses have flexuous filamentous particles mostly 730 to 790 nm long (Fig. 5.14f and g). They can be transmitted experimentally by inoculation of sap, and in nature their vectors are aphids, which transmit them in the non-persistent manner. Some have moderately wide host ranges; others, especially those often transmitted through the seed (e.g. bean common mosaic and lettuce mosaic viruses), have narrow host ranges. Infected plants usually show mosaic symptoms.

Viruses transmitted by eriophyid mites such as wheat streak mosaic R/1: 2.8/*: E/E: S/Ac and ryegrass mosaic */*: */*: E/E: S/Ac have much in common with the potyviruses but have particles only about 700 nm long.

BEET YELLOWS VIRUS R/1: 4.3/5: E/E: S/Ap
AND CITRUS TRISTEZA VIRUS R/1: */*: E/E: S/Ap

These economically important viruses have very flexuous filamentous particles with a clearly defined cross-banding (Fig. 2.1e). The particles of citrus tristeza virus are about 2 μm long and those of beet yellows virus about 1.2 μm. The viruses cause yellowing and phloem necrosis in a narrow range of plants. They are transmitted by aphids in which they persist for a few days, and they can be transmitted mechanically, though only with difficulty.

Some fruit tree viruses, of which the best known are apple chlorotic leaf spot R/1: 2.3/5: E/E: S/*

and apple stem grooving */*: */*: E/E: S/*, have particles resembling those of beet yellows virus but only about 600 to 700 nm long.

2.2.2 *Viruses with isometric particles*

CUCUMOVIRUS GROUP
[R/1: 1.3/19 + 1.1/19 + 0.8/19: S/S: S/Ap]

*Cucu*mber *mo*saic virus and the related peanut stunt and tomato aspermy viruses have rounded isometric particles about 30 nm in diameter. RNA obtained from virus preparations contains four species of 0.4×10^6 to 1.25×10^6 daltons weight. The three largest species are required for infectivity and there are probably three different nucleoprotein particles. Cucumber mosaic virus infects plants of at least forty angiosperm families, and is one of the commonest viruses in cultivated plants in temperate regions. In most plants it causes mosaic symptoms, and occasionally necrosis (see Chapter 3). The virus is readily transmitted by inoculation of sap, and also by many species of aphid in the non-persistent manner. Alfalfa mosaic virus (§2.2.3) seems to have close affinities with the cucumoviruses even though its particles are of a different shape.

TYMOVIRUS GROUP [R/1: 2/37: S/S: S/Cl]

The type member of this group of about ten viruses is *t*urnip *y*ellow *mo*saic, one of the best-known viruses. These viruses have rounded isometric particles 25 to 30 nm in diameter which show a characteristic pattern of groups of protein subunits on their surfaces (Fig. 2.1h, Fig. 5.14a, and see Chapter 5). Some of the particles characteristically contain no nucleic acid and are not infective. The viruses have narrow host ranges and cause mosaics, but rarely necrosis. All are readily transmitted mechanically, and most are known to be transmitted by leaf eating beetles.

COMOVIRUS GROUP
[R/1: 2.3/34 + 1.5/28: S/S: S/Cl]

This world-wide group of about a dozen viruses includes *co*wpea *mo*saic, bean pod mottle and squash mosaic viruses. They have angular isometric particles 25 to 30 nm in diameter. Some particles contain no nucleic acid, and others contain either about 34% or about 28%. Nucleic acid from both kinds of nucleoprotein particle is needed for a plant to be infected; the genome is bipartite. Most of these viruses have narrow host ranges, and infected plants show mosaic or necrosis symptoms. All are readily transmitted mechanically, and several are known to be transmitted by leaf-eating beetles. Broad bean wilt virus has many similarities with the comoviruses, but is transmitted by aphids.

NEPOVIRUS GROUP
[R/1: 2.4/43 + 1.4 to 2.1/30 to 40 $^{\pm}$
(Σ2.8/46): S/S: S/Ne]

This group is so called because the viruses in it have *ne*matode vectors and *po*lyhedral particles about 30 nm in diameter. However they have also been called ringspot viruses because most cause ringspot symptoms (Fig. 3.1) in one or more hosts. Many cause damage of economic importance to perennial horticultural crops in temperate regions (e.g. grapes, hops, fruit crops), and the group includes tobacco ringspot, tomato ringspot, arabis mosaic, grapevine fanleaf and raspberry ringspot viruses. The particles of these viruses are isometric, have an angular outline, and are usually of three types, like those of the comoviruses, or four types. They have wide host ranges, and recently infected plants show ringspots, mottles and necrosis; plants that survive the early severe stage of infection may 'recover' and later show few or no symptoms even though still infected. These viruses are readily transmitted mechanically. Their vectors are nematodes of the genera *Longidorus* or *Xiphinema*; and they are also often transmitted through seed or pollen.

TOBACCO NECROSIS VIRUS R/1: 1.5/19: S/S: S/Fu

The particles of this common virus are rounded and about 26 nm in diameter (Fig. 2.1j). It is readily transmitted mechanically to a very wide range of plants, causes necrotic local lesions in the inoculated leaves of most and usually does not become systemic. In nature, it is spread between plant roots by zoospores of the chytrid fungus *Olpidium brassicae*, and is usually confined to the roots.

SATELLITE VIRUS R/1: 0.4/20: S/S: S/Fu

This very small virus is unique in that it multiplies only in plants also infected with tobacco necrosis virus, to which it is serologically unrelated. Its particles are only 17 nm in diameter (Fig. 2.1j). It is readily transmitted mechanically and, like tobacco necrosis virus, by the zoospores of *Olpidium brassicae*.

BROMOVIRUS GROUP
[R/1 : 1.1/23 + 1.0/22 + 0.7/21 : S/S : S/*]

The viruses in this group are *bro*me *mo*saic, broad bean mottle and cowpea chlorotic mottle viruses. They have rounded isometric particles about 25 nm in diameter. These were the first isometric virus particles to be experimentally taken to pieces and reassembled. Their genome is in three pieces that are found in particles of slightly different density. Viruses of this group have narrow host ranges, and cause mosaics and mottles. They are readily transmitted mechanically, but their vectors in nature are not known.

TOMBUSVIRUS GROUP [R/1 : 1.5/18 : S/S : S/*]

This group includes *tom*ato *bu*shy *st*unt, and perhaps four other viruses. They have rounded isometric particles 30 nm in diameter (Fig. 5.14b), and differ from those of bromoviruses in size and structure (see Chapter 5), in sedimenting faster when centrifuged, and in being stable in concentrated salt solutions. The tombusviruses have wide host ranges and cause necrotic symptoms in many species. They are readily transmitted mechanically, but their vectors are not known, though some seem soil-borne.

POTATO LEAF ROLL AND SIMILAR VIRUSES
[R/1 : 2/* : S/S : S/Ap]

Potato leaf roll, barley yellow dwarf, beet mild yellowing, bean leaf roll and several other viruses have several properties in common and it is likely that, when more is known about them, they will be grouped together. They mostly cause economically important diseases of crops (especially annual crops) in temperate regions. Most have moderately restricted host ranges. Infected plants are stunted but not deformed, and their leaves yellowed or reddened, and often cupped and brittle. The particles of several of these viruses are isometric and about 25 nm in diameter, but none has been transmitted mechanically to plants. They are transmitted by aphids, and may persist in the aphid for several weeks; indeed there is an unconfirmed report that potato leaf roll virus multiplies in aphids.

VIRUSES WITH LABILE PARTICLES OF TWO OR MORE TYPES

Many economically important viruses that are common in fruit trees of temperate regions have particles of two or more kinds, and are unstable in sap unless chelating or reducing agents are added. Their particles contain 15 to 25% RNA, are rounded with a diameter of 20 to 35 nm but have an irregular outline. Many of these viruses are transmitted through pollen or seed to progeny plants but no vector has been found.

The affinities of these viruses to one another have recently become somewhat clearer. One cluster includes tobacco streak, black raspberry latent and elm mottle viruses. They have three types of particle of the same density but different size; three major and one or two minor species of RNA are extracted from unfractionated preparations of virus. Tulare apple mosaic virus has some similarities to this cluster. Two other clusters of these viruses are distinguished by serological tests and together have been called the ilarviruses (viruses with *i*sometric *la*bile particles causing *r*ingspot symptoms; see Fulton, 1968). One cluster of three includes prune dwarf virus. The other cluster includes prunus necrotic rinspot, apple mosaic and rose mosaic viruses. More information is however still needed before the extent of the affinities between these clusters can be better assessed.

PEA ENATION MOSAIC VIRUS
R/1 : 1.6/28 + 1.3/28 : S/S : S/Ap

This virus occurs in leguminous plants in temperate regions and causes mosaics and enations (small outgrowths on the underside of leaves). Its genome is bipartite. It is transmissible by inoculation of sap, and has aphid vectors, in which it persists for weeks. The particles of this virus resemble those of the ilarviruses in many of their properties.

CAULIMOVIRUS GROUP [D/2 : 4.5/16 : S/S : S/Ap]

*Cauli*flower *mo*saic virus, the best-known virus of this group is the first plant virus whose particles were shown to contain DNA. Both this and the serologically related dahlia mosaic virus have rounded isometric particles about 50 nm in diameter (Fig. 2.1k). They have restricted host ranges, cause mosaic symptoms, and are readily transmitted both mechanically and by many species of aphids, in which they persist for only a few hours. Cauliflower mosaic virus occurs in all continents.

CLOVER WOUND TUMOR AND SIMILAR VIRUSES
[R/2: Σ10 to 16/11 to 22: S/S: S,I/Au]

Clover wound tumor, rice dwarf, rice black streaked dwarf and maize rough dwarf viruses share several characters. They have isometric particles about 70 nm in diameter which contain several pieces of double-stranded RNA, and which resemble the particles of the animal-infecting reoviruses and cytoplasmic polyhedrosis viruses both in appearance and composition (see Chapter 16). These viruses have narrow host ranges in plants and cause dwarfing and malformation. All are transmitted by leafhoppers or plant-hoppers, in which they may persist for life; both clover wound tumor and rice dwarf viruses have been shown to multiply in their vectors.

TOMATO SPOTTED WILT VIRUS (R)/*: */*: S/*: S/Th

This thrips-transmitted virus has a worldwide distribution and causes mosaics and necrosis in plants of many families. It is readily transmitted mechanically, but is very unstable in sap. Its particles are isometric, about 80 nm in diameter (Fig. 2.1i), contain lipid, and are unlike those of any other plant virus though they resemble the particles of some viruses of animals (see Chapter 16).

2.2.3 *Viruses with bacilliform or bullet-shaped particles*

ALFALFA MOSAIC VIRUS
R/1: 1.1/16 + 0.8/16 + 0.7/16: U/U: S/Ap

This widespread virus has bacilliform particles of at least four distinct lengths, the largest one about 58 nm long, and all about 18 nm wide (Fig. 2.1f). These different particles contain the three different parts of the RNA genome of the virus. The virus is readily transmitted mechanically, and also by aphids in the non-persistent manner (Chapter 13) to a wide range of plants, most of which develop mosaic or ringspot symptoms. This virus seems to have affinities with the cucumoviruses.

CACAO SWOLLEN SHOOT VIRUS
/: */*: U/U: S/Cc

This virus has bacilliform particles about 28 nm in diameter (Fig. 2.1c) and of various lengths, mostly about 100 to 150 nm. It has mealy bug vectors and persists in them for a few days, is very unstable in sap and only mechanically transmitted

with difficulty. It has a limited host range and causes mosaic and stem swelling symptoms. It is common in West Africa and Trinidad, where it, and the control measures used against it, cause great losses of cacao beans. Other viruses with particles of similar morphology have been found, but have not yet been studied in detail.

RHABDOVIRUS GROUP
[R/1: 4/2: U/E: S,I,V/Ap, Au, Di, O]

As originally described the rhabdoviruses, which include vesicular stomatitis and rabies viruses, were a group of mammalian viruses with complex bacilliform particles about 50 to 100 nm across and 200 to 300 nm long, and of complex structure (Fig. 5.14i, j). These particles have an outer membrane containing lipid and protein, and an inner helically-constructed core containing RNA and protein; they readily break into shorter bullet-shaped particles. Viruses with similar particles have also been found in plants, fish (trout Egtved virus) and insects (drosophila sigma virus), and are grouped with rhabdoviruses for the time being although the affinities of the viruses within this group need further study. The best-known plant viruses with particles of this type are the aphid-transmitted lettuce necrotic yellows virus, R/1: 4/*: U/E: S,I/Ap (Fig. 2.1b) and the leafhopper-transmitted potato yellow dwarf virus, R/1: 4.3/0.4: U/E: S,I/Au. Unlike most other plant viruses, their particles contain several different protein species. Only a few of these plant viruses can be transmitted mechanically, and most have restricted host ranges but some are important economically.

2.2.4 *Viroids*

Plant viroids are virus-like pathogens that do not produce characteristic nucleoprotein particles. They are readily transmitted by mechanical inoculation, but early reports of transmission by vectors are now discounted. The infective agents appear to be heterogeneous single-stranded RNA about 50×10^3 to 125×10^3 daltons in weight (Diener, 1972). The viroid causing potato spindle tuber disease has been most fully studied; it is seed-borne and it may be identical with the agent causing citrus exocortis disease (Semancik, Magnuson and Weathers, 1973). Chrysanthemum stunt and other stunting diseases may also be caused by viroids.

2.3 Conclusions

It is clear from these brief descriptions that plant viruses are a very heterogeneous collection, but it is fortunate for virologists that many of them fall into definite groups, the members of which share most of their characters. Ways of classifying viruses, and the reasons for not, at present, devising hypothetical hierarchical trees, or taxa 'higher' than the groups we have described, will be discussed in Chapter 12.

The descriptions in this chapter cover about 100 of the better-known plant viruses, and there are perhaps 100 more that we have not described. There are also at least a further 500 transmissible plant pathogens thought to be viruses but whose nature is not yet established beyond reasonable doubt. Many of these will doubtless be found to be viruses, though others may turn out to be new types of pathogens, some of which may be similar enough to viruses to be grouped with them.

Chapter 3

Effects of viruses on plants

3.1 General features

A wide range of abnormalities is caused by viruses, and almost any attribute of the plant may be changed by virus infection, though perhaps the most noticeable effects are changes in colour such as yellowing (*chlorosis*), death of tissue (*necrosis*) and changes in growth form (see the reviews listed in the appendix).

Most plant viruses infect many of their hosts systemically, that is, they spread from the site of infection to other parts of the plant. Not all tissues or cells are necessarily invaded however, and the meristematic tissues of root and shoot, for instance, seem sometimes to escape. Viruses usually persist in a systemically infected plant all its life, so that when plants propagate vegetatively (e.g. by tubers, runners, etc.) the progeny are infected. However when a virus-infected plant produces true seed, this is generally virus-free.

Viruses vary in their ability to infect different numbers of species of flowering plants. At one extreme, tobacco rattle virus can infect at least 400 different species from more than 50 families (Schmelzer, 1957), whereas the only plants known to be susceptible to bean pod mottle virus are legumes. However, most viruses that have been studied carefully are found to infect at least 30 different species. It is not known what determines the host range of a virus. If the virus infects one particular species, it will probably infect other species of the same genus, though we cannot yet predict which other species or other genera will be susceptible. The host range of a virus and details of the symptoms it causes are specific characters of the virus, and these characters are often used, together with others, to identify the virus.

The external appearance of plants is affected by viruses. The effects are, of course, related to changes within the plant. Symptoms are variable and greatly depend on the host plant and how long it has been infected, the virus strain and the environmental conditions. For example, symptoms that are particularly noticeable in plants that have been growing in cool bright conditions since infection may become masked in hot poorly-lit conditions. The time after infection before symptoms appear depends on the virus, the plant and the environmental conditions too; it is commonly a few days or weeks in herbaceous plants though it may be one or more years in woody ones.

The symptoms caused by some viruses are so slight that they are inapparent, even though the plant may contain a large amount of virus, as for example when tobacco is infected with the masked strain of TMV. At the other extreme, the symptoms may be so severe that infected plants are killed, as for example when the Malling Jewel variety of raspberry is infected with raspberry ringspot virus (a nepovirus). A given virus tends to cause the same type of symptom in most of the plants it infects. Many of the viruses that can be transmitted by inoculation of sap usually cause mosaic or necrotic symptoms, whereas several of those that can be transmitted by vectors but not by inoculation of sap cause generalized leaf yellowing.

The course of development of a virus disease in a herbaceous plant may be illustrated by describing the effect of tomato black ring virus (a nepovirus) on an experimentally infected tobacco plant kept in a glasshouse at about 20°C (Fig. 3.1). The lower leaves, about 5 to 15 cm long, can be infected by rubbing them with virus-containing sap. Small spots of dead tissue, known as *local*, or *primary*, *lesions* appear in these leaves about 5 days later. During the next 10 days these lesions enlarge slightly, the tissues around them become *chlorotic* and the lesions may become surrounded by two or three successive rings of dead tissue (*necrotic ringspots*). The next leaves to show symptoms are the tip leaves (*systemic symptoms*). These symptoms first show about 8 days after inoculation as mottling and pale patterns of lines and rings, which may be necrotic. These patterns usually develop in the three or four youngest leaves; leaves formed subsequently look almost normal, although they too contain the virus. The plant is now chronically infected and is said to have *recovered*, a phase which is typical of ringspot diseases. Thus recovery from virus infections in

Fig. 3.1 Tobacco plant cv. Xanthi-nc systemically infected with tomato black ring virus (a nepovirus), showing lesions in inoculated leaves (I), ringspot patterns on older systemically infected leaves (S), and normal-looking younger systemically infected leaves.

plants is quite different from that in animals, where the virus is often completely eliminated from the body of the host.

3.2 Effects on the appearance of plants
3.2.1 *Necrosis*

The most drastic effect of viruses on their host plants, though not the commonest one, is to kill cells, tissues or the whole plant. Many viruses, when rubbed on the leaves of particular plant species, induce necrotic lesions to form in the inoculated leaves, though this does not necessarily stop the spread of virus through the plant. The size and appearance of the lesions depend on the virus (Fig. 3.2) and plant involved, and on the environmental conditions. Necrosis is not confined to chlorophyll-containing tissues, and chlorophyll-less leaves, hypocotyls, or even roots can develop necrotic lesions similar to those in green leaves. In naturally infected plants, sectors

of systemically infected leaves may be killed (Fig. 3.3), and necrotic streaks may appear on the petiole or stem. The extent of the symptoms may indicate the extent of virus infection; for example, the 'spraing' symptoms caused by potato mop-top virus in potato tubers (Fig. 3.4) are arcs of necrotic tissue which develop at the boundary of the virus-infected region at the time (Harrison and Jones, 1971b).

3.2.2 *Effects on the morphology of the plant; amount and pattern of growth*

Most viruses seem to restrict the growth rate of their hosts, which are therefore stunted, but some cause abnormal growth patterns, either by preferentially restricting the development of some parts of the plant, or even by promoting additional growth.

Leaf shape may be altered, either subtly as in grapevines infected with grapevine fanleaf virus (a nepovirus) or more obviously as in tomato plants infected with cucumber mosaic virus, which produce leaves with little or no lamina, giving the so-called fern-leaf or shoestring symptoms (Fig. 3.5). Viruses may induce additional growth; for example, raspberry ringspot virus in cherry and tomato black ring virus in cucumber induce leafy enations to form on the undersides of leaves (Fig. 3.6). Extra growth may also be induced in stems of herbaceous plants; for example, tumours are caused by clover wound tumor virus both in roots (Fig. 3.8) and shoots of *Melilotus* spp., especially when infected tissues are wounded. Less commonly, stem form is affected. For example grapevine fanleaf virus is associated with fasciation in grapevine stems, and cacao infected with cacao swollen shoot virus has swellings in its branches and roots (Fig. 3.7). Various parts of the plant may be misshapen as, for example, the tubers of potatoes infected with potato spindle tuber viroid. The symptoms may be even more unexpected; for example, *Ceiba pentandra*, a West African forest tree which normally has spiny branches, produces spineless ones when infected with cacao swollen shoot virus.

Flowering and seed production may be decreased or even stopped. For example, when broad bean (*Vicia faba*) becomes infected with bean leaf roll virus all flowers absciss, no more are produced, and ovules already fertilized produce small seeds. The effect of viruses on the yield of crop plants is discussed in Chapter 15.

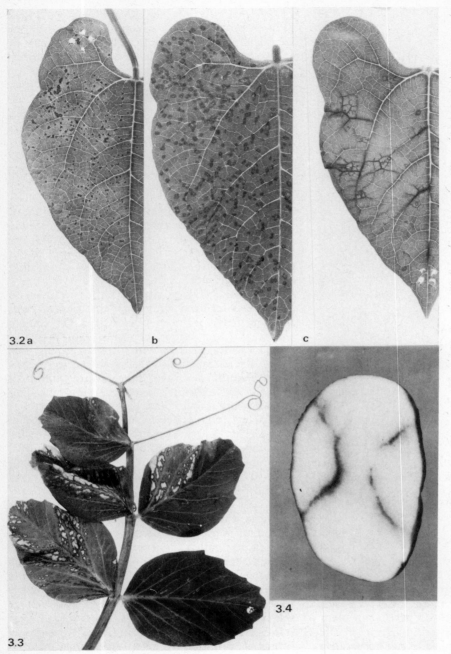

3.2 a   b   c

3.3

3.4

Fig. 3.2 (top) Local lesions in inoculated half-leaves of French bean (*Phaseolus vulgaris*). **a,** tobacco rattle virus; **b,** alfalfa mosaic virus; **c,** tobacco necrosis virus.
Fig. 3.3 (lower left) Necrotic sectors in a pea leaf naturally infected with pea early-browning virus, a tobravirus.
Fig. 3.4 (lower right) Potato *(Solanum tuberosum* cv. Arran Pilot) tuber cut to show brown internal arcs ('spraing') caused by potato mop-top virus.

(see facing page)

Fig. 3.5 (top left) Fern-leaf symptom in tomato systemically infected with cucumber mosaic virus, a cucumovirus
Fig. 3.6 (top right) Leafy enations on the underside of a cucumber leaf systemically infected with tomato black ring virus.
Fig. 3.7 (lower left) Swelling of cacao branchlet induced by cacao swollen shoot virus. (Courtesy R. E. Kenten.)
Fig. 3.8 (lower right) Roots of sweet clover (*Melilotus* sp.) infected with clover wound tumor virus. (Courtesy L. M. Black.)

3.2.3 *Effects on the colour of the plant*

Often one of the first signs of systemic virus infection is vein clearing in the youngest leaves; the veins become yellow or translucent and leaves produced subsequently may show a mosaic, mottle or general yellowing.

In mosaic diseases the leaves develop a patch-work of discrete, and usually unchanging, dark and light green areas (Fig. 3.9). The colour differs not only in the two principal dimensions of the leaf, but also in the different layers of tissue within the leaf. In leaves infected with tymoviruses, the colour differences are caused by the clumping and degeneration of the chloroplasts;

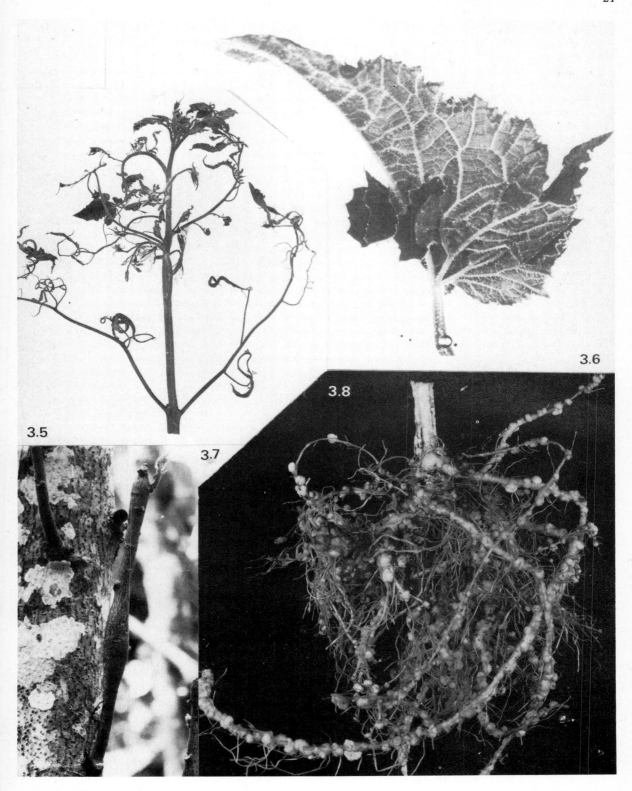

3.5

3.6

3.7

3.8

in the dark green areas the chloroplasts are distributed normally, whereas in the lighter green areas they are clumped, and in the lightest areas they coalesce and degenerate. It has been shown for at least four viruses that the dark green areas contain much less virus than the lighter areas. The

3.9

3.10

3.11

3.12

mosaic patterns caused by turnip yellow mosaic virus in systemically infected leaves are similar to the mosaic patterns in the leaves of naturally variegated tobacco varieties, and virus isolated from bright yellow areas of the mosaic and used to infect other plants gives, initially at least, only bright yellow symptoms, whereas virus from pale yellow areas gives only pale yellow symptoms (Reid and Matthews, 1966; Chalcroft and Matthews, 1967). These workers suggested that each stock of the virus is a mixture of mutable symptom variants, and that individual cells in the leaf primordia are each infected with only one symptom variant. Hence each patch in the mosaic is the clone of cells derived from one primordial cell, and is 'labelled' with one particular symptom variant of the virus. Thus the size and position of the patches of the mosaic presumably indicate the number of leaf primordial cells exposed to infection, and their subsequent growth and division.

Many of the mosaic-causing viruses and some others affect the pigmentation of flowers. The famous 'broken' flowers of tulips are infected with tulip breaking virus (a potyvirus), their petals becoming flecked or longitudinally streaked where the amount of anthocyanin pigment is abnormally small or large (van Slogteren, 1971; Fig. 1.1). Similar symptoms are produced by turnip mosaic virus (another potyvirus) in wallflower (*Cheiranthus cheiri*). A few mosaic-inducing viruses also affect pigmentation of the seed testa (Fig. 3.11).

Other kinds of pattern are caused by other viruses. The nepoviruses and potato mop-top virus are among those that can produce ringspots and line patterns in systemically infected leaves (Fig. 3.10), several other viruses cause yellowing of the veins (vein chlorosis) and still others, for example the beet yellow-net agent, induce yellowing on each side of the veins (yellow vein-banding).

Leaf rolling and yellowing viruses (e.g. potato leaf roll or bean leaf roll viruses) cause general yellowing and, sometimes, reddening of the leaves. The oldest leaves usually show the most obvious symptoms, and typically contain large concentrations of sugars and are sweet to the taste. Such leaves often become cupped or rolled (Fig. 3.12) and those of fleshy plants (e.g. sugar beet infected with beet mild yellowing virus) become turgid and brittle.

3.3 Effects on the histology and cytology of plants

The detailed structure of virus-infected plants has been much studied, initially using the light microscope and more recently by phase contrast and electron microscopy (see reviews listed in section 2 of the appendix).

The effects of hyperplasias, such as tumours or enations, and of necrotic reactions are of course shown in the histology of the plant, but other changes can also be seen. Many viruses affect the vascular tissues. For example, the xylem elements of grapevine infected with grapevine fanleaf virus contain characteristic lignified strands known as endocellular cordons, which are uncommon in virus-free plants. Barley yellow dwarf and other leaf yellowing viruses affect vascular tissue, causing tyloses to form in xylem elements, the phloem cells to die or degenerate, and callose to accumulate on the phloem sieve plates. In sugar beet, tobacco and tomato plants infected with beet curly top virus, the phloem degenerates, hyperplasia of phloem and xylem follows, and supernumerary sieve tubes are produced (Esau, 1941; Fig. 3.13). Different viruses may cause different kinds of tissue to become hyperplastic. For example, the tumours caused by wound tumor virus are the result of abnormal meristematic activity of phloem parenchyma cells, whereas the stem swellings caused by cacao swollen shoot virus in cacao are the result of overproduction of xylem tissue.

When the fine structure of the infected cell is studied, many changes may be seen. Some viruses drastically affect particular organelles. For example, tymoviruses induce marginal vesicles to form in chloroplasts (Fig. 3.14), and infection with some isolates of tobacco rattle virus (a tobravirus) greatly modifies the mitochondria and aggregates them to form so-called *inclusion bodies* (Harrison, Stefanac and Roberts, 1970; Fig. 3.15). However the most characteristic change is the presence of virus particles. These are usually confined to the cytoplasm, though with some

---

Fig. 3.9 (top left) Mosaic in leaf of *Trifolium incarnatum* infected with ononis yellow mosaic virus, a tymovirus.

Fig. 3.10 (top right) Necrotic line-pattern in tobacco leaf infected with potato mop-top virus.

Fig. 3.11 (lower left) Seeds of soybean cv. Iwate-Wasekurome. Top, from healthy plant; bottom, from plant infected with soybean stunt virus, strain A (a potyvirus). Seeds provided by N. Iizuka.

Fig. 3.12 (lower right) Potato (*Solanum tuberosum*) plant showing secondary symptoms of infection with potato leaf roll virus.

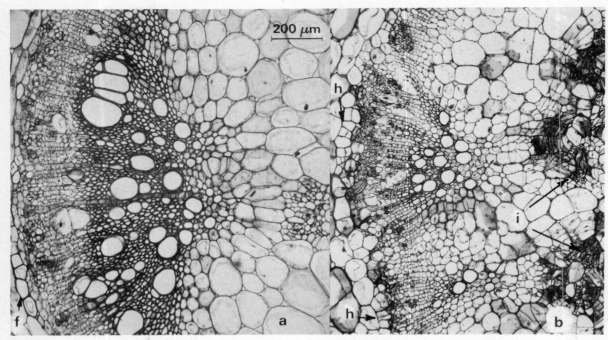

Fig. 3.13 Transverse sections of tomato stems showing vascular tissues. **a**, from healthy plant and **b**, plant infected with beet curly top virus. In **a** the external phloem is fringed by fibres. (f); in **b** by hyperplastic cells (h). **b** shows hypoplastic xylem and abnormally large internal phloem (i). (Courtesy K. Esau; from Esau, 1941.)

viruses they may be found in other parts of the cell; the particles of some strains of TMV are only found in the cytoplasm, whereas those of other strains (e.g. U5 strain) can also be found in the chloroplasts or nuclei. Particles of some rhabdo-viruses, such as potato yellow dwarf virus, are characteristically found in the perinuclear space, that is, the space between the inner and outer nuclear membranes. Nucleocapsids of the virus are found in the nucleus and it seems that they acquire their outer membrane when they bud through the inner nuclear membrane into the perinuclear space.

Virus particles may be scattered in the cytoplasm or aggregated in various ways; the spherical particles of tomato spotted wilt virus are found in membraneous bags, the elongated particles of beet yellows virus occur in bundles which are aggregated end to end to give banded inclusions (Fig. 3.16), and those of tobacco rattle virus may be found packed around mitochondria (Fig. 3.17).

Various types of larger cytoplasmic inclusions are found, and not all of these contain virus particles. Some however consist almost ex-clusively of virus particles, for example the crystal-line plates found in tobacco plants infected with type strain of TMV (Fig. 3.18) or the more rounded plates in plants infected with the rib-grass strain of TMV; these plates consist of layers in which the rod-shaped particles are arranged side by side with their long axes almost at right angles to the face of the layer (Fig. 3.18). The type of inclusion formed is characteristic of the parti-cular virus, but one virus may induce the forma-tion of several quite different types of inclusion, even in the same cell; thus the tomato aucuba mosaic strain of TMV can form hexagonal crystalline plates, spindle-shaped paracrystals, or flat angled-layer aggregates (Warmke, 1968, 1969). Plants infected with some viruses that have isometric particles may contain true three-dimensional crystals of virus particles (Fig. 3.19) or, as with arabis mosaic virus (a nepovirus), the particles may form concentric spheroidal layers. Inclusions consisting of the protein normally

Fig. 3.14 Electron micrograph of section through chloroplasts from a tobacco plant infected with belladonna mottle virus, a tymovirus. Note the abnormal marginal vesicles (V).

Fig. 3.15 Electron micrograph of a thin section of an inclusion body (I) formed by aggregation of abnormal mitochondria in a *Nicotiana clevelandii* cell infected with tobacco rattle virus, isolate CAM/DF. The nucleus (N) is adjacent. (From Harrison, Stefanac and Roberts, 1970.)

found in virus particles are also known, such as the tubular structures occurring in plants infected with some strains of turnip yellow mosaic virus

(Hitchborn and Hills, 1968).

Other cytoplasmic inclusions contain various types of material in addition to virus particles.

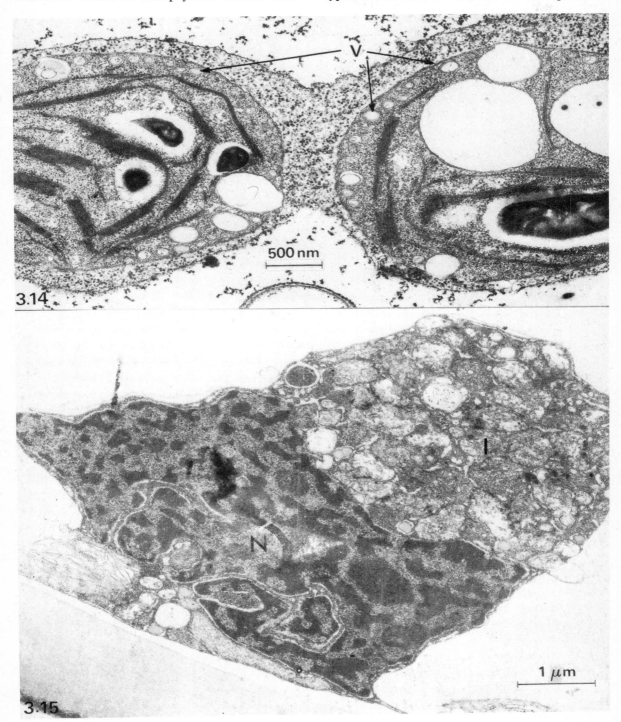

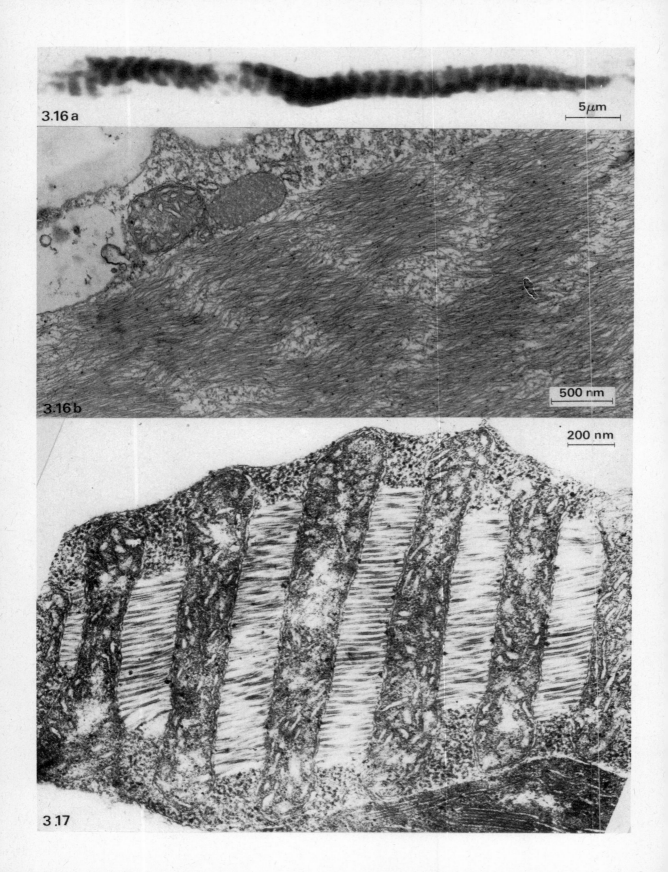

3.16 a

5 μm

3.16 b

500 nm

200 nm

3.17

Fig. 3.16 (top left) Banded inclusion bodies from phloem parenchyma cells of *Beta vulgaris* infected with beet yellows virus. **a,** Photomicrograph of whole inclusion body; **b,** Electron micrograph of part of an inclusion, showing that each band is composed of numerous filamentous virus particles lying side by side. (Courtesy K. Esau; from Esau, Cronshaw and Hoefert, 1966.)

Fig. 3.17 (bottom left) Electron micrograph of the elongated particles of tobacco rattle virus, strain CAM, around and between a group of mitochondria in a *Nicotiana clevelandii* leaf cell.

Fig. 3.18 (below) a, Light micrograph of two imperfect hexagonal crystalline plates in a cell infected with TMV. **b,** A diagram to show the arrangement of virus particles in the component layers of the plates. (Courtesy R. L. Steere; from Steere, 1957.) **c,** Electron micrograph of a section through a TMV plate crystal; the layers are 300 nm thick. (Courtesy R. G. Milne.)

For example, the amorphous 'X-bodies' in TMV-infected cells contain some virus particles but are mostly a mixture of host cell components and many flexuous tubules of unknown composition (Esau and Cronshaw, 1967). Amorphous inclusions are also seen in tissues infected with other viruses, and are readily stained with dyes, such as phloxine (Christie, 1967). By contrast the caulimoviruses induce the formation of refractile round inclusions, which consist of many virus particles embedded in a dense matrix (Fig. 3.21), and contain protein, RNA and DNA. Some inclusions contain few or no virus particles. These include pinwheel inclusions (Fig. 3.22), which are

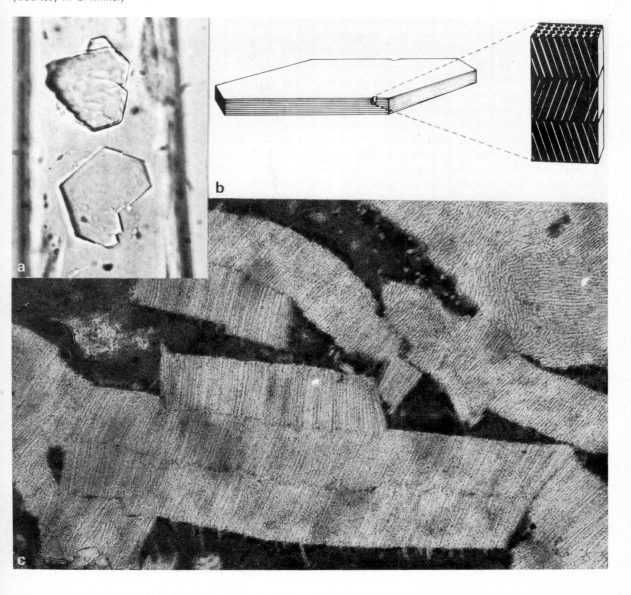

characteristic of many viruses with filamentous particles 700 to 900 nm long such as the potyviruses (Edwardson, 1966; Edwardson, Purcifull and Christie, 1968) and wheat streak mosaic virus, and which are made of protein characteristic of the virus but serologically unrelated to that in the virus particles (Hiebert *et al*, 1971).

A few viruses produce inclusions in the nucleus, as for example the crystalline proteinaceous inclusions in tobacco infected with tobacco etch virus (a potyvirus); these seem free of nucleic acid and do not contain virus particles (Fig. 3.20).

The distribution of inclusion bodies in a plant seems often to reflect the distribution of virus. Those of TMV are found in most tissues, whereas the banded crystalline inclusions caused by beet

yellows virus (Fig. 3.16) occur mainly in the phloem, and the amorphous inclusions in tumours infected with clover wound tumor virus occur only in the pseudo-phloem.

The behaviour and appearance of living infected cells has been studied by light microscopy (Sheffield, 1931; Bald, 1966). In living cells cytoplasmic streaming increases after infection, chloroplasts may clump and coalesce, and in cells around a developing lesion the nucleoli are often much swollen. Sheffield (1931) followed by cinematography the development of the X-bodies induced by a strain of TMV. She found that proteinaceous granules appeared in the cytoplasm soon after infection; these were carried round and fused to form larger bodies, which often became

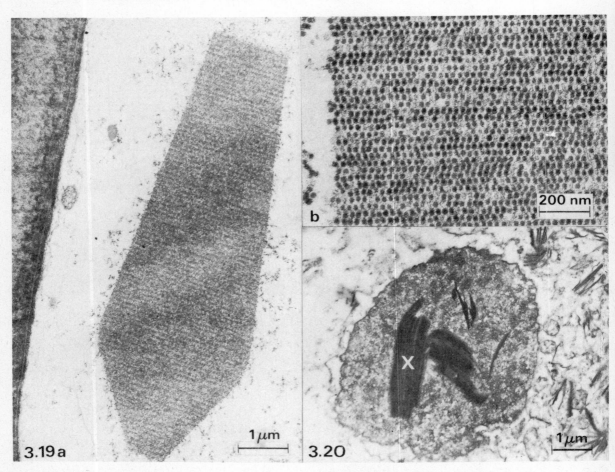

Fig. 3.19 (left and top right) Electron micrographs of **a,** Section of a crystal composed of particles of eggplant mosaic virus, a tymovirus, in the vacuole of a *Nicotiana clevelandii* cell, **b,** Part of the same section at a greater magnification.

Fig. 3.20 (lower right) Electron micrograph of intranuclear crystals (X) in a tobacco leaf cell infected with tobacco etch virus, a potyvirus. (Courtesy M. Rubio-Huertos; from Rubio-Huertos and Hidalgo, 1964.)

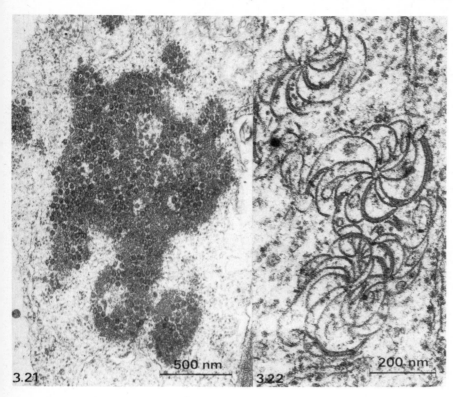

Fig. 3.21 (left) Electron micrograph of a section of an inclusion body, produced by cauliflower mosaic virus (a caulimovirus) in a leaf cell of *Brassica campestris*. The virus particles are embedded in amorphous material.
Fig. 3.22 (right) Electron micrograph of a section through pinwheel inclusions in a tobacco leaf cell infected with potato virus Y.

rounded and sometimes developed vacuoles. Finally these X-bodies broke down and the crystalline inclusions in them were liberated into the cytoplasm.

3.4 Effects on host physiology

Most of a plant's physiological processes can be affected by viruses in one way or another, either directly or indirectly. Some metabolic changes induced by virus infection are those usually associated with senescence. Thus infection of to-bacco with cucumber mosaic virus increases the amounts of alcohol-soluble amino, amide and ammonia nitrogen, and RNA- and DNA-phos-phorus, whereas other organic phosphorus compounds decrease in concentration (Porter and Weinstein, 1960). The total nitrogen content of tobacco leaves is little affected by infection with TMV, and the amount of protein nitrogen is only slightly increased, but the virus can account for up to three-quarters of the protein in the leaf and is made at the expense of normal leaf proteins (Holden and Tracey, 1948; Bawden and Kassanis, 1950).

Several viruses are known to affect chloroplast activities, and some of these changes are specific to particular viruses. For example, changes occur during the period of maximum virus replication in leaves infected with turnip yellow mosaic virus, but not some other viruses (Bedbrook and Matthews, 1973), and thus perhaps reflect the special effects on chloroplasts that accompany the replication of turnip yellow mosaic virus. These particular changes are temporary and include diversion of the products of photosynthesis into organic and amino acids instead of sugars, and an increased activity of phosphoenol pyruvate carboxylase and aspartate amino transferase. Other known virus-induced changes in the amount and activity of chloroplast components seem less specific and are not usually detected until after the main period of virus replication, and include decreases in the rate of photosynthetic carbon fixation, the amount and activity of F1 protein (ribulose diphosphate carboxylase), and the amounts of chlorophyll and $68S$ (chloroplast) ribosomes. In contrast, the concentration of $83S$ (cytoplasmic) ribosomes may be unaltered.

The leaves of plants affected by several yellow-ing virus diseases, such as those caused by beet yellows or beet mild yellowing viruses, contain increased amounts of glucose, fructose and sucrose (Goodman, Watson and Hill, 1965). Apparently the primary cause of this accumulation is not that the sugars are not translocated, but that in infected plants there is a greater 'resistance' to translocation in the petiole, and a greater sugar gradient between the leaves and the roots is needed for the same amount of translocation to occur (Watson and Watson, 1953).

Another seemingly general effect of virus infection is to decrease both the rate at which starch accumulates in infected leaf cells and also the rate at which it is translocated from the leaves during the night. Hence virus/plant combinations that do not produce visible lesions may be shown, by staining decolourized leaves with iodine solution, to form *starch lesions*, which contain either more or less starch than the surrounding tissue, depending on the time of sampling.

The respiration rate of plants is usually increased by infection, sometimes within hours of inoculation, but large increases (up to 50%) typically occur shortly before symptoms appear. The amount by which respiration increases is largest for viruses or virus strains that cause severe symptoms and where the symptoms are only slight no change may be detected. When necrotic local lesions are produced by TMV in *Nicotiana* leaves, the increase seems to result mainly from the uncoupling of respiration from oxidative phosphorylation (Merrett and Bayley, 1969), although with some other viruses there is evidence that infection activates enzymes of the pentose phosphate pathway (Dwurazna and Weintraub, 1969). When necrotic local lesions form in inoculated *Nicotiana* leaves, the upper non-inoculated leaves become more resistant to infection; the lesions produced by inoculating them with TMV are fewer and smaller than those induced in previously virus-free plants. The development of this resistance is correlated with an increased peroxidase activity in the non-inoculated leaves. It is thought that the enhanced peroxidase activity increases the rate at which quinones can be produced and cause necrosis at infection points, and hence that the spread of virus is restricted sooner (Simons and Ross, 1971b).

Woods noted as early as 1900 that the oxidase systems of tobacco were deranged by TMV infection, but such effects on oxido-reductase enzymes seem not to be specific, and Novacky and Hampton (1968) found that five different viruses all caused an increase in peroxidase activity in various plants. However, they found no novel enzymes or isozymes, and concluded that the changes were quantitative, not qualitative. Other enzyme systems of the host are also altered by virus infection. Although there is so far no unequivocal evidence of a virus-coded enzyme in infected plants, such as the RNA-dependent RNA polymerase found in virus-infected bacterial cells, there are some promising candidates (see Chapter 11). The RNA polymerases found in particles of lettuce necrotic yellows and wound tumor viruses, in cucumber plants infected with cucumber mosaic virus and in barley plants infected with brome mosaic virus may well be virus-coded (see Chapters 9 and 11). The activities attributed to the polymerases obtained from infected plants are much increased by infection but it is not yet known whether the enzymes occur in small amounts in virus-free plants.

The viruses causing enations, or changes in the pattern of growth, doubtless affect the growth hormones. Thus it is interesting that the stunting effects of some viruses can be at least partially reversed by applications of gibberellic acid (Maramorosch, 1957; Chessin, 1958).

Many attempts have been made to devise simple histological or biochemical tests that can be used to detect virus infection. Such tests include staining for callose formation, and colour reactions for sugars or polyphenols, but in general they have had limited success because quite different abnormal conditions can have metabolic effects on plants resembling those of virus infection. However, although not precise enough for some purposes, the Igel Lange colour test for callose accumulation in phloem sieve tubes has been used on a large scale in several countries to judge the health of stocks of potato tubers. Slices from the tubers are stained with 1% resorcin blue and the callose plugs counted using a light microscope. Many of the tubers from plants infected with potato leaf roll virus during the preceding growing season have an increased number of stained plugs in young sieve tubes near the cambium at the end of the tuber attached to the mother plant (heel end), although those at the extreme heel end should be discounted, and very

---

Fig. 3.23 Virus-like symptoms that are thought not to be caused by viruses. **a,** Raspberry leaf with yellow blotches caused by the raspberry leaf mite (*Eriophyes gracilis*); **b,** leaf chimaera of holly (*Ilex aquaefolium*); **c,** tomato plant grown in soil containing traces of the hormone weedkiller 4-chloro-2-methylphenoxyacetic acid; **d,** leaf from a sugar-beet plant deficient in manganese, showing brilliant interveinal chlorosis.

a

b

c

d

recent infections are often not detected (de Bokx, 1967).

3.5 Virus-like symptoms with other causes

Not all virus-like symptoms in plants are caused by viruses. Many transmissible diseases with virus-like symptoms seem to be associated with mycoplasma-like organisms (see Chapter 17). Some effects of bacterial infection may also be confused with virus infection. Bacteria such as *Agrobacterium tumefasciens* and *Corynebacterium fascians* cause plant tumours, and *Pseudomonas*

aptata, which is common on the leaf surface of various plants, can be transmitted by rubbing inocula on leaves and will give virus-like lesions (Yarwood *et al*, 1961). Some eriophyid mites and aphids cause feeding damage resembling virus symptoms, and some kinds of nutrient imbalance have effects similar to virus diseases (Fig. 3.23a, d). There are also a multitude of genetical abnormalities and chimaeras which give virus-like changes in the colour, distribution of pigments and growth form of plants and, finally, there are the deformities produced by hormone weed-killers (Fig. 3.23b, c).

Chapter 4

Experimental transmission

4.1 General features

Viruses are by definition transmissible, and in this chapter we describe some of the ways in which plant viruses can be transmitted experimentally. Virus spreads from cell to cell within a plant by way of cytoplasmic connections called plasmodesmata, and is carried from one part of the plant to another in the vascular tissue. Thus viruses are passed on from mother plants to their vegetative progeny, and they are also passed to other susceptible plants whose cells come into intimate contact with infected cells, as happens at graft unions. Grafting is one of the oldest experimental methods for transmitting viruses, and in appropriate circumstances is the most successful. Indeed many pathogens that are thought to be viruses have as yet been transmitted only by grafting.

To infect an intact plant other than by grafting, infective particles must be introduced through a wound in the outer cell wall of the plant. This outer wall is normally impermeable to viruses and infection does not occur when undamaged plants are gently sprayed with virus preparations. However, protoplasts can be obtained by treating plant tissue with enzymes to remove the cell walls, and become infected when virus particles or virus nucleic acid molecules are added to the special medium in which the protoplasts are suspended (Takebe and Otsuki, 1969).

In nature the commonest method of transmission is by vector organisms, although viruses are sometimes transmitted when shoots of infected and healthy plants rub together, and occasionally when their roots come into contact and perhaps form a graft union. A few viruses are transmitted to healthy plants that are pollinated by virus-containing pollen. All these natural mechanisms are exploited by virologists, who usually transmit the viruses either by rubbing plants with preparations of infective particles (*mechanical transmission*), or by using the organisms which transmit the viruses in nature (*vector transmission*). In addition, some plant viruses multiply in their vectors, and can also infect lines of cultured cells obtained from the vectors.

Progress in the study of viruses, their particles, properties and behaviour depends to a large extent on finding a way of transmitting the viruses from host to host. Four types of plant host are desirable: one in which a stock culture of the virus can be maintained for long periods; a second in which characteristic symptoms are produced, and which can be used to aid detection and diagnosis; a third which is a source of large amounts of virus; and a fourth for assaying infectivity, preferably by counting visible primary infections (e.g. local lesions in inoculated leaves). Different species will usually be used for these different purposes; for example, alfalfa mosaic virus can be maintained for long periods in lucerne, propagated in large amounts in tobacco plants, assayed in inoculated French bean leaves (Fig. 3.2b), and distinguished from most other viruses by the symptoms produced in plants of tobacco, French bean and *Chenopodium amaranticolor*. How these different kinds of host are used will become evident in this and later chapters. Here we will discuss in more detail the ways of transmitting viruses experimentally.

4.2 Inoculation of plants by the experimenter

4.2.1 *Graft transmission*

Grafting is an ancient horticultural practice in which the cut surfaces of tissues of different plants are brought into intimate contact so that they can establish a union. Usually grafting is done by transplanting a distal part of one plant (the *scion*) on to the root-bearing portion of another (the *stock*). When either the scion or the stock is infected, the virus usually invades the healthy partner, and may cause visible symptoms.

The first reports of what we now know was the graft transmission of viruses are from the seventeenth century when Dutch tulip growers discovered that the ability to produce the much-prized broken flower colours could be passed from one tulip to another by grafting the bulbs together. Viruses are also unwittingly and efficiently transmitted by horticulturalists when they multiply vegetatively propagated plants by grafting scions onto rootstocks. These practices are largely responsible for the widespread and frequent occurrence of viruses in woody plants such as apple, pear, cherry, plum, grapevine and citrus.

Viruses are graft-transmitted most efficiently when the stock and scion unite perfectly, which happens only when their cambia are in contact and their tissues are compatible. Various factors seem to control compatibility. For example, taxonomically distantly related species are mostly incompatible, and monocotyledons cannot be grafted in the usual sense, although bulb grafting, as already mentioned, may be possible. Virus infection of one of the grafted partners is another possible reason for failure to obtain a graft union. Another more complex phenomenon induced by grafting is found in citrus, where many of the commercial varieties are unaffected by citrus tristeza virus unless they are grafted; citrus tristeza virus transmitted by aphids to sweet orange or sour orange trees causes few or no symptoms, but kills trees consisting of a sweet orange scion on a sour orange stock (Bennett and Costa, 1949).

However it is not always necessary to have a good graft union for virus transmission, and viruses can be transmitted by means of approach grafts between such dissimilar species as *Chenopodium amaranticolor* and grapevine (Cadman, Dias and Harrison, 1960), a combination which does not form a true graft union but produces callus tissue on the grafted surfaces. Conversely a good graft union does not guarantee virus transmission. For example, tobacco rattle virus in potato is less frequently transmitted by grafting than many other potato viruses, perhaps because it is often only partially systemic in tissues.

There are many different ways of grafting (Garner, 1958), and some of those most useful for transmitting viruses are shown in Fig. 4.1. The two commonest are called detached-scion grafting and approach grafting. With *detached-scion grafting*, it is better to transmit the virus from scion to stock than from stock to scion because this will distinguish virus diseases from physiological disorders, such as nutrient deficiencies, which might be induced in a normal scion grafted on to an affected stock. Using this method, some types of scion are better than others for some types of plants: shoots with wedge-shaped ends for

herbaceous plants (Fig. 4.1c); shield buds, or occasionally bark patches, for woody ones (Fig. 4.1a); detached shoots in tubes of water (bottle grafts) for raspberry plants (Fig. 4.1b); part of the leaflet petiole of runnerless strawberry plants; and cores of tissue taken with a cork borer for potato tubers. In *approach grafting*, both partners are grown on their own roots. Fig. 4.1d shows the type of approach graft used to transmit grapevine fanleaf virus from *Chenopodium amaranticolor* to grapevine, and Fig. 4.1e shows a technique popular for grafting strawberry runners. In all kinds of grafting the tissues are usually held together until union has occurred, by raffia or grafting tape bound around the joint. With herbaceous plants in particular, the grafted surfaces should be kept moist, and water loss minimized by removing unnecessary leaves from the scion and keeping it, or the whole grafted plant, in humid conditions.

Viruses that reach a large concentration and are generally distributed through plant tissues may cross the graft junction in as little as two days, but viruses occurring in smaller concentration or having a restricted distribution in the tissues may take a week or more, and viruses in woody plants out of doors may not be transmitted for several months. Leaf symptoms usually appear first in young leaves at the tip of the shoot of the initially virus-free partner. This may take only one or a few weeks in herbaceous plants, but may need an intervening dormant period and take a year or more in woody ones. However symptoms are usually produced when the test partner is pruned to induce shoots to grow in leaf axils near the site of the graft.

4.2.2. *Transmission using dodder*

In principle, transmission by grafting and transmission with dodder are similar, since both

Fig. 4.1 Types of graft used for transmitting viruses. **a,** Shield bud graft, used with woody plants; the scion bud is inserted under the bark of the stock. The sliver of wood on the underside of the bud shield may be removed before insertion. **b,** Bottle graft, used with raspberry; the stem of the leafy scion is kept in a tube of water. The tube can be wrapped in aluminium foil (not shown) to minimize algal growth in the water. **c,** Wedge graft, used with herbaceous plants such as potato; the graft is covered with crepe rubber (not shown). **d,** Spliced approach graft, used for example to transmit viruses from *Chenopodium amaranticolor* to grapevine. The shoots of rooted plants are pared to expose their cambium and the cut surfaces bound together. **e,** Tongue approach graft, used with strawberry runners; the graft is bound with crepe rubber (not shown). Drawings based on those of Garner (1958).

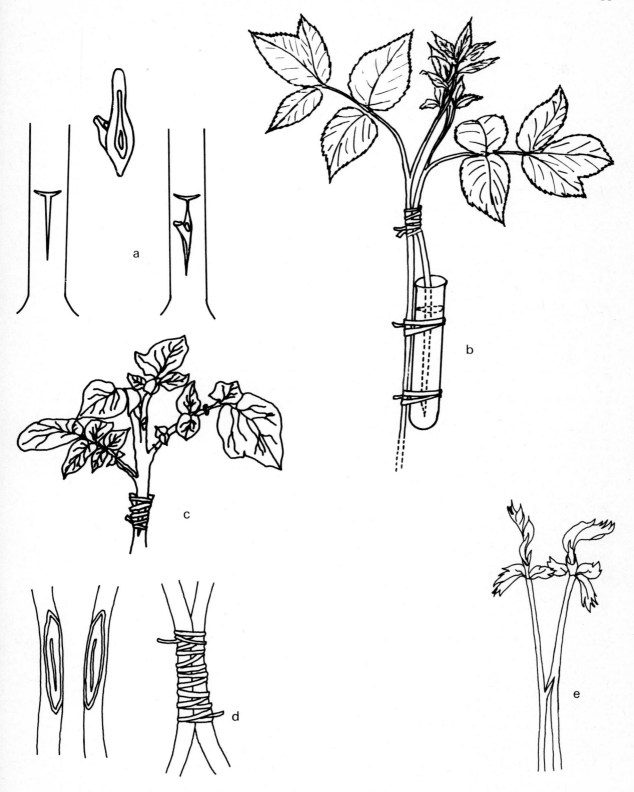

methods are a way of obtaining intimate contact between the cells and vascular systems of different plants. Dodders (*Cuscuta* spp.) are parasitic plants with slender straggling stems which, where they touch the stems of host plants, produce root-like haustoria that connect with the vascular bundles of the host. Thus a single dodder stem can be used to join two plants (Fig. 4.2), and as Bennett (1940) first showed, some viruses will move through the dodder bridge from one plant to the other. About twenty species of *Cuscuta* have been used to transmit viruses, but those with wide plant host-ranges are the most popular. Of these, *C. campestris* transmitted twenty-four out of thirty-nine viruses tested and *C. subinclusa* transmitted twelve out of twenty-three (Schmelzer, 1956). Usually, viruses which multiply in the dodder itself, for instance cucumber mosaic and tobacco rattle viruses in *C. campestris*, are more efficiently transmitted than those which are merely transported passively. Dodder is mostly used for transmitting viruses that cannot be trans-

Fig. 4.2 Transmission of virus by dodder. **a,** The dodder was established on the virus-infected sugar beet plant (left), and then trained onto a healthy tobacco plant (right). **b,** Dodder established on a *Vicia faba* plant. **c,** Necrosis produced in a *V. faba* stem near the site of inoculation of cucumber mosaic virus via dodder. (**b** and **c** Courtesy K. Schmelzer; from Schmelzer, 1956.)

mitted mechanically or by vectors, and for transmitting viruses between plants which will not form a graft union, as for example when they are too distantly related.

Dodder seed can remain viable for at least ten years, and germinates best when placed on the soil surface among seedlings of a host plant. The dodder seedlings may be used directly, or alternatively a dodder plant that is established on a virus-free host may be used as a source of shoots. The dodder seedlings or shoots are placed in the axils of the plant to be tested, and when the dodder

Fig. 4.3 (top) Manual inoculation. Before inoculation the leaves are dusted with 500-mesh carborundum (in small jar at right with neck covered by muslin). The virus-containing inoculum in the mortar (foreground) is then gently wiped over the leaf surface using a forefinger dipped in the inoculum.

Fig. 4.4 (lower) Air-brush inoculation of maize. The inoculum, containing corundum abrasive, is sprayed onto the plants, using a pressure of 4.2 to 5.6 kg/cm² (Courtesy C. C. Wernham; from MacKenzie, Anderson and Wernham, 1966.)

is well established on the suspect plant, the shoots are trained on to the indicator plants (Fig. 4.2a), or shoots are detached, placed in the axils of the indicators and allowed to establish there. When a virus causes necrotic symptoms in the indicator plant, its spread away from the dodder haustoria can be particularly well seen (Fig. 4.2c), but with most viruses the symptoms appear first in the young leaves at the tip of the shoot of the indicators. Some viruses, such as cucumber mosaic and tobacco rattle viruses, cause symptoms in infected dodder plants.

One possible source of confusion when working with dodder is the dodder latent mosaic virus, which can be seed-borne in *Cuscuta campestris*, and causes symptoms in several other species when they are parasitized by infected dodder (Bennett, 1944).

4.2.3 *Transmission by mechanical inoculation*

There are many different ways of inoculating plants mechanically with virus-containing liquids. The earliest plant virologists injected sap from one plant into another or pricked the test plant through a drop of inoculum, using a pin. These methods are very inefficient and nowadays most people rub the inoculum over the leaf surface, usually with a finger (*manual inoculation*; Fig. 4.3) or a pad of cotton gauze, but a spatula or even a stiff brush may be used. Some infections occur through wounds in leaf hair cells but other types of epidermal cell can also be infected.

Some plants that are difficult to infect by these methods are more readily infected when the inoculum, containing a fine abrasive, is sprayed onto the plant using a pressurized air-brush (Fig. 4.4; MacKenzie, Anderson and Wernham, 1966), and this is a relatively quick way of inoculating large numbers of plants. The success of mechanical inoculation depends on many factors, the most important of which we shall now discuss.

(a) HOST CELL FACTORS

Mechanical inoculation produces small temporary wounds in the plant through which virus particles can enter and infect the cells. Obviously the wounds must be so slight that the wounded cells do not die. However, ways of making suitable wounds are extremely inefficient, and this is probably the main reason why 10^5 or more virus particles must be applied to a leaf to cause each infection.

The viruses that are most readily transmitted by mechanical inoculation are those that reach the largest concentrations in plants and can infect the epidermis as well as other tissues. Typically, these viruses cause mosaic, mottle or ringspot symptoms. By contrast, viruses that seem restricted to the phloem or xylem are not mechanically transmissible by usual methods, probably because the epidermal cells are not susceptible. However some, such as beet curly top virus, can sometimes be transmitted by making deep pin pricks to the vascular bundles through infective sap. Other possible reasons for the failure of mechanical inoculation are that the infective particles are unstable or insufficiently concentrated in the inoculum, or that the inoculum contains inhibitors of infection. Later in this chapter we discuss special techniques for use in such circumstances.

(b) TEST-PLANT FACTORS

Some plants can be infected by mechanical inoculation much more readily than others – in general, herbaceous plants more so than woody ones, dicotyledons than monocotyledons, and young seedlings than older plants. Some species are particularly easy to infect and are susceptible to many viruses, and some families, such as the Solanaceae and Leguminosae, seem to contain more of these species than other families. Among the most useful species for detecting, culturing or assaying a wide range of viruses are:

Chenopodiaceae	*Chenopodium amaranticolor* Coste and Reyn.
	C. quinoa Willd.
Cucurbitaceae	*Cucumis sativus* L. (cucumber)
Gramineae	*Triticum aestivum* L. (wheat)
	Hordeum vulgare L. (barley)
Leguminoseae	*Phaseolus vulgaris* L. (dwarf, or French, bean)
	Vicia faba L. (broad bean)
	Vigna sinensis L. (cowpea)
Solanaceae	*Nicotiana tabacum* L. (tobacco)
	N. clevelandii Gray
	Petunia hybrida Vilm.

The susceptibility of a plant to infection depends greatly on its physiological state. Plants grown in shade conditions, with delicate, pale green, juicy leaves are more susceptible than those grown in bright sunlight, which have tough dark green leaves with a smaller water content. Also, the chances of infecting a plant can be altered dramatically by changing the environmental conditions immediately before or after inoculation. When plants are kept in warm air immediately before inoculation their susceptibility increases, but the same treatment after inoculation usually decreases the number of infections (Table 4.1);

Table 4.1 Effects of keeping French bean plants in darkness or at 36°C on susceptibility to infection with tobacco necrosis virus

Period when treatment applied[1]	Number of lesions per leaf		
	Plants in darkness at about 20°C (Bawden and Roberts, 1948)	Plants at 36°C in daylight (Kassanis, 1952) Exp. 1	Exp. 2
2 days before inoculation	111	46	—
1 day before inoculation	98	29	—
Untreated control	26	2	69
1 day after inoculation	32	—	0
2 days after inoculation	18	—	—

[1] At other times the plants were in daylight at about 20°C.

the time of treatment needed to get the maximum response is longer, and the size of the response is larger, for initially resistant than for initially susceptible plants. Darkening the plants before inoculation usually increases their susceptibility (Table 4.1; Bawden and Roberts, 1948), but the effect of darkening is complex, and a short period of light after the dark period and before inoculation increases susceptibility even more (Helms and MacIntyre, 1967). Not surprisingly there are diurnal changes in susceptibility, presumably in response to changing temperature, light intensity and period of illumination. For example, in Auckland, New Zealand, some plants are most susceptible to infection in the late afternoon and least susceptible near the end of the night (Matthews, 1953).

Other treatments that can increase or decrease the number of infections obtained, depending on their timing and the condition of the plants, include dipping inoculated leaves briefly in hot water (50°C), rinsing off the inoculum with a stream of tap water, drying the leaves after inoculation and allowing the leaves temporarily to wilt. These and other effects were reviewed by Yarwood and Fulton (1967).

Leaves are the most convenient, but not necessarily the most susceptible, part of the plant to inoculate mechanically. Yarwood (1960) found

that other parts (e.g. hypocotyl, root, etc.) were sometimes more susceptible than leaves and that, when different parts were inoculated, the virus spread through the plant at different rates and caused somewhat different symptoms.

(c) INOCULUM FACTORS

Abrasives Gentle abrasive powders, such as carborundum (silicon carbide), corundum (aluminium oxide) or Celite (diatomaceous earth), are often used to increase the number of infections produced by inocula, usually by an amount equivalent to a 20 to 50-fold increase in virus concentration (Fig. 4.5a). Carborundum and corundum are usually dusted on the leaves before inoculation because they soon sediment in aqueous suspension, but Celite, which settles less rapidly, is generally mixed in the inoculum. The action of abrasives is presumably to increase the number of infectible wounds produced at inoculation, but it may be more than this. The effect of the abrasive depends on its size and on the amount used; for example 500 to 600 mesh carborundum gives excellent results when dusted on leaves before inoculation or mixed (about 50 to 100 mg/ml) in the inoculum. When abrasives have been used, the inoculated leaves are liable to wilt and must be kept for a few hours in a humid atmosphere free from draughts.

Ionic content of the inoculum The type, amount and proportion of different ions in the inoculum greatly affect its infectivity. For instance phosphate, especially at 0.02 to 0.1 M in the range pH 7.0 to 8.5, usually greatly favours infection. However different viruses and hosts respond differently to these factors, and the optimal conditions have to be found for each combination of virus and host. The infectivity of most viruses is inactivated at low pH values; thus the sap of plants such as grapevine (pH 3.5) may inactivate the particles of many viruses unless infected leaves are ground in alkaline solutions.

Inhibitors of infection It is difficult or impossible to transmit viruses from some plant species by mechanically inoculating their sap to susceptible test plants of certain other species. The commonest reason for this is that the sap of the source plant contains inhibitors of infection. These inhibitors do not inactivate virus particles permanently but in some way prevent infection. They include compounds of many kinds. Some of the

Fig. 4.5 Effects of inoculum components on infection. **a,** *Nicotiana glutinosa* leaf manually inoculated with TMV; left half-leaf, previously rubbed with Celite; right half-leaf, previously rubbed with water. (Courtesy B. Kassanis; from Kalmus and Kassanis, 1945.) **b,** *Phaseolus vulgaris* leaf manually inoculated with tobacco necrosis virus. The inoculum applied to the right half-leaf contained sugar-beet sap, a powerful inhibitor of infection. (Courtesy B. Kassanis.)

most potent are proteins or polysaccharides, such as are found in leaf sap of sugar beet, *Chenopodium* spp., *Phytolacca* spp. and *Dianthus* spp. (Fig. 4.5b). These inhibitors have little effect on infection of the species from which they were obtained (Table 4.2). With viruses whose particles occur in large concentration, the effects of inhibitors can be avoided by diluting the inoculum (Table 4.2), because the effect of the inhibitor usually decreases more quickly with dilution than does virus infectivity. Alternatively the virus particles can be purified to separate them from the inhibitor, or assay species used that are not affected by the inhibitor. Some species, such as cucumber, *Chenopodium amaranticolor* and *Gomphrena globosa* tend to be less affected by inhibitors than others. Thus although it is difficult to transmit alfalfa mosaic virus by mechanical

inoculation of sap from *C. amaranticolor* to *Nicotiana* spp. because there are inhibitors in the *C. amaranticolor* sap, the virus can be readily transmitted first to *Gomphrena globosa* and then from *G. globosa* to *Nicotiana* spp.

Virus inactivators Sap from the leaves of many woody plants contain tannins, which in certain conditions combine with and precipitate virus particles, and hence prevent infection. Loss of infectivity can be avoided by grinding the leaves in pH 8 to 9 buffer, or in nicotine (Table 4.3) or caffeine solutions, because the combination is greatly decreased in alkaline conditions. Alternatively a protein, such as hide powder, which apparently competes with the virus particles for the tannin, can be added to the extraction medium; this method enables cacao swollen shoot virus to be transmitted from cacao to cacao. Another way of obtaining infective virus inocula from tannin-rich plants is to prepare the inoculum from tissues in which the concentration of tannin is relatively small, such as corollas or young roots.

Plant sap usually contains powerful oxidase enzymes, and the particles of some viruses are

Table 4.2 Importance of the assay-host species in determining the effect of inhibitors of infection occurring in plant sap (Gendron and Kassanis, 1954)

Assay host	Composition of inoculum	Dilution of inoculum mixture (in water)		
	(CMV = tobacco sap containing cucumber mosaic virus)	Undiluted	1/10	1/100
Datura tatula	1 vol. CMV+1 vol. *D. tatula* sap	18[1]	14	2
	1 vol. CMV+1 vol. *B. vulgaris* sap	0	0	1
	1 vol. CMV+1 vol. water	17	12	3
Beta vulgaris	1 vol. CMV+1 vol. *D. tatula* sap	2	8	3
	1 vol. CMV+1 vol. *B. vulgaris* sap	55	32	1
	1 vol. CMV+1 vol. water	81	38	5

[1] Figures are mean numbers of cucumber mosaic virus lesions produced per half leaf.

inactivated by products of oxidase action. This kind of inactivation (Chapter 10) can be prevented by adding reducing agents (e.g. dithiothreitol, mercaptoethanol, thioglycollic acid, cysteine hydrochloride) or chelating agents (e.g. sodium diethyldithiocarbamate) to the extraction medium. Further details are given in Table 4.3.

The nucleic acid of some so-called 'defective' viruses is inadequately protected, so that when leaves that contain these viruses are ground, the leaf nucleases and other inactivating agents in the sap quickly inactivate the virus nucleic acid, in some instances almost instantaneously. However, much infectivity can be preserved by taking suitable precautions. Bentonite clay may be added to the extraction medium to adsorb the nucleases, or the leaves may be ground in alkaline buffer (pH 9.5), in which the enzymes are inactive. Alternatively the sap may be deproteinized during its release from the plant by grinding the leaves at 2°C in dilute buffer plus an equal volume of water-saturated phenol, and centrifuging the resulting slurry to separate the less dense aqueous phase which contains the virus nucleic acid, from the denser phenol phase and leaf debris. The aqueous fraction is shaken successively with several aliquots of cold diethyl ether to remove residual phenol, a stream of nitrogen is bubbled

through it to remove any remaining ether, and the preparation is then ready to inoculate. The results are often dramatic (Table 4.3). The phenol either denatures or dissolves the various inactivating agents, and it also extracts nucleic acid from intact nucleoprotein virus particles. In addition it denatures proteinaceous inhibitors of infection.

4.3 Transmission using vectors

The vectors of plant viruses include insects, mites, nematodes and fungi, but any one virus is usually transmitted by only one or a few species which, with few exceptions, are members of the same taxonomic group.

Vectors are used in the laboratory not only to transmit viruses to plants, but also in studies of the behaviour of the virus in the vector (Chapter 13). In Chapter 7 we discuss the analysis of vector transmission experiments, and here we outline how the tests are done.

4.3.1 *Culturing vectors*

To ensure that virus-free vectors are available for experiments it is advisable to culture them in virus-free conditions and to check the vector cultures regularly to confirm that they are virus-free. The vectors are best cultured on plants immune to the virus being studied and also to any other likely contaminant viruses.

Many species of vectors reproduce best on vigorously growing plants, and therefore need subculturing to fresh plants at regular intervals to keep them in good condition. Insects and mites should be cultured on plants in well-ventilated cages (Fig. 4.6a, b), because in unventilated cages they are apt to drown in the water that condenses, and moulds grow freely. Many species of aphids are *parthenogenetic* and *viviparous*, and multiply rapidly. Wingless (*apterous*) forms, which are more readily handled than winged (*alate*) ones, usually predominate when the insects are not crowded and the plants are kept in continuous light. Other insects, such as leafhoppers and beetles, multiply more slowly than aphids and can be more difficult to culture.

Some root-feeding nematode vectors, especially the parthenogenetic species, complete their life cycles in a month or two in glasshouse conditions and can be cultured satisfactorily. The soil must be kept moist but neither dry nor water-logged, for the nematode species known to be vectors

Table 4.3 Some substances that preserve the infectivity of unstable or defective viruses when added to the leaf extraction fluid

Substance[1]	Virus	Source and assay species	Increase in infectivity of inoculum	Mode of action
Sodium diethyldithio-carbamate (0.01 M)	Cucumber mosaic virus	*Nicotiana tabacum/ Chenopodium amaranticolor*	×5–500	Chelates copper and inhibits poly-phenoloxidase, thus preventing the formation of *o*-quinones, which are virus inactivators
Thioglycollic acid (0.1%)	Cucumber mosaic virus	*Nicotiana tabacum/ Chenopodium amaranticolor*	×10–500	Reduces *o*-quinones
Nicotine (1%)	Raspberry ringspot virus	Raspberry (*Rubus fruticosus*)/ *Chenopodium quinoa*	×40–400	Removes tannin and prevents combination between virus and tannin by keeping pH above 8.0
Phenol (water saturated) at 4°C+equal vol. 0.02 M phosphate buffer, pH 7.4	Defective strain of tobacco necrosis virus	*Phaseolus vulgaris/ P. vulgaris*	×40–200	Removes or inactivates leaf nucleases and other inactivators of virus RNA, which is found in the aqueous phase
Bentonite (2.5%) in 0.06 M phosphate buffer, pH 8	Defective strain of tobacco necrosis virus	*P. vulgaris/P. vulgaris*	×20–200	Adsorbs inactivators of virus RNA, especially nucleases
GPS buffer (0.1 M glycine, 0.05 M K_2HPO_4, 0.3 M NaCl, pH 9.5)	TMV-RNA	*Nicotiana tabacum/ Datura stramonium*	×200 or more	Inhibits leaf nucleases

[1] Typically, extracts are made by grinding 1 g leaf in 5 ml fluid.

cannot survive in air-dry soil, and are adversely affected in water-logged soil.

Vector fungi are best cultured on the roots of plants grown in moist sterilized sand which is easy to remove when preparing zoospore suspensions, but if the cultures do not thrive in these conditions, mixtures of peat and sand, or sterilized soil, may have to be used. Precautions should be taken to avoid contaminating the cultures by water splash, and new cultures should be started at regular intervals.

Vector species that multiply very slowly (e.g. large nematodes such as *Xiphinema diversicaudatum*) or species that, for unknown reasons, cannot be cultured satisfactorily (e.g. some leafhoppers), are usually collected when needed, from naturally infested plants or soil.

4.3.2 *Handling vectors*

Some of the tools used for handling vectors are shown in Fig. 4.6c. These are used to transfer individual organisms being tested as vectors to or from virus source, or test (bait) plants.

Aphids usually are moved on the tip of a squirrel-hair artist's paintbrush moistened with saliva. First the abdomen of the aphid is lightly tapped, to make it withdraw its stylets from deep in the leaf, and then it is picked up on the brush and gently deposited on a fresh leaf. When working with viruses whose particles are in large concentrations in leaves, the brush may become contaminated with virus-containing sap and, to avoid touching the plant, it is best to put the aphid on a small piece of paper placed on a leaf of

Fig. 4.6 Apparatus used for culturing and handling vectors. **a,** Cage of transparent plastic used for culturing aphids; note gauze-covered ventilation holes. **b,** Small cages used to confine planthoppers on cereal plants. **c,** Tools used to pick up vectors. Top, dental pulp-canal file with bent tip, used for nematodes; middle, artist's squirrel hair paint-brush reduced to one hair, used for mites; bottom, artist's paint-brush, used for aphids (scale is in cm). **d,** Small aspirator used for collecting leafhoppers. One end of the suction tube is covered with muslin.

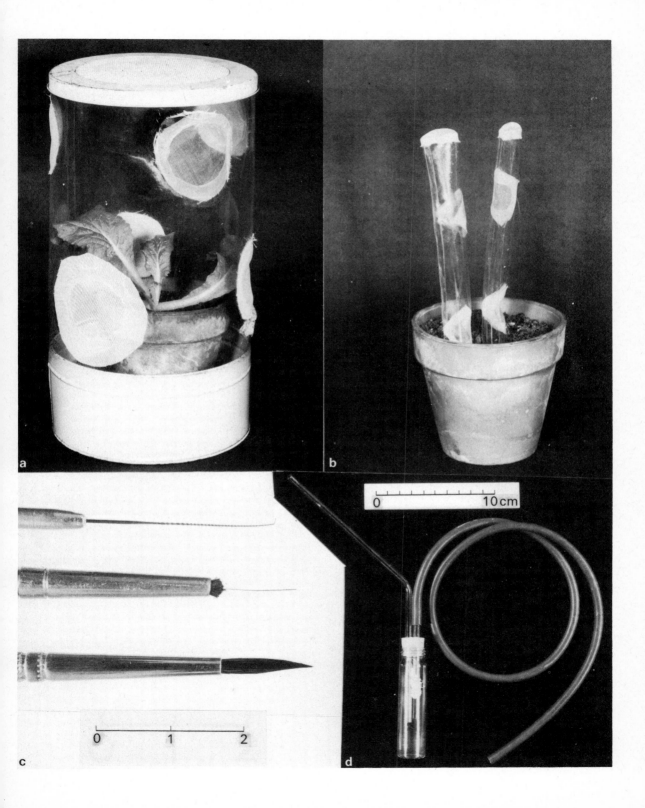

the test plant. The aphid will then walk on to the test plant. When aphids start probing they lay their antennae back along the abdomen.

Leafhoppers are too active to be moved with a paintbrush, and a small aspirator (Fig. 4.6d) is often used. Mites are very delicate, and a paintbrush trimmed to a single hair can be used for them.

Nematodes (Southey, 1965) and other soil-inhabiting vectors need somewhat different techniques. Nematodes are always handled in water and transferred to plants by watering a suspension on to the roots. The first stage in extracting them from soil is to break up the soil crumbs in a large volume of water. The nematodes can then be removed from the watery suspension by passing it through a series of sieves of decreasing pore size. Nematodes of the required species will be mostly retained by one or two of these sieves. Then, using a stereoscopic microscope, they can be picked out of the root fragments, debris and mixture of other organisms also retained by the sieves, using either an eyelash stuck to a mounted needle or a dental pulp canal file with a bent tip (Fig. 4.6c) and put in a dish of water. Alternatively the nematodes and debris caught on a particular sieve can be transferred to a piece of paper tissue or muslin supported just below the surface of water in a funnel or small dish. After several hours, many of the nematodes will have moved through the tissue into the water below and the suspension can be drained off. Nematodes can also be extracted using a special apparatus called an elutriator, which fractionates soil suspensions by allowing them to settle through a gently upward-moving stream of water and delivers different fractions at different outlets. The nematodes can then be picked out of these fractions.

When dealing with fungi, it is difficult to handle individual organisms, and suspensions of zoospores are usually used. Suspensions of *Olpidium* zoospores are best obtained by steeping fungus-infected roots that have been kept drier than usual for the preceding day or two to restrict zoospore liberation. These roots are steeped in water or some other suspending medium (e.g. 5% Hoagland's solution) for 15 to 30 minutes. The concentration of zoospores can then be counted with the aid of a haemocytometer, and the suspension concentrated, if necessary, by gentle centrifugation. Zoospore suspensions must be kept cool and handled with care, or the zoospores soon lose their motility and their ability to transmit virus and infect plants.

4.3.3 *Acquisition of virus by vectors*

When vectors are used merely as a convenient way of transmitting virus from infected to healthy plants, feeding times that are optimal for transmission are usually used for virus acquisition and inoculation (Chapter 13). For other types of experiments, however, special techniques may be used to control the way in which the vector acquires and inoculates the virus.

The simplest of these is perhaps that used to make the zoospores of the fungus *Olpidium* infective with tobacco necrosis virus. The virus and zoospore suspensions are merely mixed and kept at 5 to 15°C for a few minutes before putting the suspension around the roots of suitable bait plants. *Olpidium* zoospores are fragile, and they remain motile and able to infect roots for only about an hour when kept in distilled water or tapwater, but in some experiments have survived for two days or more in dilute neutral phosphate buffer or 5% Hoagland's solution. Success in transmission depends greatly on the strain of *Olpidium*, the strain of virus and the species of bait plant used (Kassanis and Macfarlane, 1965).

Some insect vectors can acquire viruses by feeding on virus-containing liquids through a membrane. For example, beet curly top virus can be acquired by the beet leafhopper (*Circulifer tenellus*) feeding through an animal membrane (Carter, 1927). Aphids have been reared without access to plants by allowing them to feed through a membrane on nutrient solution (Dadd and Mittler, 1966; Dadd and Krieger, 1967), and they can acquire some viruses by feeding on infective plant sap or preparations of virus particles. Fig. 4.7 shows one convenient way of supporting the fluid, membrane and aphids; the membrane most

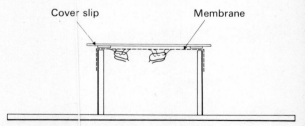

Fig. 4.7 Cage for feeding aphids on virus-containing liquids. One end of the cylindrical cage of opaque plastic (thick lines) is closed and the other covered by fully stretched Parafilm M (broken line). A drop of virus preparation is placed on the membrane and covered with a glass cover slip so as to make a thin layer of liquid accessible to aphids whose mouthparts penetrate the membrane.

commonly used is fully stretched Parafilm M (Marathon Corp., Menasha, Wisconsin, U.S.A.). The aphids feed best when 10 to 20% sucrose is added to the preparations. Methods involving feeding through a membrane have however been used successfully with relatively few viruses.

Some plant viruses persist for long periods in their vectors, and the vectors may not be able to inoculate them to healthy plants immediately after acquisition. Such viruses seem to be imbibed with food, then to pass from the gut into the haemolymph of the vector, and hence to its saliva (some multiplying en route), and thus to test plants. The vectors of some of these viruses (for example aphids, leafhoppers and beetles) can be made infective by injecting preparations of virus particles into their haemocoele. The insects are anaesthetized with carbon dioxide at about 4°C, and the virus preparation injected, usually through an abdominal intersegmental membrane, using a very small glass pipette (diameter about 30 μm). The inoculum should be free of bacteria, and the amount injected is usually less than 1% of the body weight of the insect (e.g. about 5 ml/aphid). After injection the insects are kept in cool humid conditions for a few hours to recover, before being transferred to test plants.

There are no reports that vectors other than insects can transmit plant viruses after acquiring them by artificial feeding or injection.

4.3.4 *Test or bait plants*

Vectors that have acquired virus are tested on suitable test or indicator plants, often called bait plants in work with soil-inhabiting vectors. The dose of virus carried by a vector is small, and transmission is usually obtained most readily when the test plants are as young and susceptible as possible. Indeed, germinating seeds are used in some tests; for example, germinating mung bean (*Phaseolus aureus*) seed can be used to test the transmission of tobacco necrosis virus by *Olpidium*, and germinating cacao beans are usually used to test the transmission of cacao swollen shoot virus by mealy bugs.

The symptoms of many of the viruses transmitted by vectors feeding on the leaves or shoots are readily seen, but some viruses transmitted by soil-inhabiting vectors rarely move from roots to shoots, or may not invade the shoots quickly. Where such viruses are mechanically transmissible it is therefore important, after an appropriate period, to test the roots of bait plants

for virus by inoculating their sap to indicator plants.

4.4 Infection of cultured cells

Many viruses will infect not only intact hosts but also cells cultured from their hosts, but whereas the use of cultured vertebrate cells has revolutionized the study of viruses of vertebrates, the use of cultured plant and insect cells has only recently begun to have an impact on the study of plant viruses. There are several reasons for this difference. First, not until recently were methods for growing continuous lines of cells of vector insects developed (reviewed by Grace, 1973); secondly, for many years, only plant cells with intact cell walls could be obtained, and efficient ways of infecting these were not then known, whereas vertebrate viruses are acquired by, and infect, susceptible vertebrate cells very readily; and thirdly, the plant cells (callus) used in the early experiments produced very little virus compared with the amounts obtained from intact host plants, whereas with viruses of vertebrates the reverse is true.

4.4.1 *Plant cells*

In the early experiments with plant cells (reviewed by Kassanis, 1967), callus tissue or callus cells from several species were cultured on solid media or as suspensions in liquid media. Many of the cell lines that could be subcultured in-definitely were originally obtained from tumours on plants infected with the crown gall bacterium (*Agrobacterium tumefasciens*). Although little of interest about plant viruses was discovered from these early experiments, they led to the important technique of meristem tip culture, which is used to obtain virus-free plants from infected ones (see Chapter 15).

It is possible to obtain clones of callus cells that produce very friable masses of cells. Clones of this type obtained from tobacco can be infected with TMV simply by dispersing the cells in the presence of TMV particles using a 'Vortex' mixer (Murakishi *et al*, 1971). By this technique 50 to 90% of the cells in a suspension can be infected, and yield about 10^7 virus particles per cell in 100 hours.

Enzymically separated leaf cells with cell walls still intact have also been used in plant virus studies (Jackson *et al*, 1972). However, recent experiments with leaf cell protoplasts (cells without cell walls, Fig. 4.8) show the greatest potential for studies of plant virus biochemistry. Such protoplasts can be obtained by treating tobacco leaf cells with pectinases and cellulases, and may

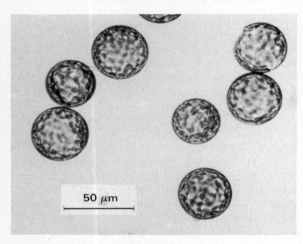

Fig. 4.8 Protoplasts obtained from the palisade cells of tobacco leaves by treating leaf tissue with pectinase, and then treating the separated cells with cellulase. (Courtesy I. Takebe.)

live for several days (Takebe and Otsuki, 1969; Aoki and Takebe, 1969). They are susceptible to infection with intact TMV particles in the presence of poly-L-ornithine or with TMV-RNA and yield about 10^6 virus particles/cell in about 20 hours. The other viruses that have recently been cultured in tobacco protoplasts include cowpea chlorotic mottle, cucumber mosaic, pea enation mosaic, tobacco rattle and potato X.

4.4.2 *Insect cells*

Although primary cultures of insect cells (i.e. cells grown from explanted tissues, but surviving for limited periods) have been studied for more than thirty years, it was not until 1962 that Grace established a continuous line of insect cells (Grace, 1962; 1973), and not until 1967 that the first continuous line of cells of a plant virus vector, the leafhopper *Agallia constricta*, was cultured (Chiu and Black, 1967; Black, 1969). The cells were originally obtained from tissue from week-old *A. constricta* eggs which were placed in a complex medium containing salts, sugars, proteins (including foetal calf serum) and antibiotics. Other

leafhopper cell lines have since been established, and primary cultures of aphid cells have been grown in similar media (Tokumitsu and Maramorosch, 1966; Peters and Black, 1970).

The primary cell cultures and the continuous lines of cells can be infected with various plant viruses (Black, 1969; Chiu *et al*, 1970; Peters and Black, 1970). They are infected, like the cells of vertebrates, merely by adding virus particles to the medium in which the cells are growing; virus infection is detected by staining with fluorescent antibody to the virus (see Chapter 8 and Fig. 13.9), or by electron microscopy.

4.5 Contamination

There is always the possibility that virus cultures are contaminated by other unknown or unwanted viruses. By taking sensible precautions this possibility can be minimized but is unlikely to be eliminated altogether. When working with mechanically transmissible viruses, pestles and mortars must be autoclaved, glassware treated with strong detergent or acid after use, and hands carefully washed and not allowed to touch other objects (towels, taps, doorknobs, etc.) before touching plants. Plants inoculated with different preparations must not be allowed to touch one another, and no plant should be touched except when being inoculated. Glasshouse ventilators should be screened to keep out airborne vectors and the glasshouses should be fumigated regularly with insecticides to kill insects that have entered inadvertently (often on the clothing of staff). Contaminations with nematode-transmitted viruses are rare unless unsterilized soil is used for growing plants, but those with fungus-borne viruses (such as tobacco necrosis virus) are commoner, and because some vector fungi thrive in glasshouse conditions, soil must not be re-used for potting. Either heat-sterilized soil, chemically-sterilized soil or mixtures of sand and peat should be used for growing experimental plants.

Once contaminations have occurred, contaminating viruses can often be removed by exploiting differences in the properties or behaviour of the original virus and the contaminant. We cannot go into details of all the possible methods here, but they should mostly be self-evident from the differences between viruses described in later chapters.

Chapter 5

The composition and structure of the particles of plant viruses

The history of the discovery of plant virus particles was outlined in Chapter 1. Most are spherical or tubular shells of protein which enclose the infective nucleic acid, and protect it when it is outside host cells.

5.1 The components of plant virus particles

The particles of different plant viruses contain different amounts of nucleic acid (Table 2.2). The nucleic acid constitutes from 15 to 45% of isometric virus particles, though some of these viruses in addition produce nucleic acid-free isometric particles. By contrast rod-shaped helically constructed particles contain about 5% of nucleic acid and bacilliform lipid-containing particles about 1%. The remainder of the particle of most viruses is largely protein, but with viruses whose particles contain lipid, this may amount to at least 20% of their weight. The particles of some plant viruses contain enzymes such as the RNA transcriptases of clover wound tumor (Black and Knight, 1970) and allied viruses, and of lettuce necrotic yellows virus (Francki and Randles, 1972).

The other universal but frequently forgotten component of virus particles is water. It constitutes about 50% of the weight of virus crystals, and 10% to 50% of the weight of virus particles in suspension. However it is usual to disregard the water when describing the composition of virus particles, and to express the amounts of nucleic acid, protein, etc. in terms of the non-aqueous components only; we follow the same practice in this book.

In addition to the major constituents, virus particles contain varying amounts of metallic ions, and up to 1% of the weight of some is polyamine. The particles of turnip yellow mosaic virus, for example, contain enough polyamine to neutralize one seventh of the phosphate groups in their nucleic acid.

5.2 Nucleic acid: composition and structure

The structure, composition and properties of the nucleic acids of plant viruses are studied by the techniques used for other nucleic acids (Davidson, 1972). Nucleic acids consist of a long backbone of alternating sugars (ribose or deoxyribose) and phosphate residues with one of five different purine or pyrimidine bases attached to each sugar residue (Fig. 5.1). Two purines, adenine (A) and guanine (G), are found in plant virus nucleic acids and three pyrimidines, cytosine (C) and either uracil (U) (in those containing ribose) or thymine (T) (in those containing deoxyribose). Unusual bases, such as the methylated bases found in the T-even bacteriophages and in transfer RNA, have not been found.

Until recently, all plant virus nucleic acids that had been studied were found to be ribonucleic acid (RNA). However Shepherd, Wakeman and Romanko (1968) found that cauliflower mosaic virus particles contain double-stranded deoxyribose nucleic acid (DNA), and other caulimoviruses presumably also contain DNA.

The particles of most plant viruses contain a linear piece of single-stranded RNA, although in those of clover wound tumor, rice dwarf and other related viruses the RNA is double-stranded and the bases are paired (Gomatos and Tamm, 1963). The two strands are coiled helically around one another; every guanine in one strand is linked by three hydrogen bonds to a cytosine apposed to it in the complementary strand, and similarly every adenine is linked by two hydrogen bonds to a uracil (Crick and Watson, 1953a, b). The double-stranded DNA in cauliflower mosaic virus also consists of two complementary base-paired strands, but here each adenine is paired with thymine instead of uracil; in addition, some of the molecules in preparations of the DNA obtained from particles appear circular in the electron microscope (Fig. 5.2) (Shepherd and Wakeman, 1971; Russell et al, 1971).

About 2×10^6 daltons of nucleic acid are found

in the infective particles of most plant viruses but some viruses contain other amounts ranging from 0.4×10^6 (satellite virus) to 15.5×10^6 daltons (clover wound tumor virus) (Table 2.2). Usually the nucleic acid in the infective particles is in one piece, but in some it is in more than one. The

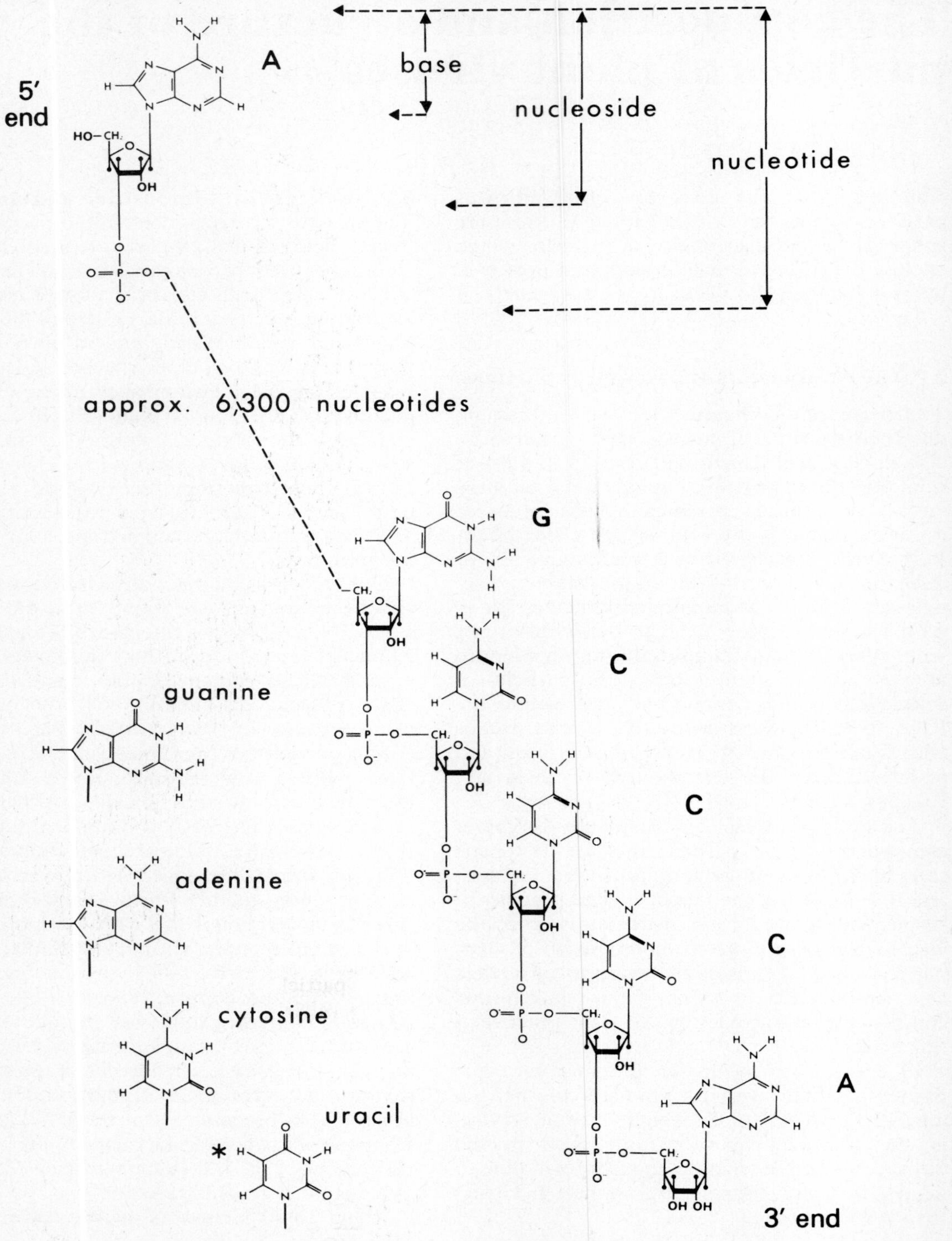

base

nucleoside

nucleotide

A

5' end

approx. 6,300 nucleotides

G

C

guanine

C

adenine

C

cytosine

A

uracil

*

3' end

infective particles of clover wound tumor virus, for example, contain twelve pieces of seven different sizes ranging from 2.5×10^6 to 0.4×10^6 daltons (Kalmakoff, Lewandowski and Black, 1969).

Preparations of some plant viruses contain only one type of particle, but preparations of others contain two or more types, which typically contain different amounts of nucleic acid. In several instances these different types of particles are not infective on their own, or they cause aberrant sorts of infection; all the types of particle containing nucleic acid may be necessary for the virus to produce its usual range of particles (Chapters 6, 11 and 12). In such viruses the genome seems to be in two or more pieces and the parts are packaged separately. Methods of extracting nucleic acids from virus particles are described at the end of this chapter, and the ways of measuring their sizes in Chapter 9.

The nucleic acid of a virus carries information which is translated into various specific proteins, and these mediate the effects of the virus. The information in the nucleic acid is encoded in the sequence of its bases (Table 12.2), and this sequence determines the sequence of amino acids in, and hence the properties and behaviour of, the virus-specified proteins.

The sequence of bases in the nucleic acid and hence the proportions of the four bases (the base ratio), are characteristic of each virus. Base ratios are estimated by carefully hydrolysing the nucleic acid by acid or alkali into its constituent nucleotides, nucleosides or bases, separating these by chromatography or electrophoresis, and estimating the amount of each base. The commonest method for estimating the amount of base in a solution relies on the fact that purine and pyrimidine bases strongly absorb ultraviolet radiation of around 250 to 280 nm wavelength, and the different bases have different and characteristic absorption spectra, which can be measured in a spectrophotometer and used to determine the amount of each base. As might be expected,

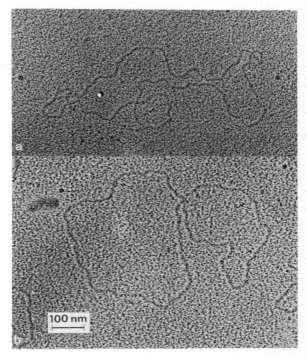

Fig. 5.2 Circular forms of the DNA from particles of cauliflower mosaic virus. (Courtesy E. A. C. Follett.)

related viruses have similar base ratios, whereas viruses with very different base ratios share relatively few properties (Fig. 5.3; Chapter 12).

Determining the sequence of bases (primary structure) in a nucleic acid is a daunting problem. It is done by breaking the nucleic acid into pieces using combinations of specific enzymes or chemicals, or by isolating newly synthesized pieces; these pieces are then separated and broken into smaller pieces to determine their base sequence. Usually the original nucleic acid is, on different occasions, broken in different ways and thus in different places so that overlapping sequences are obtained. Then with judgement, luck and the help of an anagram enthusiast the base sequence is derived. This is not yet known for any plant virus, though work has started on the RNA of TMV (Lloyd and Mandeles, 1970; Mundry, 1968). Special methods have however been used to establish large parts of the sequence of the RNAs of the bacteriophages R17, MS2 and Qβ, and the complete base sequence has been determined for a few small RNA molecules such as the transfer RNAs and small ribosomal RNAs.

There is also some information on parts of the base sequence and on the pattern of the sequence

◄ ─────────────────

Fig. 5.1 Diagram illustrating the structure and arrangement of the molecules at the 5′ and 3′ ends of the RNA of TMV. Also the structure of the bases found in plant virus RNAs; guanine and adenine are purines, cytosine and uracil are pyrimidines. In DNA there is deoxyribose instead of ribose, and thymine instead of uracil; thymine is uracil with a methyl group replacing the hydrogen marked with an asterisk. The carbon atoms of the sugar are numbered clockwise, right to left, 1′ to 5′. Base+sugar = nucleoside; nucleoside+phosphate = nucleotide.

for some plant virus RNAs. Firstly the terminal nucleotides of several are known, and these are of interest because they perhaps contribute to the starting signals used when the RNA is being transcribed into its complementary strand or when it is being translated into protein. Each phosphate residue in the backbone of a nucleic acid is between two sugar residues and is joined to the 3′ carbon of one and the 5′ carbon of the other. Thus the nucleic acid strand is polar; the sugar residue at one end has a potentially free 5′ hydroxyl group and is called the 5′ end, and that at the other a potentially free 3′ hydroxyl, and is called the 3′ end (Fig. 5.1). It seems that the nucleic acids are translated into protein by ribosomes which move from the 5′ to the 3′ end of the nucleic acid. By contrast, they are transcribed into a complementary strand by a polymerase which moves from the 3′ to the 5′ end of the original strand, which is sometimes called the *template*. The newly synthesized complementary strand is made starting at its 5′ end; the template and complementary strands thus have opposite polarity, and are said to be antiparallel. The base at the 5′ end of the nucleic acid from the particles of at least four plant viruses is adenine; with two of these, TMV (four strains) and turnip yellow mosaic virus, the terminal sugar has a free 5′ hydroxyl (Ap . . .), but with the other two, tobacco necrosis and satellite viruses, the terminal ribose is diphosphorylated, and incidentally the terminal trinucleotide has the same sequence (ppApGpUp...) (Lesnaw and Reichmann, 1970). By contrast, several small bacteriophages that contain RNA (f2, MS2, R17, Qβ etc.) have guanosine 5′ triphosphate (pppGp...) at the 5′ end. However the 3′ end of TMV-RNA seems to be the same as that of phage RNAs and transfer RNAs (...CpCpCpA), but that of satellite virus is ...CpCpC or ...CpCpCp (Horst, Fraenkel-Conrat and Mandeles, 1971), and both RNA species of tobacco rattle virus have the 3′ terminal sequence ...GpCpCpC (Minson and Darby, 1973a). By contrast, both RNA species of cowpea mosaic virus have long polyadenylate sequences, averaging 200 residues, at their 3′ ends (El Manna and Bruening, 1973).

Another clue which confirmed that each virus has a unique sequence in its RNA came from experiments in which the nucleic acids were hydrolysed by specific enzymes, such as either pancreatic or T1 ribonuclease, and the resulting oligonucleotide fragments separated chromatographically so that their relative amounts could be

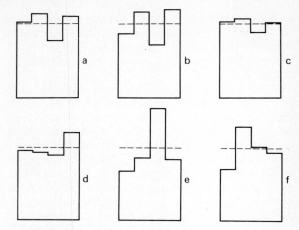

Fig. 5.3 Means of molar base ratios (left to right, G : A : C : U; broken line 25%) of the RNAs of some plant viruses grouped according to their base ratio. **a,** Tobacco mosaic (12 strains), tobacco rattle, alfalfa mosaic, the bromoviruses (3). **b,** The comoviruses (6), broad bean wilt. **c,** Tomato bushy stunt, turnip crinkle, carnation ringspot, carnation mottle, southern bean mosaic, sowbane mosaic, satellite virus. **d,** The nepoviruses, cucumber mosaic. **e,** The tymoviruses. **f,** The potexviruses. (Data from many sources.)

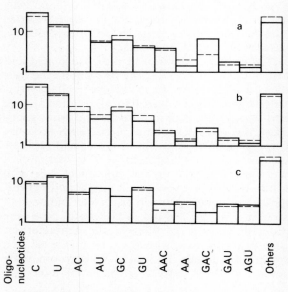

Fig. 5.4 The amounts (moles %; unbroken line) of pyrimidines, and of oligonucleotides with a 3′ terminal pyrimidine, obtained by treating with pancreatic ribonuclease the RNAs from the particles of **a,** turnip yellow mosaic virus, **b,** wild cucumber mosaic virus and **c,** TMV. The amounts expected, had the pyrimidines occurred at random in the RNAs, are shown by the broken line. Note the similarity between the histograms for the two tymoviruses, and the difference between these and the histogram for the tobamovirus (**c**). (Data from Rushizky and Knight, 1960, and Symons *et al*, 1963.)

estimated. Pancreatic ribonuclease, for example, specifically hydrolyses the linkages between all 3' phosphoryl-pyrimidine nucleotides and the adjacent nucleotide, thus the sequence . . . pGpGpCpCpApUpApGp . . . is hydrolysed into the fragments . . . pGpGpCp, Cp, ApUp, and ApGp. . . . Fig. 5.4 gives the pattern of pyrimidine oligonucleotides obtained with three different viruses, showing that although related viruses give similar patterns the two tymovirus patterns are quite unlike that of the tobamovirus.

Yet another indirect way for comparing sequences is to determine how frequently each of the four bases occurs next to each of the others. This is known as nearest neighbour, or base doublet, analysis. The technique has been much used with viruses of animals and bacteria, and their hosts, but has only recently been used with a plant virus, cauliflower mosaic virus (Russell *et al*, 1971). In this technique a complementary copy of the nucleic acid is produced *in vitro*, using an enzyme from the bacterium *Escherichia coli* and all the necessary nucleic acid precursors, which include the four nucleotide 5' triphosphates. One of these four is radioactively labelled with ^{32}P in the innermost phosphate, which is the phosphate that becomes incorporated into the nucleic acid. In this way a radioactive complementary copy is obtained and if, for example, the guanosine triphosphate was labelled (shown in the figure as $\star$), then the copy will contain a radioactive phosphate on the 5' side of every guanosine. For example the sequence 5'.pApGpCpTpCpApGp.3' will be converted into a duplex molecule, as shown in the figure.

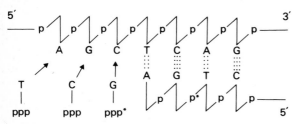

The resulting radioactive strand is then hydrolysed with suitable enzymes to give the nucleotides, and in this way the radioactive phosphate is transferred from the 5' position of the labelled nucleotide to the 3' position of the nearest nucleoside on its 5' side. For example the radioactive strand shown above gives the nucleotides Cp, Tp$\star$, Gp, Ap$\star$, Gp, Cp, and Tp. The radioactivity of the four nucleotides is then measured to give an estimate of the relative numbers of the four

possible nearest neighbours. The experiment is repeated supplying in turn each of the four nucleotide 5' triphosphates suitably labelled. From the results the relative proportions of the sixteen possible doublets can be measured. It has been found that the doublet pattern is characteristic of the nucleic acid (Fig. 5.5), and is quite

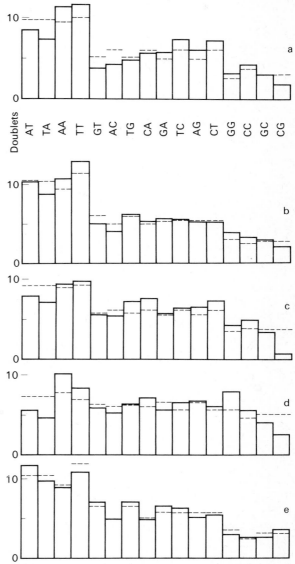

Fig. 5.5 The amounts (moles %; unbroken line) of nearest neighbour pairs of nucleotides in **a,** cauliflower mosaic virus DNA, **b,** cauliflower nuclear DNA, **c,** the nuclear DNA of a mammal (BHK 21 cells), **d,** encephalomyocarditis virus RNA (a mammalian enterovirus; Chapter 16) and **e,** vaccinia virus. Note the similarity of histograms **a** and **b**, and also **c** and **d**, and the difference between **e** and either **c** or **d**. (Hay and Subak-Sharpe, 1968; Russell *et al*, 1971.)

different from the pattern that would be obtained if the nucleotides were in random order. However, the doublet patterns of small viruses of mammals are similar to those of their hosts (Fig. 5.5c and d) whereas those of large viruses of mammals are not, suggesting that small viruses use, and therefore are closely adapted to, the host's machinery for transcribing and translating nucleic acid, whereas the large viruses perhaps carry information for modifying the machinery of the host (Hay and Subak-Sharpe, 1968).

When the sequence of bases in the nucleic acid is translated into the sequence of amino acids in a protein, three nucleotides code for each amino acid. Thus TMV-RNA, which is a chain of about 6400 nucleotides, could code for the sequence of about 2100 amino acids. The only known protein of TMV is that found in the particle. This is a single polypeptide consisting of a chain of 158 amino acid residues, and only accounts for about 7% of the information carried by TMV-RNA. The position in TMV-RNA of the gene for this protein is not known for sure, although Kado and Knight (1966; 1968) obtained some evidence that it may be in the half near the 5' end of the molecule. There is better evidence for the position of the coat protein gene in some of the viruses with divided genomes (Chapter 12).

The side chains of a long-chain molecule like a nucleic acid can bond in various ways with other parts of the same molecule or with other nucleic acid molecules to give the nucleic acid secondary structure. The secondary structure of nucleic acids seems to depend mainly on two types of bonding. Firstly the bases, which are large flat aromatic ring molecules, usually stack, and the stacks are stabilized by hydrophobic bonds between adjacent bases. Secondly, in suitable conditions, most of the bases pair with, and are hydrogen-bonded to, other bases.

X-ray diffraction studies show that the two molecules in the double-stranded RNAs of plant viruses are arranged like those in the duplex molecules of DNA, as a regular double helix (Tomita and Rich, 1964; Sato et al, 1966). Base pairing may also occur within one molecule of single-stranded RNA. The strand may fold back on itself in several places and many of the bases may thus bond to other bases in the same molecule, to give small helical base-paired regions separated by less ordered regions. For example, in neutral 0.1 M phosphate buffer at room temperature, 60% of the bases in TMV-RNA are base-paired (Doty et al, 1959).

Base-paired nucleic acids have different physical and chemical properties from single-stranded nucleic acids. They absorb about one third less radiation of wavelength 260 nm than their two constituent strands. Their base pairing is most complete and most stable at low temperatures and in solutions containing many ions. As temperature increases or ionic strength decreases, the amount and stability of base pairing and stacking decrease, the strands begin to separate and the absorption of ultraviolet radiation increases. The increase both in ultraviolet absorption and in strand separation starts at 20°C with partially paired single-stranded nucleic acids and is complete at 90°C. With completely paired double-stranded nucleic acids it occurs over a much smaller temperature range and the so-called *melting temperature* (T_m; Fig. 5.6), which is between 85 and 100°C in 0.15 M salt, depends on the base ratio, because this determines the relative number of triple-bonded GC and double-bonded AT base pairs. When a completely double stranded nucleic acid is melted, and

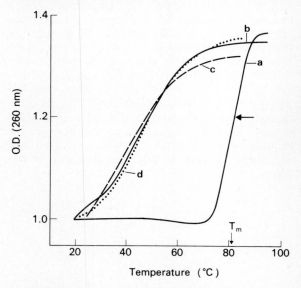

Fig. 5.6 Hyperchromic effect of slowly increasing the temperature of solutions of different nucleic acids.
a, Double-stranded RNA from the particles of rice dwarf virus; **b,** rice dwarf virus RNA after being kept previously for 10 min at 100°C and then suddenly cooled; **c,** transfer RNA; **d,** ribosomal RNA. (After Miura, Kimura and Suzuki, 1966.)

then allowed to cool slowly, the strands realign and the bases pair as before; this is known as reannealing, and is the basis of the nucleic acid *hybridization* experiments used to establish the similarity in sequence of different nucleic acids. If the heated strands are cooled quickly, the bases pair in random fashion and the product behaves like single-stranded nucleic acid (Fig. 5.6). Completely double-stranded RNAs such as those of clover wound tumor and rice dwarf viruses resist ribonuclease (0.1 μg/ml), do not react with formaldehyde (i.e. have no free amino groups), and in concentrated caesium sulphate solutions they have densities around 1.61 g/ml compared with about 1.65 for single-stranded nucleic acids (Shatkin, 1968).

The nucleic acid in virus particles which, like those of TMV, are helically constructed, is rigidly held as an extended helix by the protein subunits, and thus its bases cannot pair while in the particle. However in viruses with isometric particles the nucleic acid is not held as rigidly, and in particles of turnip yellow mosaic virus, for example, at least two-thirds of the bases are paired.

In virus particles the phosphate residues in the nucleic acids form ionic bonds either with cations, with polyamines, or with basic groups in the protein shell of the particle. Nucleic acids do not form covalent bonds with proteins, though they may with single molecules of specific amino acids (Chapter 11).

5.3 Protein: composition and structure

As already mentioned, the only well-characterized plant virus proteins are the coat proteins. However the pinwheel inclusions of two potyviruses have been partially purified and their proteins shown to be serologically unrelated, either to one another or to the particles of the two viruses (Hiebert *et al*, 1971; Hiebert and McDonald, 1973). Most plant virus particles are cylindrical or spherical shells, built of one kind of protein subunit, enclosing and protecting the nucleic acid. However the more complex particles of some plant viruses contain more than one species of polypeptide; for example, those of potato yellow dwarf virus contain four major types of polypeptide (Fig. 5.7), as also do the morphologically similar particles of rhabdoviruses of vertebrates. Another feature of complex virus particles that contain lipid is that one or more of the proteins are usually glycosylated; these are usually in the outer lipid-containing layer of such particles. Thus the three major proteins in particles of tomato spotted wilt virus are all glycosylated, and two of them seem to be in the

outer layers of the particles (Mohamed, Randles and Francki, 1973).

The composition of virus protein is estimated by hydrolysing carefully purified protein in a

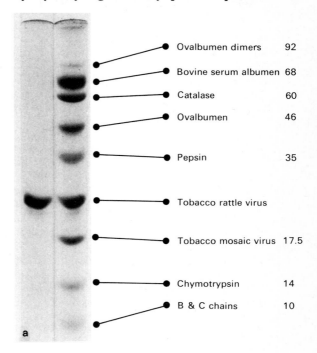

Ovalbumen dimers	92
Bovine serum albumen	68
Catalase	60
Ovalbumen	46
Pepsin	35
Tobacco rattle virus	
Tobacco mosaic virus	17.5
Chymotrypsin	14
B & C chains	10

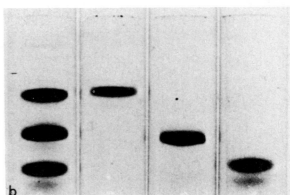

Fig. 5.7 Electrophoresis of SDS (sodium dodecyl sulphate)-treated proteins in polyacrylamide gels; a way of estimating the size of a protein. Electrophoretic migration is from top to bottom. **a,** The protein of tobacco rattle virus electrophoresed in gels alone or together with several standard proteins. The molecular weights of the standard proteins are given in thousands. (Courtesy J. M. Carpenter.) **b,** The four major proteins from the particles of potato yellow dwarf virus run in gels together and separately; their molecular weights (top to bottom) are 78, 56, 33 and 22 × 10³; the largest is a glycoprotein. (Courtesy D. L. Knudson; Knudson and MacLeod, 1972.)

6 N solution of HCl at 108°C, and then separating and estimating by chromatography the relative amounts of the resulting amino acids, usually with the aid of an automatic amino acid analyser. Different amino acids are liberated at different rates, and some of the more labile amino acids such as tyrosine may decompose. Samples of the protein are therefore hydrolysed for different periods of time so that time-dependent errors can be eliminated by extrapolation. A few amino acids are however best estimated by other methods.

Plant virus proteins are not unusual either in composition or in structure (Fraenkel-Conrat, 1968b; 1969; Kaper, 1968); they have proportions of different amino acids similar to those in most other proteins, though perhaps more serine and threonine (Harris and Hindley, 1965) (Fig. 5.8). One unusual feature is that the N-terminus of most of these proteins is acetylated, but those of satellite virus and the U2 strain of TMV have a free N-terminal amino group. The amino acid compositions of the particles of different viruses are different, although those of related viruses are mostly similar (Fig. 5.8 and Chapter 12).

From the relative molar amounts of the different amino acids in the protein, and assuming the virus particles contain only one species of protein, it is simple to estimate accurately the theoretical *minimum* number of amino acid residues in the protein (Gibbs and MacIntyre, 1970a). The *actual* number of amino acids in the protein should be a multiple of the minimum number, and so other less accurate methods must be used to estimate the size.

The approximate size of the protein is usually estimated by splitting the polypeptide chain into peptides using trypsin, which only hydrolyses peptide bonds on the carboxyl side of arginine and lysine residues, or using cyanogen bromide, which hydrolyses only on the carboxyl side of the rare amino acid methionine. The peptides are separated chromatographically and counted, and it is assumed that in the whole molecule there is one more peptide then the number of residues of the type that were selectively attacked. This information, together with the known amino acid composition of the protein will give an estimate of the size of the protein. However this method has its pitfalls, and has given misleading results in the past.

Another and much easier way to measure the total size of the protein is by electrophoresis in polyacrylamide gels (Fig. 5.7); the protein is disaggregated in a buffer containing the detergent

sodium dodecyl sulphate and its electrophoretic mobility in the gels is compared with the mobilities of other proteins of known size. It is thought that the detergent covers the protein uniformly, obliterating the charges on individual amino acids, so that the mobility of the protein depends solely on its size, but some proteins behave anomalously (Koenig, 1972; Ghabrial and Lister, 1973).

The size of the proteins in plant virus particles ranges from about 150 to 600 or more amino-acid residues (Fig. 5.7). However, estimating such sizes may be complicated by degradation during purification of the particles. For example, Koenig *et al* (1970) found that the protein from rapidly purified potato virus X particles had a molecular weight of 29.8×10^3 whereas when virus was purified more slowly, or stored, or treated with trypsin it decreased to 24.0×10^3. A similar though

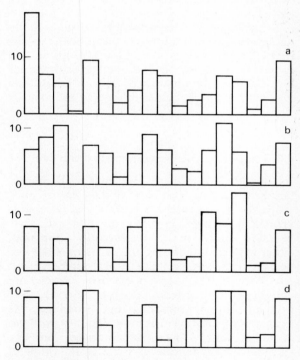

Fig. 5.8 The amino-acid composition of the coat proteins of four plant viruses; **a,** brome mosaic (20 300 daltons); **b,** cucumber mosaic (25 000 daltons); **c,** turnip yellow mosaic (20 400 daltons) and **d,** tobacco mosaic (17 500 daltons). Each histogram shows, from left to right, the amounts (moles %) of alanine (A), arginine (R), aspartic acid (D) and amine (N), cysteine (C), glutamic acid (E) and amine (Q), glycine (G), histidine (H), isoleucine (I), leucine (L), lysine (K), methionine (M), phenylalanine (F), proline (P), serine (S), threonine (T), tryptophan (W), tyrosine (Y) and valine (V). The symbols in parentheses are used in Fig. 5.9.

smaller loss occurs when the particles of cowpea mosaic or bean pod mottle viruses are stored, or treated with proteases. This results in a loss of three C-terminal amino acid residues per subunit and a change in the electrophoretic mobility of the particles (Niblett and Semancik, 1969).

The same strategy is used for finding the sequence of residues in proteins as in nucleic acids; the proteins are cleaved in different ways using specific enzyme and chemical treatments, and the fragments analysed. The terminal amino acids can be removed individually and identified, or an automatic 'sequenator' can be used; this determines the sequence of residues in small peptides by sequentially removing the N-terminal amino acid. To determine the sequence of amino acids in a protein is much easier than determining the sequence of nucleotides in a nucleic acid because there are about twenty different amino acids but only four nucleotides. However it is still a major task, and more direct new methods using mass spectrometry and computers to determine

the composition and sequence of peptides are being developed.

The sequence is known of the amino acids in a few plant virus coat proteins. For example, it is known for several strains of TMV (Fraenkel-Conrat, 1969), and for turnip yellow mosaic virus (Stehelin, Peter and Duranton, 1973). Fig. 5.9 shows the amino acid sequence of three strains of TMV; two (type and dahlemense) resemble each other more than the third (ribgrass). The sequence in these TMV proteins is obviously similar and it is different from that found in other plant and bacterial virus proteins. The differences that occur in the protein of different TMV strains are not randomly distributed throughout the polypeptide. Some parts of the polypeptide seem much less mutable than others; residues 87–94 and 113–122 are the same in all the strains and mutants studied. The less mutable regions are probably parts of the sequence whose detailed structure is important for function, so that changes in these regions give proteins that are

```
                              1
type                     -  S Y S I T T P S Q F V F L S S A W A D P
dahlemense  N-termini    -      T S                   V
ribgrass                 -      N   N S N   Y Z Y F A A V       E

                                                              50
t     I E L I N L C T N A L G N Q F Q T Q Q A R T V V Q R Q F S Q
d         L V     S S                          T              E
r     T P M L   Q   V S     S Q S Y       A G   D T   R Q   A N


t     V W K P S P Q V T V R F P D S D F K V Y R Y N A V L D P L V
d             F   S           G D V Y                          I
r     L L S T I V A P N Q       T G   R     V N S     I K


                              100
t     T A L L G A F D T R N R I I E V E N Q A N P T T A E T L D A
d           T                         Q S
r     E   M K S     P         T Z E E S R   S A S Z V A


t     T R R V D D A T V A I R S A I N N L I V E L I R G T G S Y N
d                           V N   V
r     Q                 Z   Z L   L D     S D H G     Y M D


                      150
t     R S S F E S S S G L V W T S G P A T  -
d     Q N T     M           A     S  -       C-termini
r       A Q ——— A I   P     T A    -
```

Fig. 5.9 The amino-acid sequences of the coat proteins of three strains of TMV; the sequence of the type strain protein is given in full, but those of the other strains are given only where they differ from the type strain. The protein of the ribgrass strain is two amino-acid residues smaller than the proteins of the other two strains; the amino acids in the positions corresponding to residues 146 and 147 of the type and dahlemense strains are missing. Symbols are explained in legend to Fig. 5.8. (After Fraenkel-Conrat, 1968b.)

either unstable or unable to act as particle proteins, and such variants are eliminated by selection. TMV strains that produce aberrant proteins can be deliberately selected provided the strain is transmitted from plant to plant as nucleic acid, free from plant nucleases (Chapter 12).

The subunits of particle protein have both secondary and tertiary structure. These are determined, at least for the TMV subunit, solely by its amino acid sequence, for when TMV particle protein is denatured it can be renatured under certain conditons to give subunits indistinguishable from the untreated ones (Anderer, 1959).

Polypeptide chains usually have much secondary structure, of which the commonest type in globular proteins is called α-helix (Pauling and Corey, 1951). In an α-helix the amino-acid chain is coiled helically with 3.7 amino-acid residues to each turn of the helix, and residues in adjacent turns of the helix are linked by hydrogen bonding of the carbonyl and imino groups of their peptide bonds. Such regular structures can change the

plane of polarization of light, and hence their orientation and amount can be estimated using plane-polarized monochromatic light to measure what is called their optical dichroism, or optical rotary dispersion. Measurements of this type (using infra-red light, which is absorbed by the hydrogen bonds of the α-helix) show that 25 to 35% of the amino acids in TMV particle protein are in α-helices, whose long axes are principally at right angles to the long axis of the TMV particle. It is not known whether the particle proteins of other plant viruses are similar, nor whether these proteins also contain other types of secondary structure.

It seems that virus coat proteins, like other globular proteins such as the globins (Kendrew, 1962), have most hydrophobic amino-acid residues inside the molecule and hydrophilic residues on the surface, and this arrangement presumably gives each subunit a stable hydrophobically-bonded core. Many plant virus coat proteins contain no cysteine and, even in those that do, covalent bonds between cysteine residues seem of little importance in their structure. The CV4 strain of TMV, for example, contains no cysteine and yet forms particles as stable as those of cysteine-containing TMV strains. However cysteine

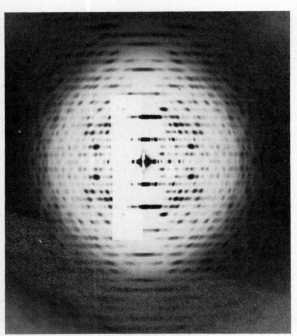

Fig. 5.10 An X-ray diffraction pattern, or fibre diagram, given by particles of the type strain of TMV. The particles were orientated to form a single paracrystal in a narrow tube, the axis of which was parallel to the symmetry axis of the pattern. A curved quartz line-focusing monochromator, and CuK$_\alpha$ radiation, were used, and part of the pattern was partially obscured with an aluminium screen so that the more intense reflections in this part of the pattern could be resolved. (Courtesy J. T. Finch.)

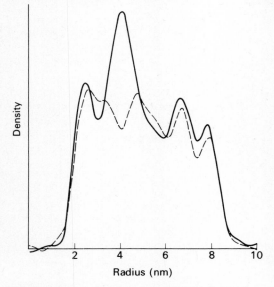

Fig. 5.11 The cylindrically averaged radial distribution of electron density in the particles of TMV; estimates obtained from X-ray diffraction studies. Note the large difference in density between the particles of TMV (unbroken line) and repolymerized RNA-free TMV protein (broken line) at a radius of about 4 nm; this indicates the position of the RNA. (After Franklin, Klug and Holmes, 1956.)

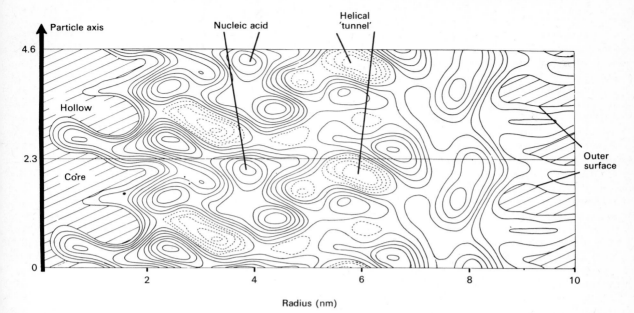

Fig. 5.12 The estimated distribution of electron density helically averaged along the basic helix of TMV particles, and obtained from X-ray diffraction studies. Diffraction patterns were obtained using TMV treated with different heavy metals to aid interpretation; the results obtained were combined in computing the electron density distribution. (Courtesy J. T. Finch; Finch and Holmes, 1967.)

may be involved in the stability of whole particles, and may contribute to the hydrophobic bonds between subunits of the particles of some viruses. Thus the particles of turnip yellow mosaic virus are degraded when treated with *p*-chloromercuribenzoate (Kaper and Houwing, 1962), a compound that reacts with free sulphydryl groups.

There have been no reports yet of the complete tertiary structure of any virus protein, though these must be imminent for TMV protein, for which many important relevant facts are known. Most important of these are the maps of the radially- and helically-projected distributions of density within the TMV particle, obtained by X-ray diffraction studies (Figs. 5.10, 5.11 and 5.12; Franklin, Klug and Holmes, 1956; Finch and Holmes, 1967). The TMV subunit must be a compact structure because its volume calculated from its dimensions (an ellipsoid, 7 nm × 2–2.5 nm) is the same as the volume of its constituent amino-acid residues (Tremaine and Goldsack, 1968). However the density maps show that there are spaces within or between the subunits, in addition to the groove containing the nucleic acid. The folded polypeptide chain in each subunit is

about 48 nm long; approximately 50 residues are in α-helices (0.146 nm/residue) and 110 fully extended (0.36 nm/residue).

Little is yet known of the structure of the coat protein subunits of other plant viruses, except that, by contrast with TMV, the C-terminal residues of the protein of several viruses (e.g. turnip yellow mosaic, broad bean mottle and tomato bushy stunt viruses) cannot be removed from the intact particles by proteolytic enzymes. This suggests that the C-termini of the subunit polypeptides are buried within the intact particles.

5.4 Lipid

There are reports of lipid in preparations of potato yellow dwarf (at least 20% of the particle weight) and tomato spotted wilt (about 20%) viruses, and it probably occurs in all large bacilliform viruses and in carrot mottle virus. Information on the composition of the lipid is limited, but that from potato yellow dwarf virus contains a larger ratio of palmitic to linolenic acid than the lipid from healthy *Nicotiana rustica* plants, the species in which the virus was grown (Ahmed *et al*, 1964).

5.5 The structure of virus particles

The way in which the component parts of virus particles are arranged is known as the quaternary structure of the particle. X-ray diffraction studies of some of the earliest purified preparations of TMV showed that the particles of TMV were rods and had a regularly repeating internal structure (Bawden *et al*, 1936; Bernal and Fankuchen, 1941), and later studies showed that the rod is a tube built of helically arranged subunits (Franklin, 1955). Similar studies with crystals of tomato bushy stunt virus (Caspar, 1956) and turnip yellow mosaic virus (Klug, Finch and Franklin, 1957) showed that the isometric particles of these viruses also have a definite internal structure of regularly arranged subunits. This confirmed earlier chemical studies, which showed that each virus particle is composed of many similar protein subunits and only one, or a very few, molecules of nucleic acid.

Crick and Watson (1956) discussed these results and pointed out that the nucleic acid in virus particles is not large enough to code for a single protein the size of a virus particle and, even if it could, this would be an inefficient and accident-prone process. They predicted that all viruses would be constructed of many small protein subunits of one or a few types, and that these subunits would be arranged either as helically-constructed tubes or as polyhedral shells with cubic symmetry (i.e. tetrahedral, octahedral or icosahedral).

Many viruses with isometric particles have now been studied by X-ray diffraction and electron microscopy, and found to have shells of protein subunits arranged with icosahedral symmetry. Caspar and Klug (1962), in proposing their 'quasi-equivalence theory', suggested that this was no accident, for the icosahedron is the polyhedron with cubic symmetry which, if constructed of identical subunits, would least distort the subunits or the bonds between them (i.e. have the minimum free energy). This same principle was used by R. Buckminster-Fuller, and later by other engineers, when designing geodesic domes such as those used to protect large radar aerials.

The subunits in virus particles are held together by various types of bonds, but not apparently by covalent bonds. The non-covalent bonds can be grouped into those that are polar (salt and hydrogen bonds) and those that are non-polar (van der Waals and hydrophobic bonds). Viruses differ in the relative amounts of these different types of bonds between their subunits. Particles of the bromoviruses quickly disrupt in molar calcium chloride solution, and it seems likely that their subunits are mainly bonded electrovalently, both to each other and to the RNA in the particles; the bonding between protein and RNA is thought to involve mainly basic amino acids and phosphate groups. By contrast many other viruses seem mainly hydrophobically bonded and are unaffected by salt; other materials such as detergents or guanidine hydrochloride are needed to disrupt them. In the particles of these viruses, bonding between protein subunits seems the major factor in their structural stability, and some of them, such as turnip yellow mosaic virus, produce particles that are protein shells free of RNA. However the protein shells are usually less stable than particles containing RNA.

How the components of virus particles come together to form particles *in vivo* is not known, but some clues about the possible stages have come from experiments on the disaggregation and reassembly of particles, and these will be described in §5.7.

We will now discuss the underlying structure of different types of virus particles.

5.5.1   *Icosahedral particles*

The geometric solid called an icosahedron (Fig. 5.13) is a regular polyhedron with 20 equilateral-triangular faces. It therefore has 12 vertices, where the corners of 5 triangles meet, and 30 edges, where the sides of adjacent pairs of triangles meet. It shows three-fold symmetry when rotated around an axis through the centre of each triangular face (i.e. when rotated it looks the same three times in each revolution). Similarly it has a five-fold rotation symmetry axis through each vertex, and a two-fold axis through the centre of each edge. Each triangular face may be thought of as containing, and being defined by, three asymmetric units (i.e. units which have no regular symmetry axes themselves); thus 60 asymmetric units are needed to construct an icosahedron. Satellite virus and the nepoviruses are so far the only plant viruses whose particles have been found to contain only 60 protein subunits (Roy *et al*, 1969; Mayo, Murant and Harrison, 1971), and most isometric virus particles contain more.

The surface of an icosahedron can be subdivided into a larger number of smaller identical triangles; such a solid, an icosadeltahedron, is exemplified by many viruses. The degree of subdivision is defined as one twentieth of the total number of triangles into which the entire surface of the icosadeltahedron has been divided; this is called its triangulation number (T). T is thus the number of triangles into which each icosahedral face has theoretically been divided. Only certain values of T are possible and they are given by the formula $T = h^2 + hk + k^2$ (where h and k are any integer). When $h = k$ ($T = 3, 12, 27, 48, 75$, etc.), or either h or $k = 0$ ($T = 1, 4, 9, 16, 25$, etc.), the triangles are arranged symmetrically on the underlying icosahedral face

(Fig. 5.13), but with other values of h and k they are in a skew arrangement; some animal viruses, for example human wart virus, have icosadeltahedral particles with skew symmetry.

The simplest icosadeltahedron has a triangulation number of 3; thus each icosahedral face has 9, and the whole polyhedron has 180 (9×20)

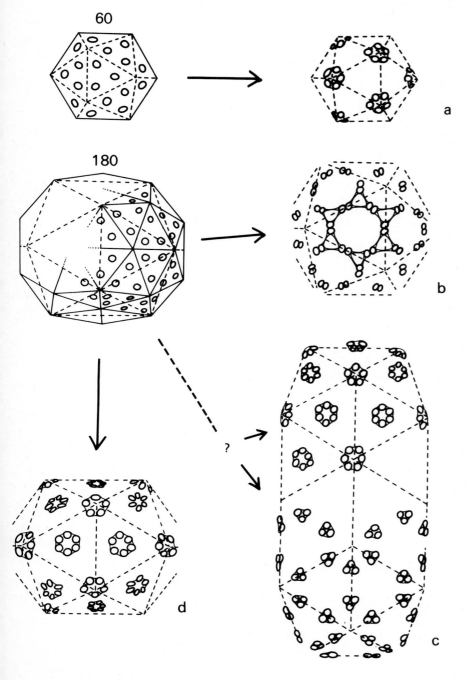

Fig. 5.13 Clustering patterns of subunits in icosahedra. *Top left* is a simple icosahedron. Its triangular faces are undivided ($T = 1$); each is defined by three subunits and thus there are 60 subunits in the entire icosahedron. These subunits may cluster in fives (**a**, pentamer clusters) at the vertices as in the particles of satellite virus (Fig. 2.1j). *Middle left* is an icosadeltahedron of 180 subunits. Each triangular face of the underlying icosahedron (broken lines) is divided into six half triangles (unbroken lines; $T = 6 \times \frac{1}{2} = 3$). Each triangular face is defined by three subunits, and these could cluster together (trimers), or may cluster with subunits of adjacent triangles either in pairs (dimers), or in fives and sixes (pentamer/hexamers). Dimer clustering gives, in one view, a 'Star of David' pattern (**b**), and this is found in particles of turnip crinkle virus (Fig. 5.14). Pentamer/hexamer clustering (**d**) is seen in particles of tymoviruses (Figs. 2.1h and 5.14a). The particles of some viruses, such as alfalfa mosaic virus (Figs. 2.1f and 5.14k) are tubular variants of an icosadeltahedron (**c**). These particles consist of two half icosadeltahedra joined by a tubular net, which has no five-fold axes. The clustering pattern in alfalfa mosaic virus particles is not known for sure; the subunits may be clustered as pentamer/hexamers (**c**, top half) or trimers (**c**, lower half)

asymmetric units. The particles of many plant viruses have this number of protein subunits, but they do not all appear alike in high-resolution electron micrographs because the subunits are clustered in different ways, and may have different shapes. The clustering pattern is charac-

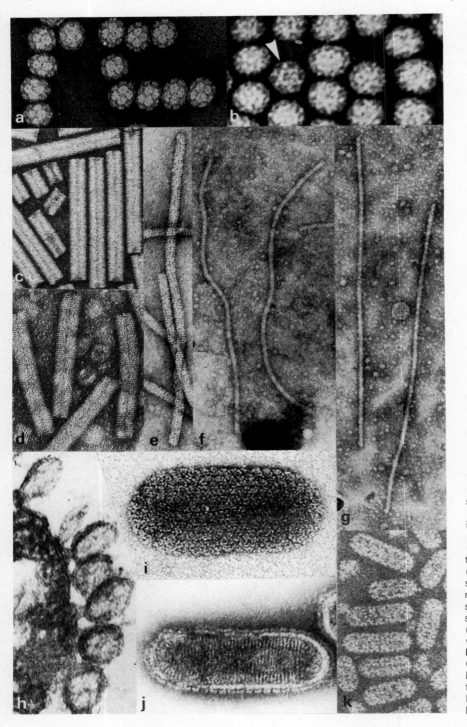

Fig. 5.14 Electron micrographs of the particles of different viruses (the maximum dimensions of the particles are given in parentheses); **a,** turnip yellow mosaic (28 nm); **b,** turnip crinkle (note the pattern on the particle indicated) (28 nm); **c,** tobacco rattle in sodium phosphotungstate (195 nm); **d,** tobacco rattle in uranyl formate; **e,** potato X (515 nm); **f,** henbane mosaic (800 nm); **g,** henbane mosaic treated with magnesium ions (900 nm); **h,** carrot mottle, sectioned at the tonoplast membrane (about 50 nm); **i,** sowthistle yellow vein in sodium phosphotungstate (230 nm); **j,** sowthistle yellow vein in uranyl acetate; **k,** alfalfa mosaic (58 nm). Courtesy of J. T. Finch **a** and **b**; I. M. Roberts **c, d** and **h**; **h** from Murant *et al,* 1969; R. D. Woods **e, f, g** and **k**; D. Peters **i** and **j**.)

teristic of the virus, and related viruses have similar patterns. Many coat proteins have a molecular weight of approximately 20 000, which is near the limit of resolution of most electron microscopes, but clusters of subunits are more easily seen. It is important to realize the difference between the chemically defined subunit of a virus particle (usually a single polypeptide) and the subunit seen in electron micrographs, which is often a cluster of chemical subunits and should be called a morphological subunit, or capsomer. It is also important to realize that the pattern seen on a particle need not reflect the way the chemical subunits are held together, for they may be bonded inside the particle in a regular 'quasi-equivalent' icosadeltahedron, yet be shaped so that their outer ends, seen in electron micrographs, appear to be clustered.

Although complex clustering patterns are possible, especially when the virus particles contain more than one type of protein, there are three basic types of clustering pattern.

(a) The three subunits defining each triangular face may cluster at the centre of the triangle (trimer capsomers).

(b) Pairs of subunits from adjacent triangles may cluster on the edges between the triangles (dimer capsomers).

(c) Subunits may cluster at the vertices of the triangles, so that where five triangles meet at the vertices of the icosadeltahedron there will be pentamer capsomers, and where six triangles meet on the faces of the icosadeltahedron, there will be hexamer capsomers.

Fig. 5.13 shows these types of clustering pattern applied to icosadeltahedra with $T = 1$ and 3. It is clear that the clustering pattern greatly affects the appearance of the particle, and that the number of chemical subunits in each capsomer can be guessed from the arrangement and size of the capsomers.

Some of the different types of icosahedral particles are shown in Fig. 2.1 and Fig. 5.14. Several examples have been studied in some detail.

(a) Satellite virus has particles only about 17 nm in diameter (Fig. 2.1j). Each particle contains 60 chemical subunits of protein ($T = 1$), and these are pentamer clustered to give 12 rather pointed capsomers.

(b) The bromoviruses have rounded particles about 25 nm in diameter, which contain 180 protein subunits ($T = 3$) arranged in 32 pentamer/hexamer capsomers. The nucleic acid, which constitutes about 20% of the particle,

seems not to fill all the space within the protein shell, for all the particles are penetrated to some extent by negative stain, and hence appear hollow. The particles of cucumber mosaic virus have the same structure and appearance.

(c) Turnip yellow mosaic virus has rounded particles about 28 nm in diameter (Fig. 5.14a) containing 180 protein subunits ($T = 3$), which group in pentamers and hexamers (Fig. 5.13) around protuberances in the underlying nucleic acid, so the particle appears to have 32 capsomers. Preparations of this virus contain many nucleic acid-free particles that are penetrated by negative stain, whereas the particles that contain nucleic acid are not. The parts of the nucleic acid molecule associated with the capsomers are perhaps better protected than other parts, for when the virus is stored at pH 10–11 its nucleic acid molecule is broken into 32 pieces (Kaper and Halperin, 1965). All the other tymoviruses have the same structure and distinctive appearance.

(d) Tomato bushy stunt and turnip crinkle viruses have rounded isometric particles about 30 nm in diameter (Fig. 5.14b) containing 180 chemical subunits of protein, clustered in pairs. The 90 dimer capsomers are arranged to give a 'Star of David' pattern (Fig. 5.13). It was claimed that there is also a single subunit of a second protein on each of the 12 vertices of the particles, but this has not been confirmed. Although the nucleic acid in the particles constitutes only about 17% of their weight, they are not penetrated by phosphotungstate negative stain. At one time it was thought that the particles of these two viruses consisted of an outer shell of 180 subunits plus an inner core of 60 similar subunits, with the nucleic acid sandwiched between, but estimates of the weight of the particles and their subunits suggest that this is not so.

Some other viruses producing a single kind of isometric particle that is stable in concentrated salt solutions (e.g. bean southern mosaic and sowbane mosaic viruses) may have a similar structure.

(e) Cowpea mosaic virus has isometric particles with angular outlines and containing 60 protein subunits of each of two kinds. One kind seems to be clustered in 12 groups of 5 subunits and the other in 20 groups of 3 (Crowther, Geelen and Mellema, 1974).

(f) Alfalfa mosaic virus (Fig. 5.14k) has unusual bacilliform particles 18 nm wide and of various definite lengths between 18 and 60 nm; all contain about 18% nucleic acid. These particles are not of course icosahedral but their rounded ends are halves of icosadeltahedra (each with a three-fold symmetry axis on the long axis of the particle). The central tubular part of the different particles consists of protein subunits arranged with two- and three-fold symmetry only. It is not clear whether the protein subunits are clustered in trimers, or in pentamer/hexamers. The arrangement of the nucleic acid within the particle is not known.

(g) Some viruses with larger isometric particles also seem to have icosahedral symmetry. These include clover wound tumor, rice dwarf and maize rough dwarf viruses, which have particles about 70 nm in diameter. Particles of maize rough dwarf virus have two layers of protein subunits. The outer, which is easily removed, apparently consists of 92 capsomers ($T = 9$ icosahedron), plus 12 spikes on the vertices. The inner layer, of 50 nm diameter, also has 12 projections and these are coaxial with the spikes on the outer layer (Milne, Conti and Lisa, 1973). Another virus with large isometric particles is cauliflower mosaic virus, whose particles are 45 to 50 nm in diameter (Fig. 2.1k) and somewhat resemble those of the papovaviruses of mammals (for example human wart virus), whose particles have a skew structure ($T = 7$) of 72 capsomers.

5.5.2 *Particles of helical construction*

The best-studied virus of this, or indeed of any type, is tobacco mosaic virus (Franklin, Klug and Holmes, 1956; Caspar, 1964; Kaper, 1968; Kushner, 1969).

TMV has straight rod-shaped particles 300 nm long and 15 nm wide. When a concentrated solution of TMV (20%) is drawn into a capillary and moved gently to and fro along the tube, the particles become orientated parallel to one another and parallel to the axis of the tube. With care a single paracrystal can thus be obtained in which the particles are randomly orientated about and along their axis, but are close packed in a regular hexagonal array perpendicular to that axis. Such paracrystals diffract X-rays, and the arrangement, spacing and intensity of the spots in the diffraction patterns obtained (Fig. 5.10) show that the TMV particle is constructed of subunits helically arranged around a water-filled axial canal. The particle structure repeats every 6.9 nm of its length, that is, in every three turns of the helix. When TMV is treated with certain mercury compounds, the mercury becomes attached to the single cysteine residue in each protein subunit. Mercury atoms diffract X-rays strongly, and using virus treated in this way it has been shown that there are 49 protein subunits in each axial repetition (i.e. $16\frac{1}{3}$ subunits in each turn of the helix, or about 2100 in each particle). The individual capsomers can just be resolved in the best electron micrographs, which show that TMV particles are right-handed helices (Finch, 1972). The capsomers are single unclustered chemical subunits, and are the basic asymmetric units of the structure. Every capsomer in the particle, except those at the ends, is in a structurally equivalent position in relation to the long axis of the particle. Such a structure is said to have translational symmetry, as contrasted with rotational symmetry, because the particle has only one axis of symmetry, the long axis; the particle can be rotated about, or moved along, this axis into equivalent positions.

TMV particles can be disaggregated into their constituent protein subunits and nucleic acid, and in certain conditions the subunits will re-aggregate to form particles, either with or without nucleic acid. A comparison of the X-ray diffraction patterns of these two kinds of particles shows that the nucleic acid molecule fits snugly into the helical array of protein subunits, at a radius of 4 nm (Figs. 5.11, 5.12 and 5.15). The optical rotatory dispersion of TMV particles (i.e. the effect of TMV on the plane of polarization of light) has been compared with that of repolymerized nucleic acid-free particles using ultraviolet light, because this is strongly absorbed by the bases in nucleic acid. These studies show that the bases, which are flat planar molecules, are orientated parallel to the long axis of the particle, so part or all of the nucleic acid strand is probably held in a fold or groove in the protein subunits. Figure 5.15 shows part of a model illustrating the main features of the TMV particle.

The particles of other viruses with straight tubular particles are similar to those of TMV but show some differences. Tobacco rattle virus (Fig. 5.14c, d) has particles of two modal lengths; some particles of all strains are about 190 nm long, and other particles are 45 nm to 110 nm long, depending on the strain. All the particles are constructed from the same protein subunits but differ in the size of the nucleic acid molecule they contain. The particles repeat every three turns of the basic helix, which has a pitch of 2.5 nm and each turn of the helix contains perhaps $25\frac{1}{3}$ subunits. The ends of each particle are clearly different; one end is somewhat rounded, the

Fig. 5.15 Photograph of a model of part of a particle of TMV. Some of the 'protein subunits' have been removed to show the position of the 'RNA'. (Courtesy A. Klug.)

other flared. Barley stripe mosaic virus has particles of at least three different modal lengths between 110 nm and 150 nm, but the particle has an axial repeat period every five turns of the basic helix, of pitch 2.6 nm.

The particles of viruses with longer, sometimes more flexuous, particles than those already described, seem all to have a basic helix with a pitch of 3.3 to 3.6 nm (Varma *et al*, 1968). They may be divided into groups by the modal lengths of their particles (Chapters 2 and 12), and also on the apparent structure of their particles.

(a) The potexviruses have flexuous particles that, when negatively stained with uranyl formate and examined in an electron microscope, show obvious cross-banding. These viruses have particles 450 to 580 nm long and include potato X (Fig. 5.14e), white clover mosaic and, possibly, narcissus mosaic viruses. Particles of potato virus X and narcissus mosaic virus have been shown by X-ray diffraction to have a basic helix of pitch 3.3 nm in dry particles and 3.6 nm in wet, and those of narcissus mosaic virus repeat every 5 turns of the basic helix.

(b) The carlaviruses have straight particles about 650 nm long, that show banding, or lines of subunits, parallel to the long axis of the particle.

(c) The potyviruses have slightly flexuous particles about 750 nm long. These show no distinct surface pattern, but optical transforms of the electron micrographs show that, like those of the preceding two groups, they have a basic helix of pitch 3.3 to 3.6 nm. In the presence of Mg^{2+} ions these particles tend to increase in length by 5 to 10% and become straighter (Fig. 5.14f, g) (Govier and Woods, 1971).

(d) Viruses with very flexuous filamentous particles include beet yellows virus, whose particles are 1250 nm long (Fig. 2.1e), citrus tristeza virus (2500 nm long) and apple chlorotic leaf spot and apple stem grooving viruses (about 650 nm long). These particles show very obvious cross-banding every 3.3 to 3.6 nm.

5.5.3 *Complex particles*

Several plant viruses with large complex bacilliform particles have been described. Maize mosaic virus (Herold, Bergold and Weibel, 1960) was the first and others include lettuce necrotic yellows (Fig. 2.1b), broccoli necrotic yellows, potato yellow dwarf (Figs. 6.4 and 6.5), and sowthistle yellow vein (Fig. 5.14i, j). The particles of these viruses are very fragile, most are 300 to 400 nm long and 50 to 80 nm wide in sections of infected tissues, but when purified usually become shorter, and often bullet-shaped. They have a tubular core which is 10 nm shorter and is narrower than the whole particle, and helically constructed, with a basic helix of pitch 4.5 nm and an axial channel. There is some evidence that the helical core may be a coiled coil. The membrane or membranes around the helical core seem to be acquired by the core when it passes through one or other of the host cell membranes, for almost complete particles are sometimes found attached by one end to a cell membrane, which is continuous with the membrane around the particles. Among the best studied of these particles are those of broccoli necrotic yellows virus (Hills and Campbell, 1968) and sowthistle yellow vein virus (Peters and Kitajima, 1970). The capsomers of the outer membrane are hexagonally close-packed, but reports differ on the size and arrangement of the hexamers. The lipid found in the particles of viruses of this kind is probably in the outer membranes.

5.5.4 *Particles of unknown structure*

There are several viruses whose particles are known but the structure of which has not been determined directly or deduced by analogy with related viruses. Some of these probably have structures differing considerably from those described above.

Tomato spotted wilt virus has rounded irregular particles (Fig. 2.1i), measuring about 70 nm diameter in sections of infected tissue. In purified preparations the particles are larger and more irregular, and some may have an irregular tail-like structure, which seems to be an evagination of the thin outer membrane of the particle and may be an artifact. These particles are unlike those of any other plant virus but resemble in some features the particles of the leukoviruses of animals (Chapter 16).

Carrot mottle virus has particles (Fig. 5.14h) resembling those of group A and B arboviruses of animals (Chapter 16). The particles are more or less rounded and measure about 50 nm in diameter. They have an ill-defined core, with an outer membrane that seems to be acquired when the particles bud from the tonoplast membrane into the vacuole.

5.6 Dissociation of virus particles into their components

Early experiments showed that protein denaturants inactivate virus particles. Some of the denaturants, such as sodium dodecyl sulphate (Sreenivasaya and Pirie, 1938), urea (Stanley and Lauffer, 1939) and phenol (Bawden and Pirie, 1940) separated the nucleic acid from the protein of the virus, but it was not until much later that Gierer and Schramm (1956) showed that the nucleic acid prepared by treating TMV with phenol was infective, and Fraenkel-Conrat and Williams (1955) recombined TMV nucleic acid and protein to form infective particles.

No universal method is known for preparing the nucleic acid or protein of viruses, presumably because the number and types of bonds between the subunits in the particles differ from one virus to another. We will mention some of the commonest methods; for further references see section 2 of the appendix.

5.6.1 *Preparation of virus nucleic acids*

Several methods have been used to obtain RNA from particles of plant viruses. A simple method,

sometimes used, is to heat the particles briefly to near 100°C. In practice a small volume of concentrated virus suspension is added to a larger volume of hot salt or buffer solution, then the mixture is cooled rapidly, and the denatured and precipitated virus protein is removed.

The particles of some viruses are disrupted by concentrated salt solutions, for example those of the bromoviruses, and also alfalfa mosaic and cucumber mosaic viruses. A suspension of the virus is dialysed against 1 M calcium chloride solution at pH 7.0, which leaves the virus protein dissolved, and the RNA reversibly precipitated.

Phenol and sodium dodecyl sulphate (SDS) are however the two chemicals most commonly used for preparing virus nucleic acids and are used either separately or in combination, often together with other chemicals. In the simplest method using phenol, the virus preparation is emulsified with an equal volume of water-saturated phenol, which either precipitates the virus protein or dissolves it in the phenol, so that after centrifuging the emulsion, the free nucleic acid is in the aqueous phase. This is washed with diethyl ether to remove the dissolved phenol. The method is improved if bentonite, a clay mineral, is added to adsorb traces of nuclease; however, in solutions containing more than about 0.02 M of salts, bentonite adsorbs nucleic acid. This method is also commonly used to extract infective virus RNA direct from infected plants. A more recent variant (Diener and Schneider, 1968) is to use a smaller amount of phenol so that there is no separate phenol phase.

In the simplest method using SDS, the detergent is added to the virus suspension, and once the virus is disaggregated, the mixture of components is fractionated by, for example, differential precipitation with ammonium sulphate. SDS is more commonly used together with phenol, particularly for preparing the nucleic acids of those viruses, such as tomato bushy stunt virus, that cannot be prepared using either chemical alone.

Various other chemicals have been used successfully; these include guanidine hydrochloride with potato virus X, ethanol with turnip yellow mosaic virus, acetic acid with TMV, and either SDS plus a mixture of proteolytic enzymes called pronase, or sodium perchlorate, with various viruses.

All the methods mentioned above have been used for isolating the nucleic acid of viruses containing single-stranded RNA, but some can also be used with viruses containing other types of nucleic acid. Double-stranded RNA has been prepared from wound tumor and other similar viruses by phenol treatment, and infective DNA has been prepared from some animal viruses using phenol and SDS combined. Other DNA-containing animal viruses and also cauliflower mosaic virus are less readily disrupted and their nucleic acid can be obtained by treating the particles with SDS plus pronase.

5.6.2 *Preparation of virus proteins*

Many methods, including most of those for isolating virus nucleic acids, will give preparations of virus proteins in one form or another, but few give *native* protein (i.e. protein with its primary, secondary and tertiary structure unaffected). A good way of deciding whether a virus protein is still native is to see if the protein will reassemble with infective RNA and form virus particles of the same appearance and stability as untreated particles. The particles of several viruses, such as TMV, the bromoviruses, and a few bacteriophages have been reassembled, but the methods used may not work with other viruses. This may be because the protein is no longer native, or because the correct conditions for reassembly have not been found.

Native coat protein of most TMV strains can be prepared by dialysing their particles against pH 10.5 buffer for several days. The particles are disrupted, the RNA hydrolysed, and a clear solution of A-protein (*alkalischer Protein*) obtained. However, the coat proteins of some TMV strains (e.g. tomato yellow aucuba and ribgrass) are sensitive to alkali, and are best prepared by dissociating the virus particles in cold 67% acetic acid. Most other methods, such as SDS or heat treatment, produce a protein preparation that will not reassemble to form virus particles. However, when TMV particles are treated with small amounts of SDS under controlled conditions the protein subunits are sequentially removed, mainly from the end of the particle that contains the 3′ terminus of the RNA (Symington, 1969). This method has been used in attempts to map the position of certain genes in the TMV-RNA molecule (Kado and Knight, 1966; 1968; §12.3.2).

Native proteins of the bromoviruses are obtained by dialysing virus preparations against molar calcium chloride, and removing first the precipitated RNA by centrifugation and then the calcium chloride by dialysis.

Fortunately, for purposes such as amino-acid analysis, it is not necessary to prepare native protein, and the protein can be prepared by many different methods; by treatment with acid, alkali, phenol, urea or guanidine hydrochloride, or by boiling with SDS plus a strong reducing agent, a popular method of preparing samples for electrophoresis in polyacrylamide gels.

5.7 Reassembly of virus particles *in vitro*

The study of the reassembly of virus particles *in vitro* is of interest because the process probably resembles that occurring *in vivo*. The particles of several simple viruses have been reassembled from their constituent parts, and this suggests that these parts contain the information needed for their assembly. However this is unlikely to be true for viruses with more complex particles, for example lettuce necrotic yellows virus, which may need other assistance, such as may be provided by the membranes of a living cell. Caspar and Klug (1962) suggested that the process of self assembly or reassembly of simple virus particles is likely to be similar to crystallization, with the particles being the structures with minimum free energy. Although work on self-assembly and reassembly has been done with many biological materials (Kushner, 1969), the most detailed and successful studies have been with TMV and the bromoviruses.

5.7.1 *Reassembly of TMV particles*

Schramm (1947) showed that TMV protein can form aggregates of various definite sizes, suggesting that the subunits can bond in more than one way, for otherwise only one type or a continuum of sizes of aggregate would be formed. The protein exists as a monomer only at low temperatures, low concentrations, low ionic strengths and high pH; conditions it is unlikely to find in the cells of infected host plants. Under less extreme conditions the protein forms aggregates which vary in size from the trimer, common in A-protein preparations, to structures as large as the TMV particle. However only in acid conditions does the protein form helical structures similar to TMV particles. The structures and sedimentation coefficients of the aggregates which predominate under different conditions of pH and ionic strength are shown in Fig. 5.16. Temperature also affects aggregation, and larger aggregates predominate as temperature and ionic strength

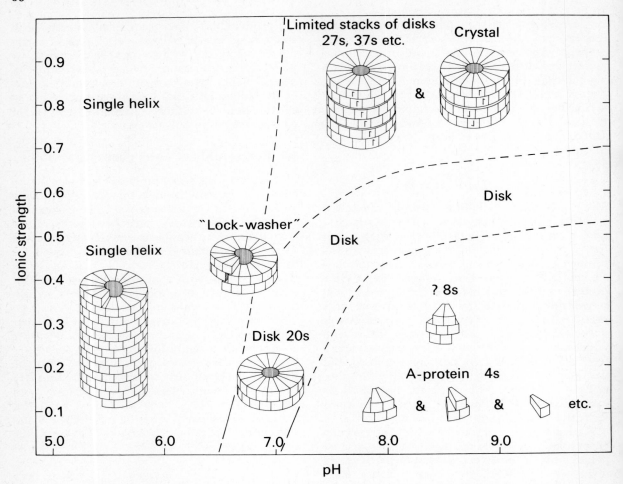

Fig. 5.16 Diagram showing which particular polymers of TMV protein predominate in solutions with different combinations of pH and ionic strength. (Courtesy A. Klug; Durham, Finch and Klug, 1971.)

increase, indicating that the aggregates are mainly stabilized by hydrophobic bonds (Lauffer and Stevens, 1968), though the influence of pH shows that electrovalent groups are also involved. Klug and his colleagues (Durham, Finch and Klug, 1971; Durham and Klug, 1971; Butler and Klug, 1971) suggest that the biologically important aggregate is the double disc made of 34 subunits. This is stable at high pH, but at low pH it polymerizes to form helices. The pH-dependent reversible conversion of disc to helix seems to depend on the ionization of juxtaposed carboxyl groups on adjacent subunits. The double discs of protein seem to be incorporated directly into virus particles, and they reassemble with virus RNA to form particles at least a hundred times more

quickly than A-protein does. It has been suggested that during reassembly the double disc is converted transiently into a two turn helix or 'lock washer'. The RNA is assumed to alter the electrochemical environment within the particle, so that the protein subunits are stable in the helical configuration even in alkaline conditions.

TMV protein will form particles of normal appearance not only with TMV-RNA or the RNA of closely related viruses, such as cucumber green mottle mosaic virus, but also with the RNA of unrelated viruses, such as turnip yellow mosaic virus (Matthews, 1966) or MS2 bacteriophage (Sugiyama, 1966), or even with artificial polynucleotides such as polyadenylic acid. However it forms particles about 1000 times more quickly with TMV-RNA, closely-related RNAs and turnip yellow mosaic virus RNA (Fritsch *et al*, 1973) than with the others, and in the particles containing some of the unrelated RNAs the nucleic acid remains susceptible to ribonuclease suggesting that the particle structure is imperfect. Guilley, Stussi and Hirth (1971) and Butler and Klug (1971) have shown that when TMV-RNA is reassembled with double discs of TMV protein, the rate of formation of particles decreases drastically when the RNA is pre-treated with traces of calf spleen phosphodiesterase, which hydrolyses RNA specifically from its 5′ end, whereas treatment with snake venom phosphodiesterase to attack the 3′ end has little effect, suggesting that the particle assembles starting at the 5′ end. Butler and Klug found that when the spleen enzyme has removed about 40 nucleotides per RNA molecule, the rate at which particles form decreases by 85%, but further removal of nucleotides does not decrease the rate much more, suggesting that there is a 5′ terminal sequence of about 40 nucleotides which has a special affinity for the first double disc of protein. In the particle, the RNA is positioned in the helix so that there are 3 nucleotides to every protein subunit; thus 40 nucleotides would represent almost one turn of the helix.

Experiments using the proteins and RNAs of different tobamoviruses show that the protein first incorporated into an assembling particle determines which protein is incorporated into the rest of the particle; when partially assembled particles of TMV-RNA and TMV-protein are supplied with the proteins of either TMV or cucumber green mottle mosaic virus, or a mixture, only TMV-protein is incorporated into the particle (Okada *et al*, 1970), but when the partly assembled particles contain TMV-RNA and the cucumber virus protein only this protein is incorporated.

5.7.2 *Reassembly of bromovirus particles*

The bromoviruses have been the focus of much rewarding work on the reassembly of isometric virus particles (Leberman, 1968; Bancroft, 1970). Incardona and Kaesberg (1964) found that brome mosaic virus particles were stable and had a sedimentation coefficient of 88S (S = svedberg units; see Chapter 9) when in buffers of small ionic concentration with a pH below 6.7, but in more alkaline buffers they sedimented at 78S, and also quickly lost infectivity, presumably because their RNA had become accessible to ribonuclease. Incardona and Kaesberg suggested that the decrease in sedimentation rate was caused by swelling, because nothing was lost from the

particles in the alkaline buffers and their effect on sedimentation rate was reversible. Two other bromoviruses have also been studied. The particles of all three swell in neutral buffers, and the swelling is almost completely inhibited by magnesium ions or other polyvalent cations including polyamines. Swollen particles fall apart when treated with ribonuclease or salt (1 M NaCl).

These and other results suggest that in acid conditions bromovirus particles are held together by hydrogen bonds between carboxyl groups in adjacent protein subunits and also by electrovalent links between the subunits and the nucleic acid (Bancroft, 1970; Johnson, Wagner and Bancroft, 1973). When the carboxyl groups are neutralized they ionize and, unless linked by a divalent or polyvalent cation, they repel one another and the particle swells exposing the RNA to possible ribonuclease attack. When swollen, the particle is held together by electrovalent bonds binding the protein subunits to the nucleic acid; however when there are sufficient counterions (e.g. 1 M NaCl) in the suspending medium to swamp the electrovalent bonds, the subunits are freed. The virus particles are reassembled by mixing the correct proportions of protein and RNA in neutral 0.5 M NaCl solution, and then slowly decreasing the salt concentration by dialysing the mixture against 0.02 M neutral buffer containing a trace of magnesium chloride; the magnesium ions are essential, perhaps because they bridge the carboxyl groups and thereby position the subunits correctly. A great variety of particles with sizes and structures different from the virus particles can be obtained by varying the composition of the reassembly mixture (Fig. 5.17).

Bromovirus proteins can also form virus-like particles with other RNAs, or even polyanions such as polyvinyl alcohol or polygalacturonic acid. With TMV-RNA, infective particles are formed which have a sedimentation coefficient of about 150S and, unlike intact TMV particles, the 150S particles are photoreactivable (see Chapter 10). Particles assembled from mixtures of proteins from pairs of different bromoviruses have electrophoretic mobilities between those of the individual viruses; individual particles presumably contain different relative amounts of the different proteins.

In the absence of RNA, protein from some bromoviruses can form hollow spherical particles (52S), analogous to the top component found in

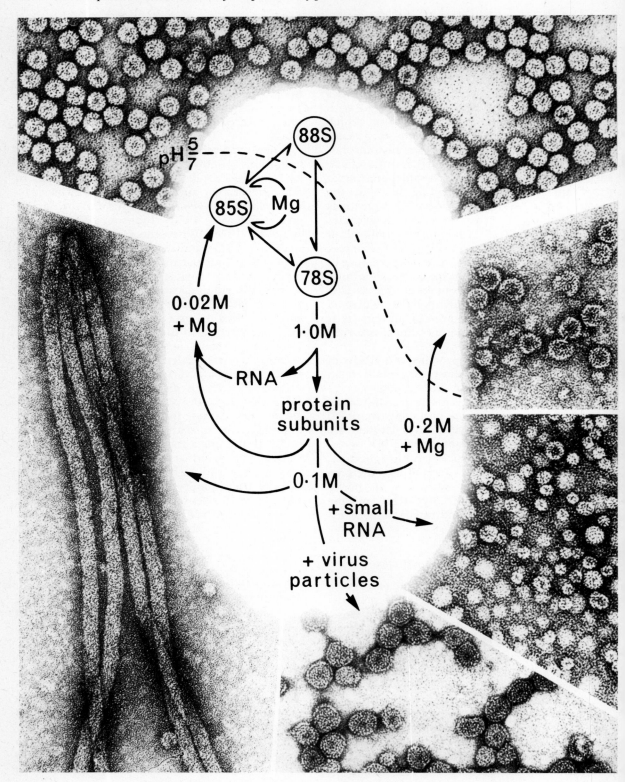

preparations of tymoviruses; however unlike those top components, the bromovirus 'pseudo-top' component particles are stable only in acid conditions, and dissociate when preparations are neutralized, even in the presence of Mg^{2+} ions.

These findings suggest that the particles of bromoviruses mainly depend for their assembly and stability on hydrogen and electrovalent bonds, whereas those of the tobamoviruses depend mainly on hydrophobic and hydrogen bonds. However with both it seems that carboxyl groups are involved in subtleties in the positioning of the subunits.

Fig. 5.17 Electron micrographs of the particles of cowpea chlorotic mottle virus (top), and some of the different particles (below) obtained in reassembly experiments. The central diagram gives the principal factors controlling the interconversions of the different forms. (Courtesy J. B. Bancroft; Bancroft, 1970.)

Chapter 6

The purification of virus particles, and some properties of purified preparations

6.1 Principles of virus purification

Beijerinck, in his classic paper of 1898, reported that TMV could be precipitated by alcohol, and that the precipitate, after drying, could be resuspended in water and was still infective. This was the first step towards purification of a plant virus. More recent work has shown a variety of ways in which the particles of plant viruses differ from the normal constituents of their host cells, and the art of virus purification is to discover and exploit these differences so as to produce fully active virus preparations as free as possible from host materials.

Before the virus can be purified it must be extracted from the cells of an infected host. The typical intact plant cell (Fig. 6.1) has a rigid cell wall surrounding the gelatinous cytoplasm, in which are suspended various inclusions and cell organelles including chloroplasts (which usually contain starch grains), the nucleus and mitochondria; also there is usually a large central, membrane-bound cell vacuole. The cytoplasm contains a variable amount of a complex convoluted lipoprotein membrane called the endoplasmic reticulum. Either attached to the endoplasmic reticulum, or free, are numerous small nucleoprotein particles called ribosomes, irregular in outline and 10 to 20 nm in diameter.

Usually, the first stage of virus purification is to crush the cells and extract the virus-containing sap of the plant. Sap is a complex mixture of the disorganized contents of the cells. It contains a wide range of enzymes capable of degrading the sap constituents, including the particles of some viruses, it is usually unstable, and frequently rather acid. Most of the larger constituents of sap, such as chloroplasts, mitochondria, starch grains and fragments of cell walls sediment quickly and can be removed by brief 'low-g' centrifugation (i.e. small centrifugal fields for a short time, such as 1000 to 10 000g for 5 to 15 minutes). Other sap constituents, such as sugars, salts and amino acids are small and soluble. Between these extremes are

sap constituents such as plant proteins, ribosomes and microsomes (small fragments of the endoplasmic reticulum); it is these that most resemble virus particles in size, composition and stability, and that are the most troublesome host materials to remove when purifying viruses (Figs. 6.2 and 6.3). The most abundant plant proteins in sap are those with sedimentation coefficients of either $4S$ or $18S$, and phytoferritin. Phytoferritin particles come from the plastids, have a core containing iron, and although only about 10 nm in diameter have a density large enough to give them a sedimentation coefficient around $60S$ (Hyde *et al*, 1963). The fraction 1 (or $18S$) protein, which in higher plants is the enzyme ribulose diphosphate carboxylase, also has particles of about 10 nm in diameter and is the major protein of the chloroplasts; it often decreases in concentration in virus-infected plants, whereas the other proteins, called the fraction 2 (or $4S$) proteins, come mainly from the cytoplasm and are usually little affected by virus infection. The several kinds of ribosomes are distinguished by their sedimentation coefficients. All contain between 40% and 60% ribonucleic acid. They disaggregate in the absence of Mg^{2+} ions; in tobacco, for example, the chloroplasts contain $70S$ ribosomes, each of which dissociates into one $35S$ and one $50S$ ribosome in the absence of Mg^{2+} ions, whereas the cytoplasm contains $82S$ ribosomes which dissociate into $35S$ and $58S$ ribosomes (Boardman, Francki and Wildman, 1965; 1966).

The object of purification is to produce a preparation containing only infective virus particles, but this is rarely, if ever, achieved for not only is it difficult to remove the last traces of host constituents, but plants infected with some viruses contain a variety of virus products (e.g. particles of different lengths or composition, and of different infectivities), and some virus particles lose their infectivity during purification. The morphology of particles is sometimes much modified during purification; Fig. 6.4 shows

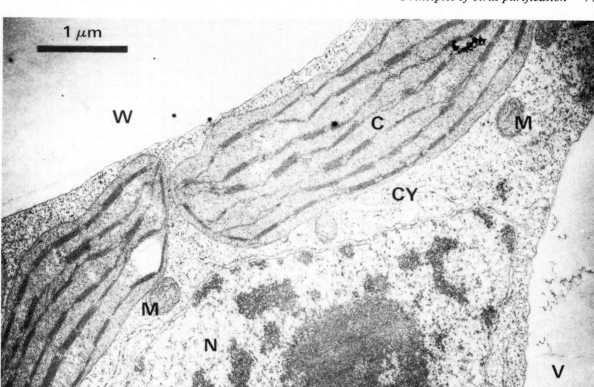

Fig. 6.1 Electron micrograph of a thin section of part of a cell from a healthy leaf of *Nicotiana clevelandii*, showing nucleus (N), nucleolus (No), chloroplast (C), mitochondria (M), cytoplasm (CY) containing ribosomes and endoplasmic reticulum, cell wall (W) and vacuole (V).

particles of wild cucumber mosaic and potato yellow dwarf viruses extracted from infected plants with minimal treatment and prepared for electron microscopy immediately. Some of the wild cucumber mosaic virus particles have two shells, whereas after purification all have only one, and they are then indistinguishable from the particles of the closely related turnip yellow mosaic virus (Fig. 2.1h and 5.14a). Potato yellow dwarf virus particles are even more drastically affected, for initially they are bacilliform, but may become greatly modified when attempts are made to purify them (Figs. 6.4 and 6.5).

In the remainder of this chapter we describe the different stages of virus purification; references to specialized reviews are given in the appendix.

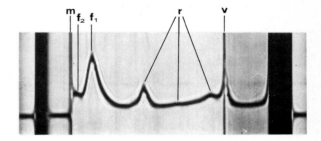

Fig. 6.2 Schlieren diagram of an ultracentrifugal analysis of untreated sap of a broad bean plant infected with broad bean mottle virus. From left to right: air/liquid meniscus (**m**); 4 S plant protein (**f2**); 16 S plant protein (**f1**); ribosomes (**r**); virus (**v**).

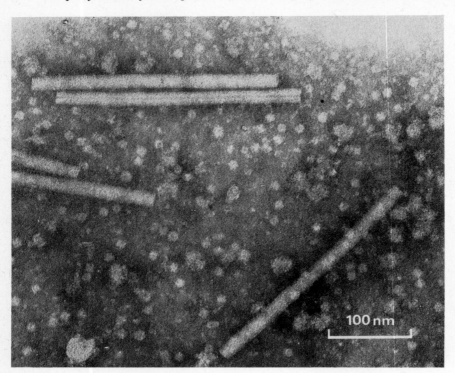

Fig. 6.3 Electron micrograph of sap from tobacco infected with the common strain of TMV; fixed with 4% formaldehyde for 2 hours before being mixed with an equal volume of 2% sodium phosphotungstate (pH 7.0), and sprayed onto the microscope grid. Note the TMV particles, the ribosomes of various sizes (about 20 nm diameter) and the (smaller) plant proteins.

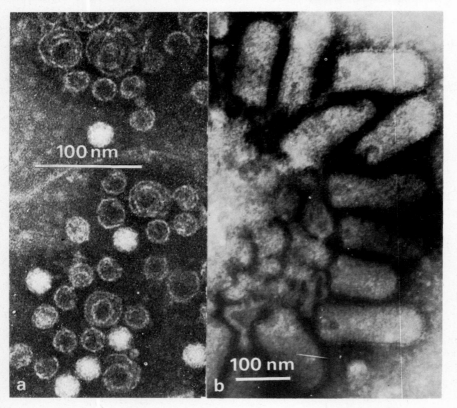

Fig. 6.4 Electron micrograph of sap of **a,** pumpkin infected with wild cucumber mosaic virus. (Courtesy G. J. Hills; Hitchborn and Hills, 1965.) **b,** *Nicotiana rustica* infected with potato yellow dwarf virus; epidermal strip of infected leaf drawn through a drop of potassium phosphotungstate (pH 6.6) on the microscope grid. (Courtesy G. J. Hills.)

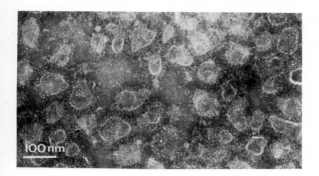

Fig. 6.5 Electron micrograph of a preparation of potato yellow dwarf virus particles purified and mounted for electron microscopy by the most suitable methods known in 1965. (Courtesy G. J. Hills; Black *et al*, 1965.)

6.2 Growing the virus

Some plants are, for various reasons, more suitable than others as sources of virus particles for purification. Virus particles attain larger concentrations in some plant species than others. Also, some plants, such as tobacco, yield sap whose normal constituents are relatively easily removed, whereas others produce sap with less manageable host constituents. Plant constituents are most easily removed from extracts when the source plants are young, adequately fertilized, and grown at 20 to 25°C in humid shady conditions.

Systemically infected leaves often, but not always, produce more virus particles than those infected only locally. Particles of viruses such as TMV remain at large concentrations for a long time in systemically infected leaves; indeed the sap of infected tobacco plants may contain 2 to 3 g of TMV particles per litre, so that they constitute 50 to 80% of the protein in the sap. The particles of other viruses, such as alfalfa mosaic (Fig. 11.3), or the nepoviruses, attain a moderately large concentration and then decrease greatly in amount. Still other viruses seem never to attain large concentrations; for example the sap of infected oat plants apparently contains only about 100 μg of barley yellow dwarf virus particles per litre, and that of most other hosts contains even less. The concentration of the particles of many viruses also depends greatly on their host's environmental conditions, especially temperature, and optimal conditions for accumulation of particles must be found for each virus by trial and error.

6.3 Extracting the virus from the plant

It is good general practice to purify viruses in the cold (4°C), because the particles of some viruses are unstable at higher temperatures. For example, whereas TMV particles remain infective for several years at room temperature, half the infectivity of sap containing those of Tulare apple mosaic virus is lost every seven minutes.

Infected leaves are crushed mechanically using, for example, a pestle and mortar, or a mincer or an electric blender, to liberate their sap. This is usually then filtered through muslin to remove cell walls and fibre, etc. The extraction of stable virus particles is often simplified by first freezing the leaves. Usually only about half the particles in the leaf are extracted with the sap and more can be obtained by grinding the fibre with water or buffer.

6.3.1 *Buffers*

For various reasons it is advantageous to add buffer to the extracted sap or, better still, to grind the plant tissue in buffer. Virus particles are more soluble and stable at some pH values than at others. Crude sap from most species is slightly acid, although that from cucurbits is slightly alkaline, and sap from other species is very acid; that from grapevine leaves is pH 3 to 4.

Buffers (Table 6.1) are solutions that contain either a weak acid and its salt or a weak base and its

Table 6.1 Some standard buffers and their pH ranges

pH range	Constituents
2.2– 3.6	Glycine, HCl
2.2– 8.0	Citric acid, disodium hydrogen phosphate
3.0– 6.6	Citric acid, sodium citrate
3.6– 5.8	Acetic acid, sodium acetate
5.8– 8.0	Mono and disodium (or potassium) hydrogen phosphate
7.2– 9.1	Tris-hydroxymethylaminomethane (TRIS), HCl
6.8– 9.2	Boric acid-borax
8.6–10.6	Glycine, sodium hydroxide
9.2–11.0	Borax, sodium hydroxide

salt, and resist pH changes most strongly when they contain equimolar concentrations of the two. Increasing the concentration of the constituents of a buffer will increase its buffering capacity, but may also affect the dissociation of the constituents and hence alter the pH. Mixtures of different buffers are effective over the pH ranges of the buffers they contain provided that the constituents do not react chemically.

Some of the amino acids in proteins have groups which ionize at characteristic different pH values; for example aspartic and glutamic acids have acidic groups whereas arginine and lysine are basic. Many of these ionizable groups are at the surface of the protein and give it a surface charge. Thus the state of ionization and charge on virus particles depend on the pH of the suspending medium. The pH value at which the total negative charge equals the total positive charge is called the isoelectric point. Because surface charge contributes to the solubility of particles, most viruses are least soluble at their isoelectric points, and many that have elongated particles are reversibly precipitated at this pH value. Most virus particles have isoelectric points around pH 4, and are most stable and soluble at about pH 7. For example, three times more alfalfa mosaic virus particles (isoelectric point about pH 4.5) are extracted when infected tobacco leaves are ground in 0.1 M phosphate buffer (pH 7.0) than when they are ground in water (pH of sap 5.8). Particles of a few viruses, such as brome mosaic and satellite viruses, have isoelectric points around pH 7, and are therefore more stable and soluble at about pH 5 than at pH 7. Plant proteins and ribosomes are most soluble and stable at pH 7, and are irreversibly precipitated at about pH 4.5. Thus with some viruses that have isometric particles, these can be extracted and at the same time freed from plant contaminants by grinding the plant in pH 4.5 acetate buffer (0.1 M).

Plant constituents are also precipitated during storage at pH 7 in high molarity buffers, which presumably damp their surface charge so that they become insoluble. Thus it is a help with some viruses, particularly those with elongated particles, to extract the virus particles into high molarity buffers in which the anion is either citrate (pH 6.5), phosphate (pH 7.0) or borate (pH 7.5). Of these ions, citrate chelates Mg^{2+} ions and thus degrades ribosomes, and phosphate tends to aggregate many viruses with elongated particles. However the precipitating effect of phosphate buffer can be used to advantage in the purification of tomato spotted wilt virus particles; plants infected with this virus are ground in a neutral buffer containing 0.1 M phosphate and 0.01 M sulphite, fibre is removed by filtration, and the extract centrifuged briefly at low-g to give a sediment. This contains most of the virus particles, which can be resuspended in 0.01 M sulphite. By contrast the particles of brome mosaic and some other viruses are disrupted by large salt concentrations, and even those viruses that can withstand high molarity buffers are mostly more stable in dilute buffers. Thus Scott (1963) found that cucumber mosaic virus was best extracted from plants in 0.5 M citrate buffer of pH 6.5, but best purified subsequently in 0.005 M borate buffer of pH 9.0.

6.3.2 *Other additives*

Ribosomes are unstable in the absence of Ca^{2+} or Mg^{2+} ions. These ions are most easily removed from sap by adding a chelating agent, such as ethylene diamine tetra-acetate (EDTA), to the extracting buffer. However EDTA must be used with care because the particles of some viruses, such as alfalfa mosaic virus, require divalent cations for their stability. The flexibility and length of the elongated particles of some potyviruses is also affected by the presence of divalent cations (Fig. 5.14f, g), but whether flexibility is correlated with stability or infectivity is not yet known.

The particles of many viruses aggregate all too readily, and many compounds have been used to try to minimize this tendency, because particles often denature when they aggregate, or may become difficult or impossible to resuspend in an undenatured form, and so sediment during low-g centrifugation. Detergents, such as polyoxyethylene sorbitan mono-oleate (Tween-80) have been used to prevent or reverse aggregation, but Brakke (1959) found that barley stripe mosaic virus was dispersed better by sodium N-methyl N-oleoyl taurate (Igepon T-73) than by nine other anionic, five non-ionic and two cationic detergents, and that Tween-80 (non-ionic) reversed the effect of Igepon T-73. An alternative, which has been successfully used with some of the potyviruses, is to add 0.5 M urea to the extracting buffer (Damirdagh and Shepherd, 1970); the urea, which in greater concentrations disrupts virus particles, seems to act by weakening hydrophobic, and perhaps hydrogen, bonds.

Plant saps usually contain polyphenol oxidases and some contain the tannins produced from oxidized phenolic compounds. Some viruses are inactivated by intermediate products of polyphenol oxidation (e.g. cucumber mosaic virus in tobacco sap), and others are precipitated by the polyanionic tannins (e.g. raspberry ringspot virus in raspberry sap). The effect of these inactivating or precipitating systems may be minimized in various ways (Table 4.3). The plant may be ground into a pH 8 to 9 buffer or 2% nicotine solution, and various reducing agents may be added (e.g. sulphite, thioglycollate, mercapto-ethanol or dithiothreitol). Copper ions, which are essential for the activity of oxidase enzymes, may be removed by chelating agents such as diethyl-dithiocarbamate (DIECA); DIECA must how-ever be used with caution as it inactivates prune dwarf and some other viruses. A further alterna-tive is to add hide or milk powders to compete with the virus for the oxidized polyphenols, and this is successful with cacao swollen shoot virus.

6.4 Fractionating the virus-containing extract

The virus particles can be separated from the other components in sap by many different physical and chemical techniques.

6.4.1 Centrifugation

The operationally important part of a centrifuge is the *head* or *rotor*, which is spun about a vertical axis, and in which the suspensions to be centri-fuged are held, usually in separate *tubes* or *cells*. The rest of the centrifuge consists of ancillary equipment which drives the rotor safely at a chosen speed and temperature. Fig. 9.6 shows the four principal types of rotor used in virus research. The one that is most commonly used is the *fixed angle rotor* (Fig. 9.6a) in which the tubes con-taining the virus preparations are held in holes drilled in the rotor at a fixed angle, ranging from $18°$ to $40°$, to the axis of rotation. During cen-trifugation the particles sediment centrifugally until they hit the outer wall of the tube; they then slide down this wall and form a pellet at the bottom.

Nowadays it is possible to centrifuge large quantities of liquid in relative gravitational fields of $100\,000g$ or more. In these conditions even the smallest virus particles sediment quite quickly; the rate at which a particle sediments is defined in svedberg units (S), and the methods for deter-mining sedimentation rates are described in §9.4. Centrifugation may be used to separate virus particles from host constituents because it is a fortunate fact that most virus particles differ from host constituents either in their sedimentation rate or their density or both (Fig. 6.6). For

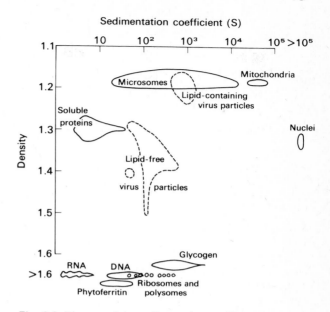

Fig. 6.6 Diagram of the sedimentation coefficient and density in a CsCl gradient of the particles of various plant viruses and some of the common host components found in plant sap.

example, constituents such as cell-wall fibres, chloroplasts etc. sediment at more than $1000S$, virus particles at 50 to $1000S$ (most at 100 to $200S$), ribosomes at 25 to $100S$, 'soluble' plant protein mostly at $4S$ and $18S$, and simple mole-cules like amino acids, sugars etc. at less than $2S$ (Fig. 6.2). Hence by choosing both an appro-priate speed and period of centrifugation or by adjusting the density of the suspending medium, it is usually possible to separate host and virus components. However, as is obvious from Fig. 6.6, this strategy is not successful with lipid-containing virus particles such as those of lettuce necrotic yellows virus (a rhabdovirus) or tomato spotted wilt virus.

Extracts are first clarified by brief low-g centrifugation for a time that depends upon the virus particles being purified. For example, satellite virus particles ($50S$) can be centrifuged for 20 minutes at $10\,000g$ to give a clarified extract containing only virus particles, ribosomes, soluble plant proteins, and smaller soluble compounds, whereas an extract containing lettuce necrotic yellows virus particles ($943S$) can not be centri-fuged at $2000g$ for more than 3 minutes without sedimenting many of the particles, and will there-fore also contain some larger host constituents such as microsomes and chloroplast fragments.

The particles of some viruses seem to be

attached to the larger cell constituents, and may be sedimented when the sap is clarified by low-*g* centrifugation. This property may lead to a loss of virus particles at the start of the purification procedure, or it may be exploited and used as the first stage of purification, provided a way of resuspending the virus particles in the sediment is known.

Once the virus particles are extracted from the plant, and suspended in a suitable buffer, they are usually sedimented from the clarified extract by long periods of high-*g* centrifugation (0.5 to 4 hours at 75 000 to 150 000*g*), which leaves most of the soluble plant proteins and smaller compounds in suspension. Further cycles of sedimentation and clarification, a technique called *differential centrifugation*, will further purify the particles, but may damage them and, should particles be of different types, will usually deplete the preparation in those that sediment the most slowly. The underlying theory of differential centrifugation is discussed by Schumaker and Rees (1972).

A most valuable fractionation of virus-containing preparations can usually be obtained by centrifuging them in density gradients (Brakke, 1960, 1964; Brakke and van Pelt, 1970a). For this technique the tubes that hold the virus preparations in the centrifuge are placed in the buckets of a swinging bucket rotor (*swing out rotor*, Fig. 9.6b). At rest the buckets, which are attached near their tops to the rotor, hang down, with their long axes parallel to the axis of rotation of the rotor, but when the rotor spins they swing out until at speed their axes are at right angles to the axis of the rotor. In this way the centrifugal force is always applied down the long axis of the centrifuge tube. Before centrifugation, the tubes are filled to about 80% with virus-free solvent to which varying amounts of a dense inert solute has been added to give a gradient of density which decreases from the bottom towards the top of the tube and which stops convective mixing. A small volume of the virus preparation, which must be less dense than the top of the gradient, is carefully layered on to the density gradient and then centrifuged. The different components of the virus preparation sediment into the gradient at different rates and, where concentrated enough, will be seen as thin light-scattering layers (Fig. 6.7). Initially the components sediment and separate at a rate determined by their relative sedimentation coefficients (*rate zonal separation*). However when centrifuged long enough each component will sediment to the bottom of the tube, or, if the

Fig. 6.7 Photograph of a sucrose density gradient into which has been centrifuged a preparation of tobacco rattle virus particles. The two main bands contain the shorter (upper band) and longer (lower band) virus particles. The gradient is of 10 to 40% sucrose and was centrifuged for 90 minutes at 21 000 rev/min (about 44 000*g*).

solution in the gradient is dense enough, it will sediment to, and float in, the part of the gradient with its own density (*isopycnic banding separation*).

A special type of centrifuge rotor, called a *zonal rotor* (Fig. 9.6d) can be used for fractionation on a larger scale than is possible in swinging bucket rotors. The zonal rotor is hollow, and in it a density gradient is accommodated in a large annular space. There are many variants of the zonal rotor, most commonly used is the pancake rotor. This rotor has a special axial seal through which the rotor is filled and emptied while spinning at about 5000 rev/min.

Rate zonal separation is commonly used for purifying plant virus particles, using sucrose (0 to 400 g/l), 'Ficoll' (polysucrose), or sometimes glycerol as the dense solute; all have the merit that they increase not only density but also viscosity, which stabilizes the gradients for handling.

Isopycnic banding separation is also used for fractionating plant virus preparations. Most plant virus particles have densities of 1.3 to 1.45 g/ml, and only those containing lipid are less dense (about 1.18 g/ml). Thus sucrose is only suitable for isopycnic banding of lipid-containing particles (60% sucrose at 20°C has a density of 1.22) and, for the others, solutions of salts such as caesium chloride (50% solution at 20°C has a density of 1.59) or potassium tartrate must be used.

Gradients of sucrose or potassium tartrate are usually prepared before centrifuging, either using a gradient-generating mixer or by layering about four solutions of different densities one above the other and allowing these to diffuse together to form a 'smooth' gradient. Gradients of CsCl may be made in the same way, but are sometimes generated in the centrifuge; the virus preparation is mixed with sufficient CsCl to give a solution of about the same density as the virus particles, this is then centrifuged at high speed often in a fixed angle rotor for 12 to 48 hours. This partially sediments the salt (until sedimentation and diffusion balance) to form a gradient, and at the same time the virus particles sediment centrifugally to, or float centripetally to, the position where their density equals that of the solution. Whereas the particles of all viruses seem stable in sucrose solutions, not all are stable in CsCl solutions.

The separated components in the centrifuged density gradient columns can be removed from the tubes in various ways, and the dense solute removed either by dialysis, or by dilution and centrifugation to recover the virus particles.

6.4.2 *Restricted diffusion: dialysis and gel chromatography*

Dialysis is a technique by which small molecules are removed or added to a virus preparation through a semipermeable membrane, by diffusion and osmosis. The most commonly used membranes are made of viscose cellulose, usually as a seamless tube, which can be sealed by knotting the ends. Virus particles cannot diffuse through the membrane but water, sugars, buffers, salts etc. can.

Agar or agarose gel chromatography is another technique for sorting substances by their sizes and diffusion rates; agarose is one of the two major constituents of agar, and is an uncharged polysaccharide. A wide glass tube is filled with chips or beads of dilute agar or agarose gel suspended in buffer (a 1 to 8% gel in the form of chips or beads, 'Sepharose', of 0.5 mm or less in diameter is suitable). A small volume of virus preparation is added to one end of the column and slowly washed through with buffer. Virus particles diffuse only slowly and not very far into the 'stationary' buffer in the gel, and therefore pass through the column more quickly than, for example, phytoferritin, which is smaller, has a higher diffusion rate, and is retarded (Fig. 6.8). The same principle is involved when columns of granulated dextrose gels, such as 'Sephadex', are used to separate small molecules, such as salts or sugars, from virus particles. The virus particles run through the column unimpeded because the pores in the gel are even smaller than in Sepharose, and totally exclude the particles from the gel.

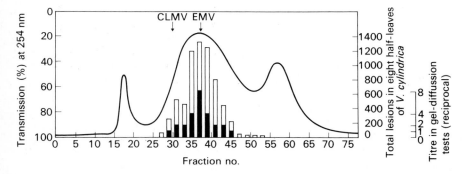

Fig. 6.8 Agarose gel chromatography of a partially pure preparation of raspberry bushy dwarf virus, which has isometric particles about 30 nm in diameter. Ultra-violet absorption (continuous line), virus antigen content (solid columns) and infectivity for *Vigna cylindrica* (open columns) of fractions obtained by diffusion chromatography in 2% agarose beads. The left-hand peak contains impurities that have large particles; the right-hand peak contains phytoferritin. EMV and CLMV indicate the elution positions of the particles of eggplant mosaic (30 nm) and cauliflower mosaic (45 nm) viruses in other experiments. (Courtesy A. F. Murant; Barnett and Murant, 1970.)

6.4.3 *Electrophoresis*

Virus particles are charged and, except at their isoelectric point, migrate in an electrical field. The particles of different viruses, or even strains of one virus, may have different electrophoretic mobilities which reflect their surface charge densities. Electrophoretic mobility is relatively independent of particle size and molecular weight. Electrophoresis has been little used to purify plant virus particles, though plant proteins have been removed from virus preparations by electrophoresing the preparation in a liquid column stabilized by a sucrose density gradient (Brakke, 1955; van Regenmortel, 1964; 1972).

6.4.4 *Liquid two-phase systems*

A water solution of two organic polymers such as dextran and polyethylene glycol, or dextran and methyl cellulose, will separate on standing into two phases, and any virus particles or proteins in the mixture are often much more concentrated in one phase than the other. This difference can be used for purifying the particles, but seems to have few advantages over the precipitation methods to be described in §6.4.6 (Leberman, 1966).

6.4.5 *Heating*

Proteins denature when heated but differ in their susceptibility to such denaturation. Particles of some viruses, such as TMV, can withstand exposure at 50 to 60°C for 10 minutes, whereas much host protein is coagulated and can thus be removed from tissue extracts by this treatment.

6.4.6 *Precipitants*

Virus particles are soluble because in suspension they are electrically charged and hydrated. They are therefore precipitated from suspension when compounds are added to the solvent to damp their charge, or dehydrate them, or both. Thus, as already mentioned, particles of some viruses are reversibly precipitated at their isoelectric points, and most plant proteins and ribosomes are irreversibly precipitated.

Inorganic salts both discharge and dehydrate, and can be used at high concentrations to *salt out* virus particles and plant proteins. Salts differ in their ability to precipitate hydrophilic colloids, and on this ability their ions can be arranged in the well-known Hofmeister or lyotropic series.

Anions have the greatest effect on precipitation, and their order of decreasing ability to precipitate is sulphate > phosphate > acetate > citrate > bicarbonate > chloride > nitrate. Cations have less effect and a less well-defined order.

Virus particles are most commonly salted out with ammonium sulphate which is very soluble and cheap. Particles of most viruses are precipitated in one-third saturated ammonium sulphate solutions, but some have very soluble particles that are not precipitated until the concentration reaches half saturation, and those of broad bean mottle virus not until three-quarter saturation. Plant proteins and ribosomes are also precipitated by ammonium sulphate, but most are at the same time denatured.

Much plant protein is precipitated when a sap extract is incubated at room temperature for 6–24 hours, especially when phosphate has been added to the extract; presumably the proteins are denatured by the salts and enzymes present. Similarly, extracts containing virus particles may be clarified by freezing, presumably because of the precipitating effect of the high concentrations of salt produced during freezing; the solvent freezes first and the solutes including proteins and salts become increasingly concentrated until their eutectic points are reached and the solution freezes.

Simple organic compounds have a range of hydrophilic and lyophilic affinities and some are of great use in virus purification. The most hydrophilic ones used are ethanol and acetone, both of which are miscible with water in all proportions, and precipitate virus particles by competing for water of hydration. They are perhaps best used at small concentrations for clarifying tissue extracts because in these conditions they precipitate and denature plant proteins and lipid-containing structures such as chloroplasts, mitochondria and endoplasmic reticulum but not the particles of most viruses. At greater concentrations they may denature virus particles. For example, sap containing turnip yellow mosaic virus particles is usefully clarified by 21% ethanol, whereas the particles are partially denatured by 30% ethanol. However the particles of some viruses are reversibly precipitated by these compounds and have been purified using them; for example, those of TMV and squash mosaic virus by 50% ethanol and of raspberry ringspot virus by 70% ethanol or 50% acetone.

Diethyl ether and *n*-butanol are frequently used as protein and lipid denaturants to clarify tissue extracts. They are less hydrophilic and less soluble in water than ethanol and acetone (7.5 g ether and 7.9 g butanol will dissolve in 100 ml water at 20°C). They are however likely to damage any virus particles containing lipids. Ether has been used to clarify saps containing filamentous virus particles, and is usually used together with carbon tetrachloride, as described below. Butanol is useful for clarifying sap containing, for example, the particles of cucumber mosaic or lettuce mosaic viruses, but it acts slowly, and the sap is therefore mixed with the butanol (8.5% by volume) and stored for several hours before centrifuging to remove precipitated plant proteins. It has little effect on ribosomes. Butanol is often used together with chloroform, as described below.

The least hydrophilic compounds used are chloroform, carbon tetrachloride and, for the very unstable particles of lettuce necrotic yellows virus, the fluorocarbon Freon 113 ($CCl_2F.CClF_2$). These three compounds are almost insoluble in water (chloroform 1.0 g, and carbon tetrachloride 0.08 g per 100 ml water, Freon 113 even less) and they do not precipitate virus particles. When they are vigorously shaken or stirred with a tissue extract an emulsion is formed, which when centrifuged at low *g* separates into three layers. The lightest layer is the aqueous phase and contains the hydrophilic components including virus particles, the densest is the organic compound and contains lyophilic materials such as fats, waxes, chlorophyll etc., and between the two is a layer of precipitated proteins. The particles of many viruses can be purified using chloroform, or butanol and chloroform together; infected plant tissue is disrupted mechanically into buffer plus a mixture of equal volumes of butanol and chloroform, the resulting emulsion is centrifuged, and the aqueous layer removed and stored overnight at room temperature, when the butanol dissolved in it denatures and precipitates some 'soluble' plant proteins. Similarly with other viruses, the infected tissue is blended into a mixture of buffer and ether, then carbon tetrachloride is added and the mixture blended further to form an emulsion, which is centrifuged to obtain the aqueous phase. Ether/carbon tetrachloride removes less plant protein than butanol/chloroform, but seems more suitable for purifying elongated virus particles.

Polyethylene glycol (PEG) is another hydrophilic organic compound fully soluble in water in all proportions, that will precipitate the particles of many different viruses. Its ability to precipitate them depends on its concentration and on the amount of salt present. For example, Hebert (1963) found that the same amount of TMV particles was precipitated from a preparation by either 4% PEG in 0.1 M NaCl or 2% PEG in 0.3 M NaCl. It is possible to separate virus particles from host constituents by using different concentrations of PEG and NaCl to precipitate the particles from sap and to elute them from the precipitate (Clark, 1968). A variant of this method is to centrifuge virus particles, which have been precipitated by PEG, through a reversed gradient of PEG stabilized by sucrose (Clark and Lister, 1971).

6.4.7 *Adsorbents*

Proteins are adsorbed from suspension by various materials. For example much of the plant protein in an extract containing virus particles is removed when it is mixed with hydrated calcium phosphate or activated charcoal and then centrifuged briefly at low *g*. Similarly, sap containing the particles of lettuce necrotic yellows virus can be clarified by mixing it with activated charcoal and DEAE-cellulose, and filtering through Celite (McLean and Francki, 1967). Filtration has also been used in the purification of the particles of oat blue dwarf virus (Banttari and Zeyen, 1969). PEG, glucose and NaCl were added to clarified sap from infected oat plants to give a solution containing 5%, 4.5% and 2% respectively, the resulting precipitate was collected by a cellulose filter, and washed with a neutral buffer containing the same concentration of PEG, glucose and NaCl. Finally the virus particles were eluted from the filter by 4.5% glucose in neutral buffer.

Bentonite, a highly charged clay material with a large surface area, will also adsorb proteins and virus particles when it is magnesium-saturated but usually not when sodium-saturated; presumably the divalent cation forms a 'bridge' between the anionic clay and the proteins. Bentonite is not easy to use because the ionic conditions must be carefully controlled if ribosomes and plant proteins, but not virus particles, are to be adsorbed (Dunn and Hitchborn, 1965), but it has been used successfully with some viruses whose particles are difficult to purify in other ways (de Sequeira and Lister, 1969). Bentonite in any form at about pH 6 adsorbs the enzyme ribonuclease, and is therefore used when extracting RNA from virus particles (Chapter 5), or when transmitting viruses from plant to plant as infective RNA (Chapter 4).

Virus preparations can be fractionated chromatographically using a variety of materials for the stationary phase such as cellulose, substituted celluloses (Venekamp, 1972) and hydrated calcium phosphate (Murant *et al*, 1969). This method has not proved very popular for general use, perhaps because it is time-consuming, and because the amount of virus preparation which can be fractionated at one time is small.

6.4.8 *Antihost serum*

Antisera prepared against normal constituents of plant sap can be used to remove host materials from virus preparations. A suitable dilution of the antiserum (Chapter 8) is incubated with the virus preparation and the resulting precipitate removed by brief low *g* centrifugation. The virus particles must then be separated from unreacted serum proteins.

6.5 Concentrating the virus preparation

Usually the virus particles are concentrated during purification. This is done most simply by resuspending sedimented or precipitated particles in a smaller volume of solvent. When preparations are dialysed to remove salt or sucrose they may become greatly diluted by osmosis unless the volume of the dialysis sac is restricted; and they may become diluted during agar gel chromatography. Solvent may be removed directly from such preparations in various ways. The preparation, in a dialysis sac, can be concentrated by the following methods:

Osmosis: the sac is surrounded by a hydrophilic compound of large molecular weight such as polyethylene glycol (20 M grade) or egg albumen.

Pervaporation: the sac is hung up to dry in an air stream in a cold room.

Vacuum filtration or pressure dialysis: the dialysis membrane is supported on a special frame, the preparation put on one side of the membrane, and the solvent molecules forced out through the membrane either by applying pressure to the preparation or a vacuum to the other side of the membrane.

Alternatively the preparation may be partially frozen and the ice removed, or dry agar or polydextran gel can be added, allowed to imbibe water and then be removed by centrifuging.

Table 6.2 Ways of purifying the particles of TMV and raspberry ringspot virus

A TMV

1. Mince systemically infected tobacco leaf and squeeze sap through muslin. Discard fibre.
2. Heat sap for 10 min at 60°C and centrifuge 5 min at 5000 *g*. Discard pellet.
3. Acidify supernatant fluid to pH 3.2. Centrifuge 10 min at 5000 *g*. Discard supernatant fluid.
4. Resuspend sediment in 0.05 M phosphate buffer (pH 7.5) and centrifuge 5 min at 5000 *g*. Discard pellet.
5. To 2 vol. supernatant fluid add 1 vol. saturated solution of ammonium sulphate and centrifuge 10 min at 5000 *g*. Discard supernatant fluid.
6. Resuspend sediment in phosphate buffer and centrifuge 5 min at 5000 *g*. Discard pellet.
7. Centrifuge supernatant fluid 2 h at ∕5 000 *g*. Discard supernatant fluid.
8. Resuspend sediment in phosphate buffer, and centrifuge 5 min at 5000 *g*. Discard pellet. The supernatant fluid is the purified preparation.
9. Store purified preparation at 4°C.

B Raspberry ringspot virus

1. Disrupt systemically infected *Nicotiana clevelandii* leaf tissue in 0.05 M phosphate buffer, pH 7.5 (2 ml/g leaf), using a mechanical blender.
2. Filter extract through muslin and add 8.5% (v/v) *n*-butanol.
3. After 20 min remove coagulum by centrifuging 10 min at 5000 *g*.
4. Centrifuge supernatant fluid 3 h at 75 000 *g* and resuspend sediment in phosphate buffer. Leave overnight at 4°C.
5. Centrifuge 5 min at 5000 *g*.
6. As 4.
7. Centrifuge 5 min at 5000 *g* and place supernatant fluid on a column of 2% agarose beads (Sepharose 2B).
8. Pool the eluted fractions containing the virus and centrifuge 90 min at 140 000 *g*.
9. Resuspend sediment in phosphate buffer and centrifuge 5 min at 5000 *g*. Supernatant fluid is the purified preparation.
10. Store purified preparation at 4°C, in the presence of 0.02% sodium azide.

6.6 Purification schemes

The methods of purification we have described can be combined in many ways, and the best combination must be found for each virus. Two examples of satisfactory combinations are given in Table 6.2. Other schemes are given by Steere (1959) and Noordam (1973).

6.7 The purified virus preparation

The ideal product of virus purification would be a pure suspension of fully infective virus particles. However purified preparations are invariably only partially pure, and despite all precautions usually contain some traces of host constituents, pigments, chemicals used in purification, and damaged virus particles, in addition to the intact virus particles. In practice the specific infectivity (i.e. infectivity per unit weight of virus particles) of the purified preparations of many viruses decreases as purity increases.

6.7.1 *Suspension medium*

The stability of the virus particles and the intended use of a purified preparation should determine the buffers used during purification. There is, for example, little point in estimating the phosphorus content of virus preparations made using phosphate buffers. However buffer solutes can be altered by prolonged dialysis, or Sephadex column chromatography, though these methods may not remove substances incorporated into the virus particles by chemical reaction. Different suspending media are obviously best for different purposes; for example preparations for serology, especially those for injecting animals, should be suspended in saline, whereas a solution of ammonium acetate, which sublimes *in vacuo*, is best for suspending virus particles for electron microscopy.

6.7.2 *Storage*

When stored in suitable ways, virus preparations will retain full activity for long periods. The particles of most viruses are best kept in suspension at 0 to 4°C with an added anti-microbial compound such as a drop of chloroform or sodium azide solution. Alternatively they may be frozen or dried while frozen (lyophilized). Whereas little infectivity may be lost when infective sap or tissue is frozen, much more can be lost if the preparation is pure, especially when it is completely free of salts and/or at the isoelectric point of the virus particles. Less infectivity is lost when sap or peptone and glucose (5 to 10% of each) is added to the purified preparation before freezing (Hollings and Stone, 1970).

6.7.3 *Components of preparations of virus particles*

Rarely do all the virus particles in a purified preparation appear identical, and this hetero-

geneity can be studied using the techniques described in Chapters 7 to 9. Preparations of TMV, for example, contain tubular particles most of which are about 300 nm long, but also others appreciably shorter or longer than this, and in

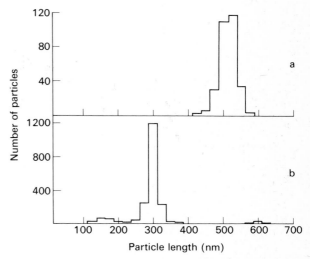

Fig. 6.9 Length distributions of virus particles. **a,** Potato virus X; modal length 513 nm (Bercks and Brandes, 1961). **b,** TMV; modal length 299 nm (Brandes and Chessin, 1965).

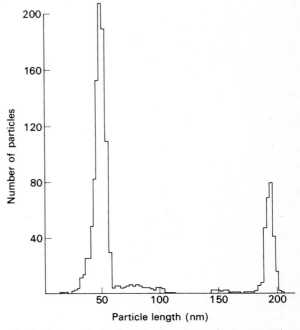

Fig. 6.10 Length distribution of particles of tobacco rattle virus, strain CAM. Modal lengths are 52 and 197 nm (Harrison and Woods, 1966).

some preparations the 300 nm particles are out-numbered by the others. Some of these shorter and longer particles seem to be produced by the breakage or aggregation of 300 nm particles during the experimental manipulations, and they are fewest when the manipulations are minimal, for example when the exudate from a cut edge of an infected leaf is examined by electron micro-scopy. Even then, some particles of different lengths are found (Fig. 6.9) and these probably also occur in intact cells.

The particles of each virus are usually of one or more characteristic sizes. Each virus with iso-metric particles has particles all of the same diameter, or of a few discrete size classes, and each virus with elongated particles has particles of the same diameter and usually mostly of one or a few characteristic lengths (Brandes, 1964; Figs. 6.9

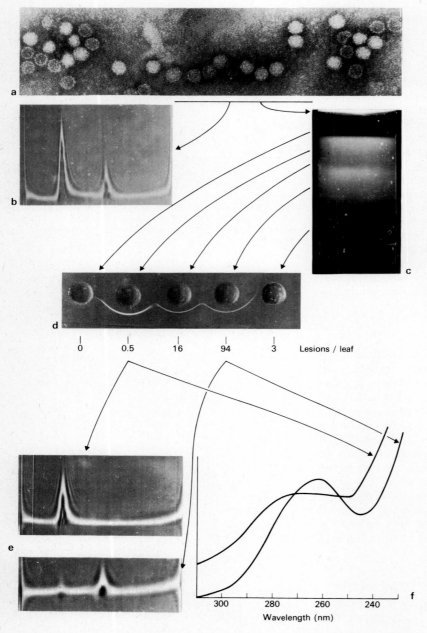

Fig. 6.11 Properties of the two components in a preparation of dulcamara mottle virus, a tymovirus. **a** and **b,** Electron micrograph and sedimentation diagram, respectively, of an unfractionated virus prepara-tion, showing the two components. **c,** Tube showing the two light-scattering zones obtained by centrifuging the virus preparation into a sucrose density gradient. **d,** immunodiffusion test showing that both zones contain the same antigen (confluent precipitation bands); figures below show that only the lower zone contained infective particles. **e,** Sedimentation diagrams of samples from the upper and lower zones, showing that the upper zone contains top component and the lower zone contains predominantly bottom component, plus a little top component. **f,** Ultra-violet absorption spectra of samples from the upper zone (protein particles) and lower zone (nucleoprotein particles).

Fig. 6.12 (opposite) Separation of components in virus preparations using an analytical ultracentrifuge. **a,** Schlieren diagram of cowpea chlorotic mottle virus particles sedimented to equilibrium in CsCl solution. Upper pattern, unfractionated virus prepara-tion containing three components, of density 1.356, 1.360 and 1.364 g/cm³; they have slightly different RNA contents but are not separated by moving

and 6.10). These lengths are called the 'normal' or modal lengths.

Particles shorter than the modal length can be separated from the others by density-gradient centrifugation or by chromatography on chips of agar gel (§§6.4.1 and 6.4.2). They are chemically similar to the larger particles but have little if any infectivity, and in general it seems likely that the modal length is the minimum infective length. This is probably because the length of the virus particle is determined by the length of the nucleic acid strand it contains (see Chapter 5), and the complete intact nucleic acid strand is essential for infection.

In preparations of turnip yellow mosaic virus there are two main kinds of isometric particle but only one contains RNA (Markham and Smith, 1949). The other kind, which is a protein shell of the same size and construction as the nucleoprotein, apparently plays no role in infection. Some properties of two such kinds of particle produced by another tymovirus are illustrated in Fig. 6.11. Empty protein coats are produced not only by tymoviruses but also by several other viruses with isometric particles, including nepoviruses and comoviruses. The different kinds of particle produced by any one of these viruses have the same size and protein coat but different densities, and different sedimentation coefficients (§9.4.3).

Most viruses with multipartite genomes (Chapters 11 and 12) produce nucleoprotein particles of two or more types. In viruses with elongated particles, such as tobacco rattle and barley stripe

mosaic viruses, the particles have the same density, but two or more modal lengths corresponding to the sizes of the RNA species they contain (Cooper and Mayo, 1972; Jackson and Brakke, 1973). Different isolates of tobacco rattle virus provide a particularly interesting example. All produce particles about 190 nm long but all also produce shorter particles whose modal length depends on the isolate (Harrison and Woods, 1966). With alfalfa mosaic virus, the bacilliform particles are of four lengths corresponding to the four characteristic RNA species, three of which are indispensable parts of the genome (Gibbs, Nixon and Woods, 1965; Bol, van Vloten-Doting and Jaspars, 1971). In these viruses, the different sized particles have the same percentage of RNA but different sedimentation coefficients.

In multipartite genome viruses with isometric particles, the nucleoprotein particles usually have the same diameter but different RNA contents, and hence different densities. Thus the three kinds of nucleoprotein particle produced by cowpea chlorotic mottle virus have the same size but slightly different densities and are separable by equilibrium sedimentation in CsCl gradients (Bancroft and Flack, 1972; Fig. 6.12), although not by rate zonal sedimentation. The three kinds of particle produced by raspberry ringspot virus have the same diameter but different sedimentation coefficients and densities, and contain 0%, 30% and 44% RNA (Murant *et al*, 1972; Fig. 6.12). However the two kinds of isometric particle of pea enation mosaic virus have different sizes and sedimentation coefficients but similar per-

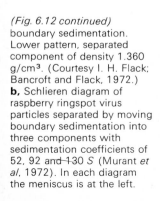

(Fig. 6.12 continued) boundary sedimentation. Lower pattern, separated component of density 1.360 g/cm³. (Courtesy I. H. Flack; Bancroft and Flack, 1972.) **b,** Schlieren diagram of raspberry ringspot virus particles separated by moving boundary sedimentation into three components with sedimentation coefficients of 52, 92 and 130 S (Murant *et al*, 1972). In each diagram the meniscus is at the left.

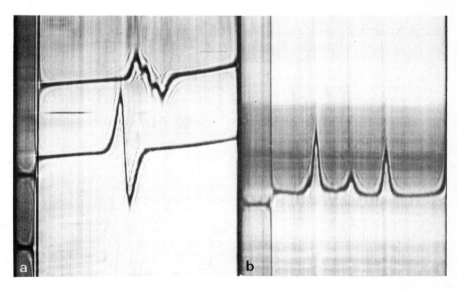

centages of RNA (Hull and Lane, 1973), and the three kinds of isometric particle of tobacco streak virus differ in size but not percentage RNA (Lister, Ghabrial and Saksena, 1972).

Preparations of virus particles may also contain monomers or small aggregates of virus protein. Thus differences between the antigenic determinants of TMV particles and disaggregated TMV coat protein were most easily found when special precautions were taken to exclude the unassembled virus protein which normally occurs in preparations of TMV particles (van Regenmortel and Lelarge, 1973).

These various components of preparations of virus particles differ in size and composition, and they can be distinguished by appropriate chemical, physical, biophysical and serological techniques (Chapters 8 and 9), and by their roles in infection (Chapter 11). There is increasing evidence that most of these components occur in infected cells and are not simply artifacts of purification. Several have special biological functions, and others may be precursors or breakdown products of intact virus particles, or by-products of virus replication.

6.7.4 *Components of preparations of virus genome nucleic acid*

Purified preparations of virus particles are the best source of the nucleic acid genomes of plant viruses (§5.6.1). However, such preparations of virus nucleic acid are in practice neither perfectly homogeneous nor absolutely pure. Nucleases are among the most important impurities, for unless suitable precautions are taken they can quickly degrade nucleic acid molecules and abolish infectivity. Virus coat protein is another characteristic impurity. However, the best preparations of TMV-RNA contain less than one molecule of TMV coat protein for ten RNA molecules, and there is no reason to suppose that this protein, or any virus component other than RNA, is necessary for infectivity. The same holds for many other viruses, but with alfalfa mosaic virus the coat protein is needed for infection unless the inoculum contains, not only the three largest species of RNA found in the particles, but also the fourth and smallest species of RNA (Bol, van Vloten-Doting and Jaspars, 1971).

Preparations of virus RNA may also be contaminated with host RNA. The predominant RNA species found in plants are single-stranded and have molecular weights of about 2.5×10^5,

and 0.56×10^6, 0.7×10^6, 1.1×10^6 and 1.3×10^6 (Loening, 1968). The 0.7×10^6 and 1.3×10^6 molecular weight species come from cytoplasmic ribosomes, and the 0.56×10^6 and 1.1×10^6 molecular weight species come from chloroplast ribosomes (§6.1). Mitochondrial ribosomal RNA is probably similar in size to that from chloroplasts. These host RNA species are useful as markers in polyacrylamide gel electrophoresis of preparations of virus RNA (§9.5.2).

Electrophoresis in polyacrylamide gels and centrifugation show that preparations of single-stranded RNA of several viruses predominantly contain RNA molecules of a specific size, which depends on the virus. Viruses with multipartite genomes produce more than one nucleic acid species, usually of different sizes but with similar nucleotide ratios (Chapter 2; Fig. 6.13). In some preparations of RNA from tobacco ringspot (Schneider, 1971) and tomato black ring (Murant *et al*, 1973) viruses an extra 'satellite' RNA is also found. In both instances this is smaller than the two species of virus RNA (Fig. 6.13), is not

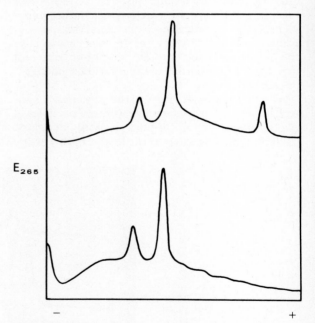

Fig. 6.13 Electrophoresis in polyacrylamide gel of the single-stranded RNA from particles of tomato black ring virus. Lower pattern, ultra-violet absorption scan showing the two essential RNA species, of molecular weight of about 2.5×10^6 and 1.5×10^6. Upper pattern, scan showing the two essential species and a third, 'satellite', species of molecular weight 0.5×10^6, which is also produced when the inoculum contains it. Migration is from left to right. (After Murant *et al*, 1973.)

needed for replication and is produced only when the inoculum contains it. Virus RNA preparations also contain variable amounts of non-infective molecules smaller than the predominant RNA species, many of which are probably derived from them.

The double-stranded RNA from particles of viruses allied to clover wound tumor virus is in several pieces which differ in size and are not linked together in preparations of the RNA. Normal isolates of wound tumor virus produce twelve such pieces (Reddy and Black, 1973a), and rice dwarf virus produces ten. These preparations are not infective (§11.3.1).

There is relatively little information about the components of preparations of plant virus DNA, that of cauliflower mosaic virus being the best studied example. Preparations of this DNA contain strands with different lengths, and molecules of the predominant length occur in both circular and linear forms, which can be detected both by electron microscopy and by analytical centrifugation (Russell *et al*, 1971; Shepherd and Wakeman, 1971).

6.7.5 *Comparative properties of virus particles and virus nucleic acid*

The nucleic acid contained in the particles of different viruses represents the virus genome at a variety of points in the replication cycle, and preparations of some virus nucleic acids are infective but of others are not (Chapter 11). Infective virus nucleic acids usually produce fewer infections than the same amount of nucleic acid in the form of nucleoprotein particles. The relative infectivity, however, depends on the virus, the assay host and its physiological state, and the method used to prepare the nucleic acid. For instance, when carefully made preparations of TMV-RNA are assayed by inoculation to *Nicotiana glutinosa*, they are usually only about 1% as infective as TMV particles containing the same amount of RNA (Gierer and Schramm, 1956), but the relative infectivity of the nucleic acid preparations can be enhanced by keeping the assay plants in darkness or at 37°C for two days before inoculation (Bawden and Pirie, 1959). By contrast, when assayed on French bean, RNA preparations from tobacco rattle virus are about 5% as infective as intact virus (Harrison and Nixon, 1959b) and those from tobacco necrosis virus may be as infective as virus particles (Kassanis, 1960).

The loss of its protein coat makes virus nucleic acid accessible to several inactivators and, in general, virus nucleic acid is much more labile than virus nucleoprotein. Some of the properties of infective single-stranded RNA and infective nucleoprotein are compared in Table 6.3.

Table 6.3 Some properties of infective virus particles and single-stranded virus RNA

	Particles of virus nucleoprotein	Single-stranded virus RNA
Visibility in the electron microscope	Easily observed using negative stain	Special methods needed (Kleinschmidt *et al*, 1962)
Infectivity per unit RNA	Usually greater than of virus RNA	Usually less than of virus particles
Sedimentation coefficient	50–1000 *S*	14–45 *S*
Inactivation by pancreatic ribonuclease (1 μg/l at 20°C in 0.015 M NaCl +0.0015 M sodium citrate)	Infectivity of most viruses unaffected	Infectivity rapidly lost
Reaction with antiserum to virus particles	Strong	None

6.8 Crystallization of plant viruses

One criterion long used by chemists to assess the purity of chemicals is their ability to crystallize. In 1926 Sumner aroused great interest when for the first time he crystallized an enzyme, urease. Thus it was natural that when, ten years later, viruses were being purified for the first time, virologists should attempt to crystallize them, for the then-known properties of viruses suggested that they were similar to enzymes. When the first crystals and paracrystals of viruses were made, much speculation was provoked, for viruses were thought to be living and to reproduce, yet the ability to crystallize was thought to be characteristic of inanimate matter. Few people at that time realized that when a number of identical objects are held in a confined space they tend to pack into a regular array that has a minimum energy.

Takahashi and Rawlins (1932) viewed preparations of TMV with polarized light and found that some containing less than 1% of virus particles did not change the plane of polarization (i.e. they were isotropic) when undisturbed, but did change the plane of polarization when flowing (i.e. they showed streaming birefringence), and from this they deduced that the virus particles were anisometric – either disc or rod-shaped. Later it was found that when a purified preparation of TMV containing more than about 10% of virus particles is stored, it separates into two layers. In the upper layer the virus particles are in a small enough concentration for Brownian movement to keep them in random orientation and thus the layer is isotropic until it is stirred, when the shearing forces in the stirred liquid align the particles and it becomes birefringent. In the lower layer the particles are close-packed and aligned in paracrystals, so that the layer is bire-fringent even when not stirred (Bawden and Pirie, 1937). Flow birefringence is also shown by con-centrated preparations of other viruses with anisometric particles, including potato X, barley stripe mosaic and tobacco rattle viruses.

True three-dimensional crystals of TMV particles often occur in virus-infected plants (Fig. 3.18), but are difficult to prepare artificially; minute ones may be prepared by co-precipitating a pure salt-free virus preparation with either cytochrome *c* or histone (Milne, 1967). Three-dimensional crystals are more readily made using isometric virus particles. The first such particles to be crystallized were those of tomato bushy stunt virus (Bawden and Pirie, 1938), and many other examples are now known. Crystals are made from virus preparations that are concen-trated and pure. They form when the particles are in conditions that limit solubility, but the most suitable conditions for crystallization differ for different viruses. The particles of some viruses crystallize when preparations are stored in the cold with added ammonium sulphate (tomato bushy stunt, tobacco necrosis, turnip yellow mosaic and southern bean mosaic viruses) or ethanol and acid (turnip yellow mosaic virus). Satellite virus particles will not crystallize out of solutions containing salts but will do so from salt-free ones, and also, like the particles of turnip yellow mosaic virus, when sedimented in the ultracentrifuge.

The crystals of virus particles differ greatly in type; some are rhombic dodecahedra (tomato bushy stunt), others are hexagonal plates (satel-lite virus) or small birefringent needles (turnip yellow mosaic). The particles of closely related viruses may consistently crystallize in different forms, and those of tobacco necrosis virus, for example, can crystallize in two forms, isotropic dodecahedra and birefringent bipyramids.

Fig. 6.14 Electron micrograph of a small crystal of tobacco necrosis satellite virus particles shadowed with metal.(Courtesy of R. W. G. Wyckoff.)

In electron micrographs these crystals usually show a close-packed structure (Fig. 6.14), but X-ray diffraction analysis shows great differences between the wet crystals of particles of different viruses; differences that are often obscured by shearing and shrinkage when the crystals are prepared for electron microscopy. The particles of tomato bushy stunt virus are arranged in wet crystals in a simple body-centred cubic lattice (Fig. 6.15a) in which each particle has eight nearest neighbours, and there are two particles, which have the same orientation, in each unit crystallographic cell (the simplest repeating unit in the crystal, and analogous to the asymmetric unit of structure in individual virus particles; Chapter 5). By contrast the particles of turnip

yellow mosaic virus are usually arranged in a simple diamond lattice (Fig. 6.15b; Klug, Longley and Leberman, 1966) so that each particle has four nearest neighbours arranged at the corners of a regular tetrahedron; each particle is orientated at right angles to its nearest neighbours so the unit cell contains eight particles.

Crystals of virus particles are not made solely to assess the purity of a virus preparation nor even for aesthetic reasons, but also as a way of holding a large number of virus particles in a regular orientation so that their structure may be deduced from the way they diffract electromagnetic radiation (Chapters 5 and 9), particularly that of 0.05 to 0.25 nm wavelength (X-rays).

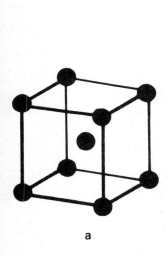

a

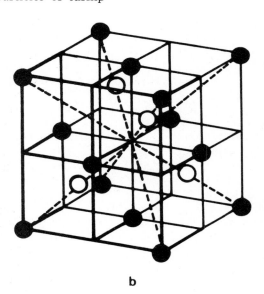

b

Fig. 6.15 Basic crystallographic unit cells of wet crystals of particles of **a,** tomato bushy stunt virus; body-centred cube, two particles per unit cell, all particles in same orientation, cube edge of length 38.6 nm. **b,** Turnip yellow mosaic virus; simple diamond lattice (based on tetrahedron), eight particles per unit cell, in two orientations at right angles, cube edge of length 70 nm. (After Klug, Longley and Leberman, 1966.)

Chapter 7

Infectivity assay

Quantitative estimates of virus content may be based on any property of the virus that can be measured. Different kinds of test are based on different properties and therefore results obtained from one type of test do not necessarily parallel those from other types. Thus the infectivity of a virus preparation depends primarily upon the amount of undamaged virus nucleic acid it contains, whereas the amount of virus detected in the same preparation by a serological test will usually depend upon the amount and state of aggregation of virus protein. Some treatments affect one property but not another; for instance, infectivity may be decreased by exposure to ultra-violet radiation, or by harsh purification methods, but neither will necessarily affect the amount of serologically active protein in the preparation. Both treatments affect the *specific infectivity* – that is, the infectivity per unit weight – of the virus. Therefore it is a sound principle when testing an unfamiliar virus, or a familiar virus in unfamiliar conditions, to assay the virus samples by more than one method. In this chapter we discuss infectivity tests, in the following chapter serological tests, and in Chapter 9 we describe some useful chemical and physical methods for studying viruses.

7.1 Basic information

In an infectivity test, host cells are exposed to virus particles, and the proportion of cells that becomes infected, and/or the time of appearance of symptoms, is recorded and used as a measure of infectivity.

If all virus particles were infective and all cells readily infectible by single particles, infectivity tests would give absolute measures of virus concentration. In practice, neither of these conditions is met, and 10^5 or more plant virus particles must usually be inoculated to infect an intact plant cell. When more than one active component is needed to cause infection the number is still larger; about 10^7 for cowpea mosaic virus (two active com-

ponents; van Kammen, 1968) and about 10^9 for alfalfa mosaic virus (three or four active components; van Vloten-Doting *et al*, 1970; Bol and van Vloten-Doting, 1973). These numbers compare with the 2 to 20 particles needed to cause infection of intact bacteria by some bacteriophages, or of cultured animal cells by some animal viruses. Indeed insects are more susceptible than plants to those viruses that infect species of both kinds. For instance, the leafhopper vector of clover wound tumor virus becomes infected when injected with only about 400 virus particles (Gamez and Black, 1967), and cultured cells of the vector *Aceratagallia sanguinolenta* are $10^{3.7}$ times more susceptible to infection with potato yellow dwarf virus than are the leaf cells of *Nicotiana rustica* (Hsu and Black, 1973).

The probable main reason for these differences in sensitivity is that rubbing leaves is an inefficient way of wounding their rigid outer cell walls so that virus particles can enter and infect. Thus when the cell walls are removed, only about 400 particles per tobacco leaf protoplast are adsorbed to cause 50% infection with cowpea chlorotic mottle virus, and, as the genome of this virus is divided among three kinds of particle, the theoretical minimum number of particles needed to cause 50% infection is 4 to 5 (Motoyoshi, Bancroft and Watts, 1973). The actual number is therefore only about 100 times the theoretical minimum. Thus despite their limitations, and the importance of other factors described in Chapter 4, infectivity tests are by far the most sensitive assays available for most viruses. Also, in many instances virus can be assayed in crude leaf extracts.

In most infectivity tests, host cells are exposed to a relatively small number of virus particles so that only a few cells are infected, each by a single dose of virus; thus the proportion of infected cells gives an estimate of the virus concentration. We called this the 'minimum dose infectivity test' by contrast with the 'multiple dose infectivity test' in which organisms are infected simultaneously by several infective doses of virus, and a dose-dependent phenomenon, such as the time of appearance of symptoms, is used as a measure of the virus concentration.

Until recently most attempts to infect, or to detect the infection of, single plant or animal cells with plant viruses failed, and therefore infection of the initially infected cells is detected by allowing the virus in them to spread to other cells, in which it produces characteristic symptoms. When, for example, a virus sample is manually inoculated to the leaf of a local lesion host, the virus spreads from each initially infected cell to surrounding cells and all die or become abnormal in colour, thus producing the local lesions; or in other circumstances starch lesions may be produced (§3.4). Some virus-infected hosts show no symptoms, and a more complex test system is needed; for example, leafhoppers infected with clover wound tumor virus by injection show no symptoms, and the virus in them must be detected either serologically (Reddy and Black, 1966), or by electron microscopy (Gamez and Black, 1968), or by putting them to feed on, and hence to inoculate, individual indicator plants. Similarly, viruses transmitted by soil-inhabiting nematodes and fungi to the roots of test plants often fail to reach the shoots, so they must be detected by inoculating sap from the roots to suitable indicator plants. It is important to be cautious when interpreting the results of indirect tests of this type because many factors in addition to the proportion of initially infected cells may affect the amount of virus detected. For example, in the leafhopper injection experiment described above, some individual leafhoppers may be infected, yet fail to transmit the virus to plants; thus if each insect is placed on one indicator plant, the proportion of infected indicator plants will be the product of the proportion of leafhoppers infected and the proportion of infected leafhoppers that transmit (Chapter 13).

7.2 Types of infectivity tests

In the earliest infectivity tests the virus preparations were injected (or pricked) into the test plants; however, later experiments (Holmes, 1929; Samuel, 1931) showed the great advantages of rubbing the inoculum over the leaves of the test plant.

Originally only hosts that were systemically infected were used as test plants, and the proportion of plants that became infected was used as a measure of the infectivity of the preparation. Such tests are time-consuming, and for accuracy each

inoculum has to be tested on many plants. This is practicable when the test plants are small and easily grown in large numbers (for example, barley for assaying barley stripe mosaic virus; Brakke, 1970) but less so otherwise. However Holmes (1928, 1929) found that necrotic local lesions developed in leaves of certain *Nicotiana* species, such as *N. glutinosa*, a few days after they had been rubbed with TMV preparations. Furthermore the number of local lesions in each leaf depended on the concentration of virus in the inoculum. Thus, whereas in previous tests the minimum infectible unit was a whole plant, in the local lesion test it is a small group of leaf cells. Holmes (1929) estimated 'that one test plant of *Nicotiana glutinosa* . . . serves the purpose for which at least several hundred *N. tabacum* plants are required . . .'. The local lesion infectivity test is therefore used in preference to the whole plant assay whenever suitable lesion-forming test plants, and virus strains, can be found. Local lesions can also develop when the virus has been transmitted by a vector. For example, local lesions of turnip mosaic virus develop in tobacco leaves infected using the aphid *Myzus persicae*, and those of tobacco necrosis virus form in roots of mung bean exposed to virus-carrying zoospores of the fungus *Olpidium brassicae*.

Plant viruses that infect their vectors can be assayed using the vector as the host. With many such viruses whole vectors are the minimum infectible units (Whitcomb, 1969), but a few, such as clover wound tumor virus, can be assayed using cultures of leafhopper vector cells and, by using fluorescent antibody to the virus, determining the proportion of cells infected (Chiu and Black, 1969; Chapter 8). Plant protoplasts could theoretically be used in the same way as cultured vector cells and there are developments in this direction (Takebe and Otsuki, 1969; Motoyoshi *et al*, 1973b).

The ways in which viruses are experimentally transmitted for such purposes as infectivity tests are described in Chapter 4, and reviews giving further details are listed in the appendix.

7.3 Inoculum concentration

When accuracy is needed in a 'minimum dose infectivity test', the virus inoculum must be used at a concentration that will only infect a small proportion of the infectible units (e.g. cells); preferably less than about 25%. This is because when a large proportion of the infectible units are infected, many of the infections are likely to result from multiple infective doses of virus, and the total number of infective doses in the inoculum will be underestimated. Thus it is often advisable to do a pilot experiment to find the best inoculum concentration to use in a full-scale infectivity test. Of course, in many infectivity tests, the total number of infectible cells is unknown, but if the inoculum is assayed at several different concentrations, and the results expressed graphically, then the effect of multiple infection is noticeable as a flattening of the 'infectivity/dilution' curve at high virus concentrations (Fig. 7.1, line *a*).

When the proportion of infectible units that are infected increases much above a quarter, so too does the proportion of infections caused by more than one infective dose of virus. It may then be useful to have a record of the time of appearance

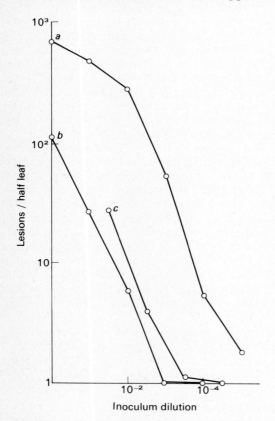

Fig. 7.1 The number of lesions produced in leaves of *Chenopodium hybridum* manually inoculated with dilutions of brome mosaic virus preparations. *a*, Infective wheat sap diluted with 0.1 M phosphate buffer (pH 7.0). *b*, Infective sap diluted with distilled water. *c*, Purified virus diluted with phosphate buffer. (Redrawn from Rochow (1959) with permission.)

of symptoms (multiple-dose infectivity test), for Holmes (1932) and others (review; Brakke, 1970) have shown that, for some viruses in some plants, the incubation period is correlated with the inoculum concentration. Whether this applies to all viruses is not known. However Whitcomb (1969) suggested that when assay is by insect transmission, the minimum interval between acquisition and inoculation of virus by the vector (latent period) is a better estimate of the amount of virus acquired than is the proportion of insects that transmit.

The optimum 'virus/infectible unit' ratio is usually obtained by altering the virus concentration. The concentration of manually inoculated (or injected) viruses can be altered simply by diluting the inoculum. However, in vector transmission tests the individual vector and individual plant are usually the minimum units, and thus the probability of transmission can be easily varied only by testing a range of numbers of vector individuals per test plant. This is only necessary and possible when, as often happens, the probability of transmission by a single vector is small.

Knowledge of the effect of dilution on the apparent infectivity of an inoculum is most important for the design and interpretation of infectivity tests. Only rarely are the numbers of lesions produced by a series of dilutions of a virus preparation proportional to the virus concentration. When the results of such a test are expressed graphically (Fig. 7.1), the line representing the results is sigmoid, provided that a large enough range of virus concentrations has been tested. The central part of such a 'dilution/infectivity' curve has the steepest slope, which may be near unity if a single virus particle is sufficient to cause infection, whereas at larger and smaller inoculum concentrations the apparent infectivity of the inoculum is less affected by dilution. Different virus preparations will be most accurately compared when their infectivities fall on the steepest part of the curve. For example in the tests giving the results shown in Fig. 7.1, differences between inocula were most accurately determined when 20 to 55 lesions per half leaf were produced. One consequence of the sigmoidal shape of the 'dilution/infectivity' curve is that the dilution endpoint, which is defined as the least the inoculum needs to be diluted to make it uninfective, is a very crude measure of virus concentration, although it is often quoted.

The 'dilution/infectivity' curve is greatly affected by many factors such as the species, variety and condition of the test plants used, the virus being tested and the constituents of the inoculum (Fig. 7.1). For example, for many viruses the slope of the steep part of the curve is less than unity; for some, such as tobacco rattle virus in *Phaseolus vulgaris*, it is close to unity, but with viruses needing more than one kind of particle to cause an infection, such as alfalfa mosaic or cowpea mosaic viruses, it may be greater than unity (van Vloten-Doting, Kruseman and Jaspers, 1968).

Anomalous 'dilution/infectivity' curves can be obtained when there are inhibitors of infection in the inoculum, for usually the effect of the inhibitor is 'diluted away' faster than the infectivity of the virus, and hence the inoculum may seem to *gain* infectivity during the first dilution steps (Chapter 4 and Table 4.2). Inoculum additives, such as phosphate ions or carborundum, increase the sensitivity of an assay, but may also alter the shape of the 'dilution/infectivity' curve.

7.4 Design of tests

The test organisms (plants or vectors) should be as uniformly susceptible as possible; they should preferably be of one age and size and they should have been reared together. Uniformity of plant susceptibility is often increased by removing the apex and all unwanted leaves, and keeping the plants in darkness for a few hours before inoculation. Most experiments on the causes of variation in the susceptibility of plants to viruses, described in Chapter 4, have used hosts that produce local lesions and viruses that are manually transmissible. Other combinations of virus/host/site of inoculation may respond differently. For example, preinoculation and postinoculation temperature seem to affect susceptibility to infection by aphid-inoculated virus differently from infection by manually-inoculated virus (Welton, Swenson and Sohi, 1964).

When all the leaves of a uniform group of plants are manually inoculated with the same inoculum of a lesion-forming virus, and the lesions that develop are counted, it is found that the numbers that develop in opposite halves of a leaf usually differ only slightly. There is a greater difference in the numbers in different leaves of the same plant, and greater still in the numbers in leaves of different plants. Thus when two inocula are compared, maximum accuracy is obtained by testing them on the opposite halves of as many leaves as possible, and minimum accuracy by inoculating them on different plants.

When whole plants or vectors are used as the minimum infectible units, it is important to test each inoculum on to a large group of test individuals, and the individuals should be allocated at random to each group. Where there is obvious heterogeneity in the test individuals, in size for example, it is best first to group the individuals according to their sizes, and then to allocate individuals from each of these size groups to each of the batches receiving different inocula. Then, when the results of the experiment have been recorded, it will be possible in the statistical analysis to estimate and discard the variation in susceptibility that is correlated with size of the test organism.

In local-lesion tests, great use can be made of experimental design to increase accuracy. Often more than two inocula are to be compared; thus there is more than one comparison to be made and several plants must be used. The inocula should be distributed among the leaves of such a group of plants so that each inoculum is compared with every other inoculum on opposite half-leaves an equal number of times. Incidentally, each inoculum should be tested on an equal number of 'right' and 'left' half-leaves because most experimenters inoculate the two halves in slightly different ways and consistently obtain different numbers of lesions on right and left half-leaves. Each comparison should also be made at least once at each leaf position on the plants, and each inoculum should occur at least once on each test plant. In this way the variation between plants, between leaves in different positions and between halves of the same leaf can be estimated and removed in the analysis of results. Figure 7.2a shows a layout of this sort, designed to compare four inocula on the two primary leaves of French bean plants. Where whole leaves instead of half-leaves are used as the minimum test unit, each inoculum should, where possible, be tested the same number of times at each leaf position and at least once on each plant. One design for this is the Latin square; Fig. 7.2b shows a Latin square design for comparing six inocula on six leaves of each of six test plants. There are a great many other possible designs, some of which give results that are easier to analyse than others (Kleczkowski, 1968).

The accuracy required and the variability of the test host determine the number of replicate test units needed in each experiment. For

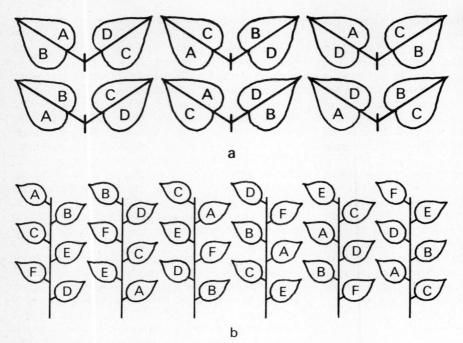

a

b

Fig. 7.2a A possible design for comparing four inocula (A–D) on opposite halves of the primary leaves of *Phaseolus vulgaris*. Each inoculum occurs opposite every other an equal number of times, occurs once on each plant and on an equal number of left and right half-leaves

b, A Latin square design for testing six inocula (A–F) on six leaves of six plants. Each inoculum occurs once on each plant and once at each leaf position, and other arrangements may be obtained by randomizing the rows and the columns in this design; for other possibilities see Fisher and Yates (1963).

example, in the experiments using *Chenopodium hybridum* that provided the data for Fig. 7.1, Rochow (1959) found that a 20% difference in inoculum concentration could be detected ($p = 0.01$) with a minimum of thirty-six half-leaf comparisons, whereas we find a similar accuracy can be obtained when comparing four inocula of tobacco necrosis virus by using only twelve half-leaves of French bean per inoculum.

7.5 Analysis of results

In the past many plant virologists analysed the results of infectivity tests making the assumption that the infection of each plant depended simply on the chance that one of a group of uniformly infective virus particles infected one of a number of uniform infectible sites, and thus at small infection levels the results would fit the Poisson distribution. This certainly is not true for some multi-component viruses, and is probably untrue for most other plant viruses. However, the infection of whole plants or vector insects has not been shown to depart from the Poisson distribution (Brakke, 1970). For local-lesion plants, Kleczkowski's (1950) results with three viruses suggest that individual infection sites differ in susceptibility so that different doses of virus are needed to cause an infection; the logarithms of the

minimum effective doses are normally distributed. However other hypotheses have been proposed (Furumoto and Mickey, 1967a, b) and discussed (Kleczkowski, 1968). No detailed analyses have been made for viruses that need particles of different kinds to give an infection, but their lesion numbers seem largely dependent upon the concentration of the least abundant component.

Most statistical tests for analysing the significance of results assume that the variate has a normal or Gaussian distribution, but Kleczkowski (1949) found that local-lesion numbers (two different virus/host combinations) had a skew distribution, and that their variance increased with their mean. He showed that lesion numbers of more than 10 per leaf or half-leaf are best transformed for analysis to values of z, using the formula $z = \log_{10}(x + c)$, where x is the number of lesions on each leaf or half-leaf and c is a constant (usually between 5 and 20) that can be determined for each experiment. For smaller numbers of lesions a more complex transformation (Kleczkowski, 1955) is required:

$$z = \log_{10} \tfrac{1}{2}[x + c + \sqrt[2]{(x^2 + 2cx)}].$$

Results of infectivity tests are expressed in various ways. One way is to assess the dilution end-point of the inoculum, but as shown above this is an inaccurate measure of concentration.

The relative infectivity of different preparations can be determined by comparing the infectivity of various dilutions of each preparation. Table 7.1 shows how the infectivity of a virus preparation can be compared with that of a standard inoculum, and hence with the infectivity of a second preparation, which has been compared with the standard using a second group of plants. The virus preparation is used at a range of dilutions, and the dilution at which it would produce the same number of lesions as the standard is estimated from a plot of 'log (adjusted lesion number)' against 'log (inoculum concentration)' (see Table 7.1). When the virus content of one of the preparations has also been determined in some other way (e.g. serology, nucleic acid content), and assuming that the specific infectivity is the same for all inocula, then the absolute virus content of other preparations can be estimated.

In tests using whole plants (or vectors) one can use a range of inoculum concentrations and estimate the inoculum concentration that would infect say 50% or 10% of the test plants; when

Table 7.1 Estimating the relative infectivities of two virus preparations by comparison with a standard inoculum

Dilution of inoculum	Total lesions in 8 half-leaves		Lesion numbers adjusted arithmetically to make standard = 1000	
	A	B	A	B
1/25	1795	—	3540	—
1/125	410	2820	799	7000
1/625	117	670	231	1664
1/3125	—	156	—	387
Standard inoculum				
	507	403	1000	1000

By plotting 'log (adjusted lesion number)' against 'log (inoculum concentration)', it is found that:
 A = 1000 when diluted 1/102
 B = 1000 when diluted 1/1100
Hence the relative infectivities of the preparations (standard = 1) are:
 A = 102
 B = 1100

the 'dilution/infectivity' curve is not known it is dangerous to extrapolate, and better to have a sufficient range of infectivity determinations to interpolate. Alternatively one can calculate some relative estimate of the infectivity either by assuming that the results fit the Poisson distribu-

tion or by the maximum likelihood estimator (Gibbs and Gower, 1960; Finney, 1964). If, for example, one assumes the Poisson distribution is applicable in an experiment in which 45 out of 200 leafhoppers infected plants after injection with 0.1 μl of a virus suspension, then the first term of the Poisson series, $P = e^{-m}$, will give the proportion that did not transmit the virus, where m is the mean number of infective doses injected into each leafhopper. Thus $m = 2.3 \log_{10}(200/155) = 0.255$. This result may be expressed in three different ways:

1 0.1 μl of the preparation contained 0.255 infective doses;
2 50% of a group of leafhoppers would be expected to become infective (50% infective dose, or ID_{50}) when each was injected with $2.3 \log_{10}(100/50)$ $(0.1/0.255) = 0.27$ μl of the preparation;
3 the inoculum contained $10^3/0.27 = 3.7 \times 10^3$ ID_{50} per ml.

In vector transmission studies, the 'infectivity' of the vector is calculated in a similar way and can be expressed as the probability that a single vector will infect a plant. Often this probability is small, and to economize on the number of test plants used in an experiment it may be best to test several vector individuals on each test plant; this 'multiple transfer' in effect increases the virus concentration in the test. From the results, the probability that single vector individuals would have infected the test plants can be calculated, again either assuming the Poisson distribution or by the maximum likelihood method. Using the latter, $P = 1 - \sqrt[n]{(1 - R/N)}$ where P is the probability of transmission by a single vector, R is the number of test plants infected out of N plants used, and n is the number of vectors tested on each plant. As n increases, however, there is a tendency to underestimate P (Gibbs and Gower, 1960).

Ways of analysing the results of 'multiple dose infectivity tests' have received much less attention. With some animal diseases the incubation period seems to be proportional to the logarithm of the inoculum concentration, but it is not known if this is true for plant viruses either in plants or in insects. Another alternative has been used by Raymer and Diener (1969), who showed that an 'infectivity index', which they calculated empirically from the cumulative total proportion of test plants showing symptoms at two-day intervals after inoculation, was a useful measure of the infectivity of preparations of potato spindle tuber viroid; this 'index' combines both incubation period and the proportion of plants infected.

In summary, an infectivity test is the most sensitive simple way of assaying a virus. By themselves infectivity tests do not give an absolute measure of virus concentration, and they ignore non-infective virus-related particles. Many factors affect the probability that a given cell will be infected by a virus particle in an infectivity test, and by suitably designing the tests the inaccuracy caused by these factors can be minimized. Where they can be used, 'minimum dose infectivity tests' are preferred to 'multiple dose infectivity tests' because they are often less time-consuming, and more advanced statistical methods are available for their analysis. Infectivity tests are used most frequently for simple purposes such as estimating the relative concentrations of virus particles in inocula, or estimating the proportion of infected plants in a sample of plants from a field crop. They are however also used to study the factors affecting infection and the mechanism of the process.

Chapter 8

Serological methods

8.1 Principles

It has been known for thousands of years that animals that recover from certain infectious diseases rarely catch the same disease again; they become immune to it. Most of these immunity reactions have a common basis. The immunizing agent, called an *antigen* or *immunogen* (in disease, the pathogen), stimulates the animal so that proteins called antibodies appear in its blood serum; these react specifically with the antigen that stimulated their production (the *homologous antigen*). Antibodies to a virus may neutralize the infectivity of that virus or precipitate it, and antibodies to a culture of bacteria may agglutinate the bacteria. Fortunately for plant virologists, the virus does not have to infect and multiply in an animal to elicit antibodies, and many plant viruses are efficient immunogens. Moreover, the reaction between the antibodies in a serum (*antiserum*) and the antigen can easily be studied *in vitro*.

Many substances cause the production of antibodies when introduced into an animal parenterally (not by intestine), either by infection or injection. The main prerequisites are that the substances have a molecular weight of at least 10^4 and a definite molecular structure, and that they are not a normal constituent of the animal being immunized. Thus most proteins and many polysaccharides can act as immunogens, though not all have equal *immunogenicity*.

Antibodies are found in the globulin fraction of the blood serum proteins (i.e. they are insoluble in solutions one-third saturated with ammonium sulphate). Globulins are present in the sera of normal animals, but after immunization new kinds appear, which differ from those in the normal (that is, *pre-immune*) serum in that they can react specifically with the immunizing antigen (immunogen).

The commonest type of antibodies in sera are γ-globulins (i.e. globulins with the smallest electric charge, and therefore the smallest electrophoretic mobility). They have a molecular weight of about 1.5×10^5, and a sedimentation coefficient of $6.5S$, and are called IgG antibodies. The other

main type are usually called macroglobulins or IgM antibodies, and have a molecular weight of about 8×10^5 and a sedimentation coefficient of $19S$. IgG antibodies are elongated in shape ($10–20 \times 3–5$ nm) and IgM antibodies are mostly spider-shaped, with five legs.

Antibodies are thought to react specifically with their homologous antigens because the two fit together in a unique way. Antibody molecules have combining sites on their surfaces, and these are thought to be complementary, presumably in shape, charge and hydrophobicity to the antigenic sites (or antigenic determinants) on the surfaces of the homologous antigens, so that in the right conditions the two can combine. IgG antibodies have two combining sites on each molecule. By contrast, antigens often have many antigenic sites whose number and variety depend on the size and complexity of the antigen. For example, each TMV particle has about 2100 similar protein subunits, each active and having the same antigenic determinants. Usually, not all the antigenic sites on each virus particle can combine with antibody at the same time, presumably because of steric hindrance; thus, even when mixed with excess antibody, each TMV particle only combines with 600 to 900 antibody molecules.

One possible result of the reaction of antigen and antibody is the formation of a visible precipitate. It is widely believed, though not proven, that the antibody molecules form bridges between antigen molecules, and thus form an aggregate which precipitates when big enough. However there is also evidence that the precipitate is formed simply because hydrophilic groups on the antigen and antibody are covered when the two combine, the resulting complex being insoluble when sufficient counter-ions are present in the medium (e.g. 0.15 M sodium chloride).

Antibodies are produced in the animal by cells of its reticulo-endothelial system, especially by cells of the lymphocytic and plasma cell series, which are common in lymph nodes, the spleen and bone marrow (Raff, 1973). Antibodies occur in all the circulating body fluids of immunized animals so that at any moment about half are in the blood serum, and the other half in the lymph and other body fluids. It is not yet known how the antigenic determinants are recognized by the animal and elicit the production of specific antibodies, but there is much current research on this problem.

8.2 Antigenicity of plant viruses

With few exceptions, the antigenically active part of a plant virus particle is its protein, particularly the external part of the protein shell which encloses the nucleic acid. The protein shell of most plant viruses is built from many apparently identical subunits, but a few plant viruses, such as lettuce necrotic yellows and potato yellow dwarf, have more complex particles with at least two quite different structural layers which seem to contain different proteins and therefore differ antigenically.

There is no definite evidence that single-stranded virus nucleic acid can act as an antigen either alone or within the virus particle. For example, TMV particles can be chemically dis-aggregated into a mixture of protein subunits and nucleic acid, and virus-like particles reconstituted from the protein subunits either with or without the nucleic acid. In serological tests the nucleic acid-free particles are indistinguishable from the original intact particles; however the nucleic acid-containing particles (whether naturally or artificially produced) are more immunogenic than the protein subunits or the nucleic acid-free particles made from them. Similarly the nucleo-protein particles of turnip yellow mosaic virus are more immunogenic than the RNA-free protein shells of the virus, although the two are indistinguishable in serological tests (Marbrook and Matthews, 1966). Perhaps the nucleic acid stabilizes the virus particles within the injected animal, and thereby enhances their immunogenicity.

Antisera specific for double-stranded RNA have, however, been produced using complexes of synthetic double-stranded RNA and methylated bovine albumin as immunogen, and these antisera have been used in work with plant viruses (e.g. Stollar and Diener, 1971). Moreover the reaction between the antigen of sugarcane Fiji disease virus, and antisera to rice dwarf and maize rough dwarf viruses, is apparently caused by antibody to double-stranded RNA (Ikegami and Francki, 1973). Serological reactions involving viruses containing double-stranded RNA may therefore be caused either by superficial similarities between the nucleic acids in the virus particles or by detailed similarities between the virus proteins.

It is, nevertheless, important to remember that, whereas infectivity tests depend on the nucleic acid of the virus, most serological tests depend on its protein, and though these two parts of the virus usually occur together they sometimes are separate, and they differ greatly in stability. Not all parts of the virus protein contribute equally to its serological activity, largely because not all parts of the peptide chain in each subunit are at the surface of the particle. In the TMV particle, the C-terminal end of the amino-acid chain of each subunit is serologically active. Antisera prepared against synthetic mono-, di-, tri-, tetra, penta- and hexa-peptides identical in sequence to the C-terminal peptide of the TMV protein will specifically neutralize the infectivity of the virus (Anderer and Schlumberger, 1966). However, small peptides of this size are not immunogenic unless coupled chemically to a large molecule such as bovine serum albumin, which was used in these tests, when they act as so-called *haptens*.

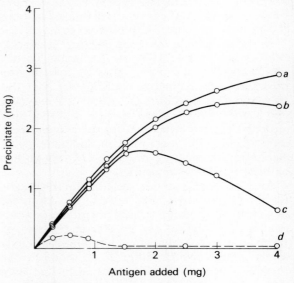

Fig. 8.1 Quantitative precipitin tests. Reaction of the particles of different strains of TMV with an antiserum to those of the type strain. (Data from von Sengbusch, 1965.) *a*, Type strain; *b*, a mutant with asparagine changed to serine at position 140 (numbered from the N-terminus in the polypeptide); *c*, a mutant with asparagine changed to lysine at position 140; *d*, the dahlemense strain, whose protein has thirty amino acids different from that of the type strain.

Von Sengbusch (1965) and van Regenmortel (1967) compared the serological behaviour of the particles of TMV strains that differ from the type strain in one or more of the amino acids in the protein chain. They found that eight of twenty amino-acid replacements altered the serological behaviour of the particles (Figs. 8.1 and 12.6). The effect of the replacements depended on their position in the chain (and hence in the particle). For example the virus was serologically altered when proline replaced leucine at amino-acid position 156 (amino acids numbered from N-terminus; 158 in all), but not by the same change at position 20, or even simultaneous replacements at positions 59 and 129, or 63 and 129. Of the twenty amino-acid replacements studied, only those at positions 65, 66, 107 and 136 to 158 affected serological behaviour, suggesting that these parts of the peptide chain are at, or near the surface of the particle.

8.3 Uses of serological tests

Serological tests with plant viruses are made for several purposes, of which the following are perhaps the most important.
(a) To index plants for infection, as for example in the large-scale commercial testing of 'seed' potato plants or tubers for potato viruses X and S.
(b) To determine the degree of antigenic relationship between the particles of different viruses or isolates.
(c) To determine the amount of virus nucleo-protein or protein in a preparation.
(d) To locate virus antigens in tissues or cells.
(e) To provide information on the structure of virus particles.

How different serological tests can be used for these different purposes will be explained in the part of this chapter dealing with serological techniques. For further information on the uses of serology in the study of plant viruses, see the reviews in section 2 of the appendix.

8.4 Preparation of antisera

A virus preparation in any state of purity can be used to make an antiserum, but as the animal will produce antibodies to all the immunogens injected, including those that are impurities, it is better to use a purified virus preparation.

Before methods of purifying virus particles were devised, the clarified sap of diseased plants was used as the immunogen preparation. However, the particles of few viruses are sufficiently concentrated in plant sap for this to be successful, and some plant saps are toxic to animals. Nowadays virus particles are mostly purified and concentrated for use as immunogens, and antisera have been prepared against the particles of at least a hundred plant viruses.

All normal healthy vertebrates seem able to produce antibodies, but most antisera to plant virus antigens have been made using the European rabbit (*Oryctolagus cuniculus*), for it is easy to handle, keep, feed and breed, and, once it is 3–6 months old, one animal produces a convenient volume of serum. Also, rabbits mainly produce precipitating antibodies. Some strains of rabbit produce higher titred antisera than others; this ability is inherited but it is not known whether a strain of rabbits that responds well to one immunogen will respond well to all. Individual rabbits vary greatly in the amount of antibody they produce, in the rate at which they produce it, and in the specificity of the antibody (van Regenmortel and von Wechmar, 1970). Thus the results obtained with a single antiserum must be interpreted with caution.

Other animals have also been used. Horses were used in the Netherlands to prepare large quantities of antiserum for diagnosing virus infection in commercial potato 'seed' stocks; however horse antisera often contain mostly non-precipitating antibodies. Antisera to plant virus particles have also been prepared in chickens, guinea pigs and frogs, but none of these animals seems as convenient as the rabbit, though the use of mice with intraperitoneally injected immunogen together with ascites tumour cells or Freund's adjuvant to stimulate serum production (Munoz, 1957) might be of value.

8.4.1 *Factors affecting the antibody content of antisera*

When a rabbit is first injected with an immunogen, antibodies usually appear in the blood within a week and are then mainly of the IgM type. The titre of the serum increases for a few days and then decreases until few antibodies can be detected. If the animal is injected again with the same immunogen, the latent period is shorter, the titre rises more quickly and reaches a higher level, and the antibodies, which are mainly the IgG type, persist in the blood longer than the IgM antibodies. These two types of response to immunization are usually called the primary response and the secondary, or hyperimmune, response. Whether there is any clear difference between them is doubtful, and the response depends greatly on the individual animal, the particular immunogen and many other factors.

Plant virus antigens vary in immunogenicity. For instance particles of TMV are more immunogenic than those of alfalfa mosaic virus, but fortunately most plant virus particles are more immunogenic than normal plant proteins. Thus the concentration and specificity of antibodies in an antiserum depend on the immunogen, and also on many other factors.

(a) *Route of injection* The immunogen may be injected either into the blood stream of the rabbit through the marginal ear vein (intravenously); into a muscle, usually that in the thigh (intramuscularly); into the connective tissue in or under the skin of the back or, more rarely, into the foot pad (intradermally or subcutaneously); or even directly into a lymph node (Goudie, Horne and Wilkinson, 1966). Formerly, clarified saps or impure virus preparations were injected into the peritoneal cavity, because toxic sap, such as that from sugar-beet leaves, can be injected intraperitoneally without killing the rabbit; this method is however now rarely used.

Circulating antibodies usually attain their maximum concentration about two weeks after an intravenous injection, and from four to eight weeks after intramuscular or subcutaneous injections.

(b) *Number of injections and amount of immunogen injected* More antibodies are produced to a given amount of immunogen when it is divided and injected in small quantities over a period of time than when it is given as one injection. For example, Matthews (1957) found that four injections each of 10 μg of turnip yellow mosaic virus given at intervals of 21 days resulted in an antiserum with four times the titre of one produced by a single injection of 1 mg of virus. Thus the titre of an antiserum is not directly proportional to the amount of immunogen injected; the response is of the diminishing returns type. Matthews found that when a rabbit was injected with a series of increasing amounts of turnip yellow mosaic virus, the first injection (100 μg) elicited the production of 0.75 mg of antibody/ml of serum, whereas each 100 μg of virus in the last

injection (100 mg) yielded only an additional 0.011 mg of antibody/ml of serum.

Experiments, such as those with influenza virus by Webster (1968a, b), have shown that at different times after injection of an immunogen the antibodies differ not only in type, but also in specificity; their specificity is related to their avidity (i.e. the dissociation constant of the antigen-antibody reaction). Antibodies of low avidity (high dissociation constant) are very specific and will only react with the homologous antigenic determinant, whereas antibodies of high avidity (low dissociation constant) are less specific and will also react weakly with slightly different antigenic determinants. After an injection of immunogen, the first antibodies found are of the IgM type and have high avidity, but these soon disappear from the serum. The first IgG antibodies formed have low avidity, but, especially after further injections of immunogen, more antibodies with a greater avidity and lower specificity are formed, together with antibodies to minor antigenic determinants of the immunogen, and also perhaps to impurities such as host proteins in a virus preparation. Therefore the early hyperimmune antiserum will be highly specific and will only react with the major antigenic determinants of the homologous immunogen. After repeated immunization the antiserum attains a higher titre, but is also less specific; it will react with related but distinct immunogens, and is more likely to have antibodies which will react with contaminants. This response to different immunization schedules has been clearly shown in work with tobacco necrosis and satellite viruses (Kassanis and Phillips, 1970).

For distinguishing between viruses, especially closely related ones, it is therefore best to use early hyperimmune sera, whereas for grouping viruses it is best to use high-titred antisera obtained after prolonged immunization.

(c) *Other substances injected with the immunogen* Virus preparations for intravenous injection are best suspended in neutral saline solution (0.85% = 0.15 M sodium chloride), with a minimum of phosphate, borate or other ions harmful to the rabbit. For injecting into a muscle or under the skin, the virus preparation is usually mixed, before injection, with an adjuvant that enhances its immunogenicity. The most commonly used adjuvant is that devised by Freund and named after him. The 'incomplete' form of Freund's adjuvant consists of a mineral paraffin (85%) and an emulsifying agent (15%, mannide mono-oleate); the 'complete' form contains, in addition, about 0.05% w/v killed and dried *Mycobacterium butyricum*. One volume of adjuvant is emulsified with one volume of the virus preparation by squirting it repeatedly from a hypodermic syringe or by using an ultrasonic vibrator. Other compounds used as adjuvants include agar, sodium alginate and phosphorylated hesperidin, but all these give inconsistent results.

An intramuscular injection of immunogen emulsified with Freund's adjuvant yields a higher titred antiserum than the same amount of antigen injected intravenously and, particularly when Freund's complete adjuvant is used, the antibody concentration in the animal decreases more slowly.

Treating an immunogen with formaldehyde may also enhance its immunogenicity; for example Hollings and Stone (1962) found that one strain of tomato black ring virus was immunogenic only when injected after exposure to 0.2% formaldehyde solution, whereas with other strains of the same virus formaldehyde treatment made no difference.

(d) *Time between injections; immunization schedules* The most efficient way to produce high-titred antisera using rabbits seems to be to inject moderate amounts of immunogen (up to 1 mg) at long intervals, using an adjuvant. Govier (1958) found that two intravenous injections of potato virus X separated by one month yielded higher titred antisera than eight times as much virus injected either as a single intramuscular injection (with Freund's adjuvant) or in a series of

twice weekly intravenous injections. Wetter (1961) found that the highest titred antisera to various viruses with elongated particles were obtained when the rabbits received first an intravenous injection and then intramuscular injections using Freund's adjuvant (Fig. 8.2). Whether this is also true for other viruses is not known, but limited experience suggests it is.

A commonly used immunization schedule is first to give an intravenous injection, followed after four weeks by an intramuscular injection into both hind legs of the rabbit, then by a second intramuscular injection after an additional two weeks; the serum is collected two to four weeks after the last injection and can have a titre in the tube precipitation test of up to 1/5000.

8.4.2 *Collection and storage of antisera*

Up to 20 ml of blood can be taken from a full-grown rabbit on each of three successive days; a fortnight afterwards, the immunogen can be injected into the rabbit again to boost antibody production, and the animal can be bled again later. However with each booster injection the specificity of the antiserum decreases, and when a very specific antiserum is needed, it may be best to kill the animal and collect all its blood (100 ml of blood, 50 ml of serum) when the antiserum has reached the minimum useful titre.

Small blood samples are usually taken from the marginal vein of the ear, either by making a small cut or using a hypodermic syringe. When all the blood in the rabbit is to be collected, the animal is either killed by breaking its neck and then bled by cutting its throat, or it is anaesthetized by injecting barbiturate, and bled from the carotid artery or by cardiac puncture.

After collection, the blood is allowed to clot, and then centrifuged to remove the fibrin, blood cells, etc. The serum should be clear and straw yellow in colour; sera from rabbits that have recently fed are often cloudy with fat droplets, which can be removed by shaking with carbon tetrachloride or chloroform; alternatively the rabbits can be starved for 12 hours before bleeding.

When stored correctly, antisera remain active for several years. Sera may be either frozen or lyophilized (i.e. dried *in vacuo* while frozen, and stored in a vacuum) or, alternatively, stored liquid at 4°C usually with added antiseptic. Glycerol (50% v/v) or sodium azide (0.02%) are the most popular antiseptics, but phenol (0.5%; 1/20 vol

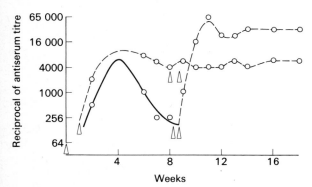

Fig. 8.2 Antiserum titre in two rabbits at different times after injections of preparations of potato virus S. Unbroken line, titre after intravenous injection, broken line after intramuscular injection, arrows indicate times of injection, circles indicate the times when serum was collected and tested. (Data from Wetter, 1961.)

of 10% emulsion added slowly while stirring), carbon tetrachloride or chloroform are alternatives. Merthiolate (sodium ethyl mercurithiosalicylate; 0.02%) has also been used but can cause anomalous results in gel-diffusion tests.

8.5 Serological techniques

8.5.1 *Basic information*

There are many ways of showing the specific combination between the particles of plant viruses and their antibodies. Most are *in vitro* methods, and many can be used quantitatively to estimate antigen and antibody concentrations, but they differ in convenience and sensitivity. Most serological tests rely on the formation of a precipitate when the virus antigen and its homologous antibodies are mixed in the presence of electrolytes. Precipitation occurs most quickly and completely at a definite antigen/antibody ratio, which is called the *optimal proportion ratio*. With either *antibody excess* or *antigen excess*, precipitation is retarded, and when one of the reactants, in particular the antigen, is in great excess, precipitation may be inhibited. The greatest dilution of the antigen to produce a visible precipitate is called the *virus antigen dilution end-point*, and the antiserum dilution end-point is called the *antiserum titre*. The largest amount of precipitate is formed when antigen and antiserum are mixed in the optimal proportion ratio; the supernatant fluid left after the precipitate has formed in such a mixture contains little or no antigen or antibody. By contrast, supernatant fluids from mixtures containing excess antiserum contain free antibody after the precipitate is removed, and similarly, free antigen is found when this is in excess.

In most quantitative serological tests various antigen/antibody ratios are obtained either by titration (i.e. aliquots of a dilution series of one reactant are mixed with aliquots of a constant concentration of the other), which gives a discontinuous series of ratios, or by allowing the reactants to diffuse together through liquid and thereby to mix in a continuous series of ratios.

To ensure that the precipitates which form in serological tests are specific, suitable 'controls' must be included. The virus preparation used to immunize the rabbit may have contained host proteins and hence the antiserum may contain antibodies against them; also, the antiserum/ antigen mixture may, for various reasons, be unstable and precipitate non-specifically. There-

fore each serological test should be, in essence, a 2 × 2 Latin square experiment in which preparations from either virus-infected or virus-free plants are mixed with either antiserum or serum from an animal that has not been immunized. The latter is best obtained before immunization from the animal which later provided the antiserum being tested, and is then called a *preimmune* serum; however it may come from another animal and is then called a *normal* serum.

8.5.2 *Chloroplast, latex and red blood cell agglutination tests*

Chloroplast agglutination is the simplest serological test used with plant viruses. It can be used only with viruses, such as TMV and potato virus X, whose particles are elongated, and sufficiently concentrated in plant sap, because these can usually react over a wide range of antigen/antibody ratios. This test is used extensively in the Netherlands and Germany for testing commercial stocks of 'seed' potatoes for virus infection. Mixtures are made from a drop of either antiserum or normal serum and a drop of sap taken from either virus-infected or healthy plants. If the virus particles and antiserum react then the chloroplasts either co-precipitate or agglutinate.

Other similar tests have been devised in which virus particles or antibody molecules are adsorbed to larger particles such as sheep red blood cells or latex particles and these are then used in agglutination experiments; the latex agglutination test (Bercks, 1967; Abu Salih, Murant and Daft, 1968a) is a fully quantitative test, and is quick, sensitive, reliable and thrifty. In a variant of the latex test, which seems to be even more sensitive for titrations of antisera (Bercks and Querfurth, 1971a), virus particles are attached indirectly to latex particles by adsorbed purified antibody molecules, and the complexes of virus/antibody/ latex tested for agglutination by antisera. The red blood cell, or haemagglutination, test also is more sensitive for detecting antigen than the conventional latex test (Abu Salih, Murant and Daft, 1968a, b), but uses more antiserum and the sensitized red blood cells cannot be stored for more than a few months, even when treated with formalin (formolized).

8.5.3 *Tube precipitation test*

The theory of this important quantitative test will be described in some detail because this test is

Fig. 8.3 Tube precipitation test with a preparation of broad bean mottle virus; antigen titration test. Tubes (6 mm diameter) contain 0.5 ml of approximately (left to right), 32, 16, 8, 4, 2 mg/l of virus antigen and 0.5 ml of antiserum diluted $\frac{1}{32}$. **a,** after 10 min at 37 C, **b,** after 20 min, **c,** after 40 min.

widely used by plant virologists, and the principles underlying its design and interpretation can be applied to most other serological tests.

In this test various dilutions of antiserum and virus antigen are mixed, incubated in small glass tubes, and the mixtures watched for a precipitate to form (Fig. 8.3). The type of precipitate ob-tained depends on the shape of the antigen; elongated virus particles produce bulky flocculent precipitates like those produced by flagellar bacterial antigens, whereas isometric virus particles or disrupted elongated particles give dense granular precipitates, like those given by somatic bacterial antigens (Figs. 8.3 and 8.4).

Usually 0.5 ml each of diluted virus preparation and antiserum are mixed in a narrow glass tube which is then partially immersed in a water bath at 37°C to promote mixing by convection. Dilute plant saps may be stable at 37°C, but un-stable plant material can often be removed from

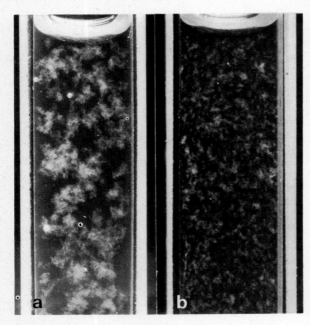

Fig. 8.4 Tube precipitation test; flocculent and granular types of precipitate. **a,** red clover vein mosaic virus particles and homologous antiserum; **b,** turnip yellow mosaic virus particles and homologous antiserum.

sap before use (provided the virus antigen is stable) by heating for 10 minutes at 45 to 50°C and centrifuging at low speed. When purified preparations of viruses with stable particles such as turnip yellow mosaic virus or TMV, are used as antigen, the mixtures can be incubated at 50°C, and precipitation will then be more rapid.

The reactants are usually diluted in neutral 0.85% saline because a precipitate will only form when electrolytes are present, and the particles of most viruses are stable and soluble at neutral pH. However those of cucumber mosaic virus may be disrupted and precipitated by neutral saline (Francki *et al*, 1966), and it is best to use 0.05 M phosphate buffer (pH 8) in serological tests with this virus (Scott, 1968).

Each test is usually either an antigen or antiserum titration. Thus in an antigen titration, 0.5 ml of each of a series of two-fold dilutions of the virus preparation is put in a separate tube and to each tube is added 0.5 ml of antiserum at constant dilution. This test gives an estimate of both the optimal proportion ratio (the mixture which produces a precipitate most quickly) and the antigen dilution end-point (Fig. 8.3). Similarly an antiserum titration will give an estimate of the optimal proportion ratio and the antiserum titre.

The complete picture of a precipitation reac-

tion is best obtained by making mixtures of every dilution of virus antigen with every dilution of antiserum and then recording the time taken for a precipitate to appear in each mixture. The results can be presented as a precipitation diagram (Fig. 8.5a and b), in which 'contour' lines are drawn to show which mixtures precipitated at similar times. Such a diagram shows that the optimum proportion ratio (thick line) for each antiserum concentration (row) is different from that (broken line) for each virus concentration (column); these are known as the α and β optimum ratios respectively, and the equivalence zone lies between them. Note that the α optimum ratio is constant and independent of the concentration of the reactants, but that the precipitation time in α optimum proportion mixtures increases as their dilution increases.

The concentration either of antibody in a serum or of virus antigen in a preparation can be estimated by titrating one or the other in standard conditions and comparing the results with those of calibration tests. The most convenient parameters to measure in a titration experiment are either the optimum proportion ratio or the dilution end-point (titre), which is the more easily and economically determined. The antiserum titre depends greatly on the antigen concentration with which it is titrated (columns), because excess antigen inhibits precipitation, particularly with viruses whose particles are isometric. Therefore the antiserum titre must be measured using either a standard concentration of virus antigen, or better still a virus antigen diluted almost to its end-point. By contrast excess antiserum does not inhibit precipitation so easily, and therefore there is a greater range of dilutions of antiserum that are suitable for measuring antigen dilution end-points.

When the precipitation diagram of a primary antiserum (Fig. 8.5a) is compared with that of a hyperimmune serum (Fig. 8.5b) collected from the same rabbit, several interesting differences can be seen.

(a) The titre of the hyperimmune serum (estimated using diluted antigen) is sixteen times that of the primary antiserum, but the virus antigen dilution end-point estimated by the two sera is almost the same. Thus the smallest concentration of virus antigen that will produce a visible precipitate is independent of the titre of the antiserum, and is usually about 1 mg/l for elongated virus particles, and 2 to 10 mg/l for isometric virus particles of about 30 nm diameter.

(b) The shape of the diagram contours in the antigen-excess region of the two sera is the same, but the region of antiserum excess inhibition is larger with the primary antiserum than with the hyperimmune serum. Matthews (1957) found that when a hyperimmune serum was diluted with normal serum the large antiserum-excess zone of a primary antiserum reappeared, and was apparently caused by a substance in the normal serum which could be removed by dialysis.

The precipitation diagrams in Fig. 8.5 are typical of those obtained with isometric virus particles. Comparable diagrams for rod-shaped or filamentous virus particles differ in the region of antigen excess, because these particles precipitate at large antigen concentrations, and their equivalence zone covers a much wider range of antigen/antibody ratios.

Precipitation diagrams show that the titre of an antiserum is best estimated by titrating the antiserum against a dilute purified virus preparation containing about 10 to 20 mg/l virus particles. When however the only antigen-containing preparation available is sap from infected plants

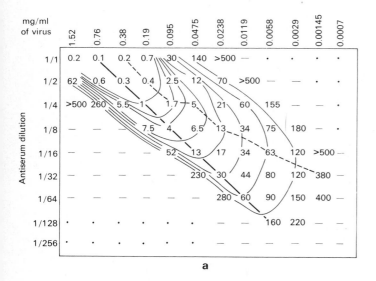

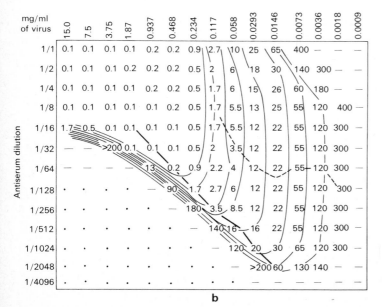

Fig. 8.5 Precipitation diagram of the reaction of a turnip yellow mosaic virus preparation with the homologous antiserum. The antiserum used in **(a)** was from a rabbit injected initially with 0.01 mg of immunogen intravenously, and after 30 days with a further 0.1 mg. The serum was collected 14 days after the second injection. Figures in the body of the diagram are the times in minutes when the precipitate first appeared during incubation at 50°C. The antiserum used in **(b)** was from the same rabbit as in **a** after further injections of 1.0 mg and 10 mg of immunogen also at 30-day intervals (Matthews, 1957).

or a partially pure virus preparation of unknown concentration, then three titrations are necessary. First the antiserum titre is determined approximately by titrating the serum against one or two dilutions of the virus preparation (equivalent, for example, to plant sap diluted to $\frac{1}{4}$ and $\frac{1}{32}$). Next the virus dilution end-point is accurately estimated using the antiserum about four times more concentrated than its approximate titre. Finally, the antiserum titre is estimated accurately using the virus preparation at four times the concentration of its dilution end-point (i.e. 4 to 40 mg/l). Control titrations should, of course, be done at the same time with a preparation from healthy plants and with normal or preimmune serum.

The precipitation test is widely used to estimate the concentration of virus antigen in different preparations by finding their dilution end-points in titration experiments (e.g. Fig. 11.3). Another important use is in virus identification, by determining whether the particles of two viruses are serologically related (i.e. have the same or similar antigenic determinants); viruses whose particles are serologically related have with very few exceptions (Bercks and Querfurth, 1971b) always been found to share most of their other characters (see Chapter 12). As already mentioned a single virus particle may carry large numbers of several kinds of antigenic determinants. Particles of two different virus isolates may have all, some, or none of their antigenic determinants in common. If particles of two isolates A and B have none in common, antiserum to A will react with A but not B, and vice versa. If particles of isolates A and B have all their determinants in common, then an antiserum to A will have the same titre when tested with either the heterologous particles (B) or the homologous ones (A). If, however, A and B share only some of their antigenic determinants, or have antigenic determinants that are slightly different, then antiserum to A will contain antibody molecules all of which can react with A, but only some of which can react with B. Therefore antiserum to A will probably have a lower titre when tested with B than when tested with A, and vice versa. The precipitation diagrams of the homologous and heterologous reactions in this last instance will resemble those of hyperimmune and primary antisera respectively (Fig. 8.5b and a). The antigenic similarity of A and B can therefore be estimated by comparing the homologous and heterologous titres of their antisera.

The antigenic similarity of the particles of two viruses can also be measured in *cross-absorption*

tests. The antibodies common to the particles of both viruses (i.e. the cross-reacting antibodies) can be removed from each serum by incubating it with the heterologous antigen, adding enough to give optimal proportions or antigen excess. The precipitates are removed by centrifugation and the serum is then termed an *absorbed serum*. Only antibodies specific to homologous particles will remain in the absorbed serum, and their concentration can be estimated by titrating it with the homologous antigen. Cross-absorption tests therefore measure the proportion of antibody in an antiserum specific for particles of the homologous virus (low avidity antibodies), whereas heterologous titrations with unabsorbed sera measure the amount of cross-reacting, heterologous antibody (high avidity antibodies). In precipitation tests, the reactants are usually titrated in a series of two-fold dilutions; therefore cross-absorption tests will be best for assessing differences between virus antigens whose antisera contain much cross-reacting antibody (more than half), whereas heterologous titrations will be best for virus antigens whose antisera have little cross-reacting antibody (less than half). Although the results given by the two tests usually agree approximately, they are rarely identical, presumably because the antigenic determinants on a virus particle differ in their ability to stimulate antibody production, and to adsorb to, or be precipitated by, antibodies of different avidity. Table 8.1

Table 8.1 Heterologus titration and cross-absorption tests with two pairs of serologically related viruses

Antiserum prepared against	Absorbed with	Test antigen	
(a) *Closely related viruses* (Dias and Harrison, 1963)		PGFV	AGFV
Portuguese grapevine fanleaf (PGFV)	—	256[1]	256
	AGFV	64	<16
American grapevine fanleaf (AGFV)	—	512	512
	PGFV	<20	20
(b) *Distantly related viruses* (Gibbs *et al*, 1966b)		APLV	OYMV
Eggplant mosaic virus (andean potato latent strain; APLV)	—	1024	16
	OYMV	1024	<2
Ononis yellow mosaic (OYMV)	—	16	512
	APLV	<2	128

[1] Figures are the reciprocals of the dilution end-points of the antisera.

gives an example of results obtained with cross-absorption and heterologous titration tests with two viruses whose particles are closely related serologically, and two whose particles are more distantly related.

The fact that a preparation of one virus will precipitate with the antiserum prepared against another should always be interpreted with caution, for this can happen when mixed or contaminated preparations are used, and must be checked by suitable cross-absorption tests. One indication of a mixed system is the occurrence of two or more optimal proportion ratios in a single titration. Mixtures of viruses that have small isometric particles can also be resolved in gel-diffusion tests which will be described later.

The micro-precipitation test is a miniaturized version of the tube precipitation test (van Slogteren, 1955). One drop of each of the reactants is mixed either on a Petri dish, which may be coated with silicone or Formvar to minimize spreading, or on a specially prepared plate with a lattice of shallow wells (e.g. a haemagglutination tray). The drops may be covered with liquid paraffin to stop evaporation. Drops containing different virus antigen/antiserum ratios are prepared, incubated, and examined for precipitates using a microscope. The method seems more sensitive than the tube precipitation test, probably because smaller floccules can be seen by microscopy, and much smaller volumes of the reactants are needed.

8.5.4 *Quantitative precipitin test*

In this test the amount of precipitate obtained from each mixture of a titration test is measured (Kabat and Mayer, 1961; Kleczkowski, 1966; von Sengbusch, 1965). After incubation, each mixture is centrifuged to collect the precipitate, which is then carefully washed in saline. The amount of precipitate is estimated by measuring either its protein or nitrogen content (see Chapter 9). Alternatively one may estimate the amount of uncombined virus antigen in the incubated mixture either using an analytical centrifuge or by centrifuging the mixture into a sucrose density gradient and measuring the concentration of virus particles in different fractions of the gradient (Ball and Brakke, 1969).

8.5.5 *Complement-fixation test*

This is a more complicated test, but is basically a tube precipitation test to which has been added a sensitive indicator system, so that serological reactions can be detected even when, for some reason, the antigen/antibody complex does not precipitate.

Bordet and Gengou (1901) found that the serum of a rabbit previously injected with sheep red blood cells would lyse the red cells when it was fresh, but not after it was kept for a few weeks or heated to 56°C for 20 minutes. The haemolytic activity of old or heated antisera was restored fully when fresh unheated serum from an uninjected rabbit was added, and the labile component of the serum needed for haemolysis was called *complement*. Complement has since been found to have many interdependent components, and those that have been characterized are proteins, some of them enzymes. Complement reacts and becomes 'fixed' during any antigen/antibody reaction, but it does not aid precipitation of the complex formed in the reaction; thus old or heated plant-virus antisera can precipitate their homologous virus antigens as readily as comparable fresh or unheated antisera. By contrast, antisera to red cells will only lyse red cells when complement is present; hence the haemolytic reaction can be used as an indicator of the presence or absence of free complement.

The complement-fixation test is done in two separate parts. First the virus antiserum, which has been heated to inactivate complement, is titrated with the virus antigen in the presence of a suitable amount of complement (e.g. a standard amount of fresh normal guinea pig serum). Then, after incubation, sheep red blood cells and their complement-free antiserum are added. In those mixtures where virus antigen and antiserum have reacted, complement will have become 'fixed' and there will be no haemolysis. In mixtures where little or no reaction has occurred, little or no complement will have been fixed and the red cells will lyse.

This test seems more sensitive than others for some uses, but not all. However because it is more complicated, other simpler tests are often preferred.

8.5.6 *Ring interface test*

This is the simplest of the tests in which virus antigen and antibody come together by diffusion. A little antiserum is put into a narrow glass tube or capillary, and the virus preparation is carefully layered on top. The virus particles and antibodies diffuse together and where a suitable antigen/

antibody ratio is reached a precipitate forms, usually within minutes (Fig. 8.6a). It is not necessary for the reactants to be at or near their optimal proportion ratio, but they must be present in sufficient concentration to give a visible precipitate. The virus preparation is always layered on top of the antiserum because any precipitate formed will sediment and is more likely to redissolve if it falls into excess antigen than if it falls into excess antiserum. To make the interface stable when dilute reactants are used, the antiserum can be diluted in saline containing 10% glycerol or sucrose.

This test is most often used qualitatively (e.g. checking for the presence of virus antigen in samples) because it is simple, quick and requires only small quantities of the reactants. However it can be used quantitatively (Whitcomb and Black, 1961), and was, for example, used very successfully to estimate concentrations of virus antigen in infected leafhoppers (Reddy and Black, 1966).

8.5.7 Gel-diffusion tests

These tests can be used only for those virus anti-

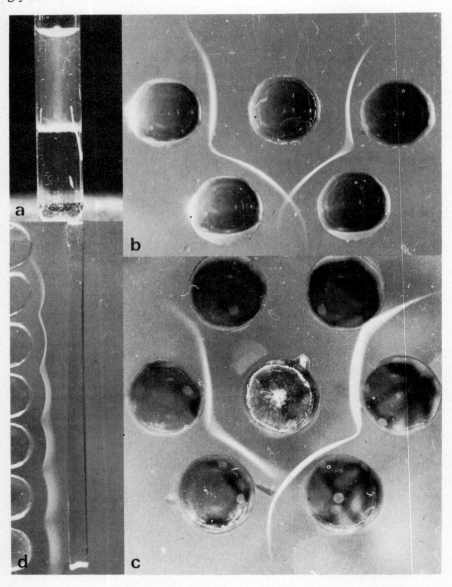

Fig. 8.6a Ring interface test with a turnip yellow mosaic virus preparation and homologous antiserum. Top layer contains antigen; bottom layer, antiserum. Note the band of precipitate at the boundary between the layers. b-d, Gel-diffusion tests: b, Confluent and crossing bands of precipitate. The two wells on the left contain a turnip yellow mosaic virus preparation, the two on the right contain a broad bean mottle virus preparation, and the top centre well contains a mixture of antisera to the two antigens. c, Confluent bands and spur. The two wells on the lower right contain an eggplant mosaic virus (Andean potato latent strain) preparation; the two on the lower left contain a dulcamara mottle virus preparation; the two at the top contain saline, and the centre well contains antiserum to eggplant mosaic virus particles. d, Gel-diffusion test with a broad bean mottle virus preparation and its antiserum, showing the effect of antigen concentration on the width and position of the band of precipitate. Long well at the right contains dilute antiserum and the circular wells contain a series of two fold dilutions of the antigen (most concentrated antigen is at the bottom).

gens that will diffuse through agar gels. The simplest type of test is named after Oudin (1946), who devised it, and is often misleadingly called the 'single diffusion' test. It resembles the ring interface test, except that the lower layer is incorporated in dilute agar gel, and can therefore contain either antiserum or, more usually, antigen. When the concentrations of the two reactants are suitably arranged, a band of precipitate forms in the gel instead of the liquid, and apparently moves through the gel away from the liquid/gel interface. A variant of this technique is the radial diffusion test. The antiserum is incorporated in a layer of agar gel in a Petri dish. Wells are then cut in the gel and filled with antigen preparations. The antigen diffuses into the gel and forms a halo of precipitate around the well. Shepard (1970) and Shepard *et al* (1971) have used this test for screening potato plants for infection by potato viruses X, S and M, which have filamentous particles. These do not diffuse readily through agar gels but antigenically active fragments which diffuse faster are prepared by treating the particles with, for example, 30% pyridine, 2.5% pyrrolidine or 1 M guanidine-HCl. This technique uses large amounts of antiserum but nearly 100 samples can be tested in a single 10 cm diameter Petri dish.

More widely used is the so-called 'double-diffusion' (in one dimension) test devised by Oakley and Fulthorpe (1953). This is like the Oudin test except that the reactants are separated in the tube by a layer of agar gel. Antigen and antibody diffuse into the agar, and where they meet in optimal proportions they form a thin band of precipitate. When a virus preparation and its antiserum contain more than one antigen/antibody system (e.g. a mixture of virus antigens or a virus and a host antigen), the different systems give separate bands of precipitate which rarely are superposed because the reactants are unlikely to have the same diffusion coefficients and be present in the same concentrations. The precipitate band is most obvious, sharpest and is stationary when antigen and antibody are present in optimal proportions. When one or other is in excess then the band will form first near the source of the dilute reactant, and will then broaden, apparently moving further away from the excess reactant.

Polson (1958) has shown that, where antigen and antibody can diffuse freely in the gel, then the position of the precipitate band at optimal proportions is related to the diffusion coefficients of the antigen and antibody by the equation:

$$Dag/Dab = Ag^2/Ab^2,$$

where Dag and Dab are the diffusion coefficients of antigen and antibody respectively and Ag and Ab are the distances of the precipitate from the menisci of the antigen and antiserum reservoirs respectively. Most antisera contain mainly IgG antibodies, which have a diffusion coefficient of 4.8×10^{-7} cm^2/sec, and hence the position of the precipitate at optimal proportions will provide an estimate of the diffusion coefficient of the antigen. In practice a series of tubes is prepared in which the antiserum concentration is kept constant, and the virus concentration varied so that the virus is in excess in some tubes and the antiserum in others. The position and width of the precipitate bands are measured and the measurements plotted graphically to find the position of the band of minimum width. The method has been used successfully to estimate the diffusion coefficients, and hence the sizes, of different antigenic components of purified virus preparations (e.g. Tremaine and Willison, 1961).

The most widely used gel-diffusion test is that devised by Ouchterlony (1948); by analogy with the other gel-diffusion tests it might be called the 'two-dimensional double-diffusion' test. Hot agar or agarose solution is poured on a glass plate or into a dish and allowed to cool to form a layer of gel; wells are then cut in the gel and filled with the reactants. The antigens and antibodies diffuse together and precipitate as in the Oakley-Fulthorpe test (Fig. 8.6b–d), but because the test is two-dimensional, several virus antigens and antisera, together with suitable 'controls', can be tested and compared in one test. When a suitably diluted antiserum and virus preparation are put in adjacent wells, a band of precipitate forms at right angles to the shortest line joining the two wells. The distances of the band from the wells depend on the same factors as in the Oakley-Fulthorpe test, and the band is straight when the antigen and antibody have the same diffusion coefficient, but when the diffusion coefficients differ the band curves around the well containing the reactant with the smaller diffusion coefficient. The Ouchterlony test is very useful for comparing several reactants at the same time. For example, we may test two virus preparations with one antiserum. The three are put in wells arranged at the corners of an equilateral triangle, or an isosceles triangle when the diffusion coefficients of the antigens are less than that of the antibodies.

At least three results are possible:

(a) When the antigens are identical, the bands of precipitate formed between each antigen and the antiserum will extend and their adjacent ends will join (Fig. 8.6b and c).

(b) When the antigens have no common determinants, though the antiserum contains antibodies to both, the bands of precipitate will cross (i.e. pass through each other) without interruption (Fig. 8.6b).

(c) When the antigens have some determinants in common, and the antiserum was prepared against only one of the antigens, the band of precipitate formed between the antiserum and the homologous antigen will join the band of precipitate between the heterologous antigen and antiserum. The antibodies specific to the homologous antigen will however diffuse through the heterologous antigen precipitate band and, where suitably placed, will precipitate with the homologous antigen. Thus the band of homologous precipitate will continue beyond the point where it joins the heterologous precipitate band to form a spur towards the heterologous antigen well (Fig. 8.6c).

Like the Oakley-Fulthorpe test, the Ouchterlony test can be used to determine diffusion coefficients or, conversely, the concentrations of the reactants (Fig. 8.6d). The many isometric virus particles about 25 to 30 nm diameter will precipitate, at optimal proportions, at 35 to 40% of the distance from the antigen to the antiserum well, whereas larger virus particles will precipitate nearer the antigen well, and the protein subunits of virus particles, and also plant proteins, will precipitate nearer the antiserum well.

Once the bands of precipitate have formed fully, the gel may be dried, treated with protein specific stain and stored. When there are doubts whether a particular band of precipitate contains virus particles, the band may be cut out, washed in saline, ground in negative stain (Chapter 9) and examined directly in the electron microscope. Bands of precipitated nucleoprotein may be distinguished from precipitated protein in several ways, such as by using acridine orange to stain for nucleic acid (Cowan and Graves, 1968).

A disadvantage of this versatile test is that it is difficult to use with most viruses that have filamentous particles, for few of these diffuse through dilute gel; however some will diffuse through 0.5% agar in borate-EDTA buffer or 0.5% agarose, and particles of others can be broken into smaller more readily diffusible pieces using ultrasonic vibration (Chapter 10), deter-

gents such as sodium dodecyl sulphate (Hariharasubramanian and Zaitlin, 1968), or compounds such as ethanolamine or pyridine (Shepard and Secor, 1969).

All gel-diffusion tests should be done at approximately constant temperatures; when the temperature varies multiple bands may form. The Ouchterlony test gives results in 0.5 to 2 days, whereas the Oakley-Fulthorpe test takes longer.

Diffusion serological tests separate mixtures of antigens and antibodies by their sizes, diffusion coefficients and concentrations, and can be combined with other sorting techniques, such as centrifugation or electrophoresis, to resolve complex mixtures. For example, a mixture of antigens may be fractionated by electrophoresing it in a sucrose density-gradient, and the composition of the fractions may be compared serologically in an Ouchterlony test. Immuno-electrophoresis is another similar technique. The antigen mixture is put in a well in a layer of buffered agar gel, a direct electric current is passed through the gel, thereby sorting the mixture in one dimension into its electrophoretic components. A long narrow trough is then cut in the gel, to one side of the well and parallel to the direction of electrophoretic movement, and filled with antiserum. The antigens and antibodies diffuse together and give separate bands of precipitate for each antigen/antibody system. This test has proved useful for distinguishing closely related strains of a virus.

8.5.8 *Immuno-osmophoresis test*

In this test (Ragetli and Weintraub, 1964) the virus and antibody are put into wells cut in a gel, and are brought together by an electric current. Conditions of pH, buffer and gel charge must be chosen so that virus and antibody are brought together. This test is very quick (10 to 30 min), can be used with filamentous virus particles as well as isometric particles, and is simple to use with particles that have a large negative charge at pH 6–8, such as those of TMV, and alfalfa mosaic and southern bean mosaic viruses. However for virus particles with either a small negative charge or a positive charge, chemically substituted agar and antibody are used to ensure that the reactants migrate towards one another (Ragetli and Weintraub, 1965).

8.5.9 *Antibody tracer techniques*

Antibodies can be used as tracers to show, for example, the position of an antigen in thin sections of infected tissue. Antibody molecules are usually not themselves sufficiently distinctive to be identified under the light or electron microscope, particularly when they are mixed with other materials. They may however be used, but with caution, to identify virus particles by electron microscopy in leaf dip preparations (Ball and Brakke, 1968), although for most work as tracers they must be conjugated with some distinctive substance. Three main kinds of tracer antibodies have been prepared, radioactively-labelled antibody, antibody conjugated with ferritin (a small iron-containing protein that has a characteristic appearance in the electron microscope), and antibody conjugated with fluorescent dyes. The first two kinds have not been much used in studies of plant viruses, but fluorescent antibodies have been used more extensively to locate virus antigens in both plant and vector tissues. The antibody is conjugated chemically with the dye, and the conjugate then used to 'stain' sections or fragments of infected and comparable healthy tissues. Fluorescein isothiocyanate is the most popular fluorochrome but others, such as rhodamine lissamine, have also been used. The position in the tissue of the 'stained' antigen is indicated by the characteristic yellow-green fluorescence of the fluorescein when it is irradiated with far-blue or ultra-violet light using a specially adapted light microscope. The light source is a mercury-vapour lamp or carbon arc, the light is filtered to remove all except far-blue and ultra-violet light, and the tissue is viewed through a pale-yellow filter to remove any transmitted far-blue and ultra-violet light. Thus the fluorescence is seen on a black background.

The fluorescein may be conjugated directly to the antibodies that react with the antigen to be stained, or, alternatively, a sheep or goat is injected with rabbit globulins and its antibodies are conjugated to fluorescein and used to stain the rabbit-antibody/antigen complex. This 'indirect' method is popular because good fluorescein conjugates are not easy to prepare, and one fluorescein-conjugated anti-rabbit serum can be used to stain any rabbit-antibody/antigen complex. In these tests it is most important to include all control treatments, because antibody conjugates may become bound to tissues non-specifically, and many tissues contain auto-fluorescent substances. Non-specific staining is minimized by taking various precautions when preparing the conjugates (Nagaraj, 1962; 1965).

Fluorescent antibodies can be used qualitatively

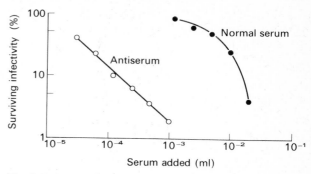

Fig. 8.7 Neutralization of the infectivity of a TMV preparation by homologous antiserum, and effect of normal serum. (Redrawn from Rappaport and Siegel, 1955.)

to see where in cells or tissues a particular virus antigen is accumulating (e.g. Sinha and Reddy, 1964; Otsuki and Takebe, 1969), or quantitatively to estimate the proportion of infected cells in a population of cultured vector cells (e.g. Chiu and Black, 1969; Chiu *et al*, 1970) or plant protoplasts (Takebe and Otsuki, 1969).

8.5.10   *Infectivity neutralization test*

When virus particles are mixed *in vitro* with their homologous antiserum, infectivity is greatly decreased. All sera, especially when fresh, decrease the infectivity of plant-virus preparations, but as a serum ages, or when it is diluted, or when IgG is separated from other serum proteins, the non-specific neutralizing ability diminishes, and the specific neutralizing ability not shown by normal serum persists (Fig. 8.7). Use of this effect has only been made occasionally; for example, Lister (1964) used it to separate a mixture of arabis mosaic and strawberry latent ringspot viruses, two viruses which share most plant hosts and whose particles have similar properties *in vitro*.

Neutralization of infectivity is widely used in work with viruses that have a higher infectivity/particle ratio than plant viruses, and especially with viruses of vertebrates. It can also be used for plant viruses in circumstances where this ratio is larger than usual, as for example when detecting infectivity by feeding aphids on virus preparations through membranes. Duffus and Gold (1969) used this technique to test the possible serological relationship between beet mild yellowing and potato leafroll viruses; Rochow (1970) used it to obtain evidence that some of the particles in plants jointly infected with two isolates of barley yellow dwarf virus are composed of the RNA of one virus isolate enclosed in the protein coat of another.

Chapter 9

Physical and chemical methods of assay and analysis

A great many techniques can be used for assaying or analysing virus particles or other virus macromolecules. Some tests are favoured more than others, usually because they are simpler or more accurate, or because they are more specific for the particular macromolecules being studied and less affected by contaminants. However it is obviously a good general practice, when possible, to assay or analyse a preparation of virus particles or virus macromolecules in more than one way, not only to obtain more comprehensive and informative results but also to check for anomalous behaviour. We have already discussed infectivity assays and serological methods; in this chapter we deal with physical and chemical techniques.

9.1 Chemical methods

The elementary composition of virus particles is about 50% carbon, 7% hydrogen, 20% oxygen, 16% nitrogen, 0.5 to 4% phosphorus, 0.2 to 1% sulphur and 2.5% ash (Bawden and Pirie, 1937; Stanley, 1939). For these analyses the particles are totally degraded and their composition estimated chemically, using for example, one of the Kjeldahl assay methods for nitrogen (Dawson et al, 1969), and the molybdate method for phosphate (Holden and Pirie, 1955). Nowadays it is more usual to estimate the different types of compounds in virus particles by specific techniques, often colorimetric, and merely to check whether the estimated total weight of compounds in the virus preparation is similar to the total weight of virus particles in it.

The only sure way to know the weight of the virus particles in a preparation is to weigh them; the virus preparation is dialysed repeatedly against a dilute buffer solution, and then measured volumes of the virus preparation and the buffer are dried and weighed. Several milligrams of virus particles are required for this method to be at all accurate, and with most viruses it is not possible to obtain such large amounts. For the same reasons the partial specific volume of virus particles (i.e., the number of millilitres of solvent displaced when 1 gram of virus is added to an infinite volume of solvent) is rarely estimated pycnometrically. Hence most estimates of virus particle concentrations are made by indirect optical methods such as those described in §9.3.

9.1.1 Proteins

The protein content of a virus preparation is usually estimated by one or other modification of Lowry's Folin-Ciocalteu phenol method (Miller, 1959; Hartree, 1972), though there are other colorimetric methods such as that described by Bramhall et al. (1969), and Dawson et al. (1969) describe still further tests. These colorimetric tests are made quantitative by comparing the colour obtained with the protein in a preparation of unknown concentration with that obtained using known amounts of a standard protein such as bovine serum albumin. It is important to remember that these tests depend on the reaction of particular amino acids. For example, the Lowry method depends principally on the tyrosine content of proteins, and hence the results obtained by this method will be unreliable if the protein being tested contains unusual amounts of tyrosine; they will also be affected by contamination with other phenols.

Some of the proteins in virus particles that contain lipid, such as those of tomato spotted wilt virus, are glycosylated (Mohamed, Randles and Francki, 1973). Glycosylation is most conveniently demonstrated after the proteins have been separated by electrophoresis in polyacrylamide gels (see §9.5) using the periodic-Schiff reagent described by Clarke (1964).

9.1.2 Nucleic acids

Virus nucleic acids differ from one another in many respects; ways of determining their size, strandedness and conformation will be described in other sections of this chapter.

The total amount of nucleic acid in a virus preparation may be estimated from the phosphorus content of the preparation (Holden and Pirie, 1955); nucleic acids contain about 9% by weight of phosphorus. This method will obviously only give useful results when nucleic acid is the only phosphorus-containing compound in the virus preparation; it will give misleading results when the preparation contains traces of phosphate buffer, or phospholipids such as occur in the particles of poxviruses of mammals, or phosphoproteins as in the particles of some poxviruses or rhabdoviruses (Downer, Rogers and Randall, 1973; Sokol and Clark, 1973).

More usually, however, the amount and type of nucleic acid in a virus preparation is estimated enzymically (§9.2), by fluorescence tests (§9.3.5) or by some colorimetric method that is specific for the sugar residues in the nucleic acid. Ribose is usually detected colorimetrically by the orcinol method (Ceriotti, 1955; Schneider, 1957), and deoxyribose by Dische's diphenylamine test (Burton, 1956; Davidson and Waymouth, 1944); both methods only detect sugars from purine nucleotides, and thus depend on the nucleotide composition of the nucleic acid. The nucleic acid content of a preparation of virus particles can also be estimated by deproteinizing it, hydrolysing the nucleic acid and measuring the optical density of the solution at 260 nm, and then assuming the hydrolysed nucleic acid has an extinction coefficient of 33 (§9.3.4). The nucleic acid can be deproteinized and hydrolysed either in N KOH overnight at 20°C followed by neutralization with perchloric acid, or in N perchloric acid for 30 min at 100°C.

Many methods are known for determining the base ratio of a nucleic acid (§5.2); some are chemical and involve hydrolysis of the nucleic acid, and other methods will be described in later sections of this chapter. Both DNA and RNA are hydrolysed when heated in acid, though DNA requires a higher temperature and lower pH than RNA and this degrades it, so enzymatic hydrolysis is usually preferred (Salivar, Tzagoloff and Pratt, 1964). By contrast RNA is readily hydrolysed, either by N-HCl at 100°C for 1 h into its purine bases and pyrimidine nucleotides, or by 0.3 N-KOH at 37°C for 16 h to yield the nucleotides. Whichever method is used, the bases or nucleotides are usually separated chromatographically (Fig. 9.1) or electrophoretically, and their concentration estimated spectrophotometrically (Markham and Smith, 1951, 1952; Wyatt, 1951).

The way in which the sequence of nucleotides in nucleic acids is determined is outlined in §5.2.

Fig. 9.1 Photograph of a contact print, made in ultra-violet light, of a chromatogram of the hydrolysed RNA from turnip yellow mosaic virus particles. The RNA was hydrolysed in acid and applied at the point indicated by the arrow. The 'spots', in order of increasing Rf (top to bottom), are of guanine, adenine, cytidylic acid and uridylic acid.

9.1.3 Other compounds

There are many techniques for the extraction, assay and analysis of lipids (Weinstein *et al*, 1969) and these have been used with, for example, the togaviruses and the ortho- and paramyxoviruses of mammals (Renkonen *et al*, 1971), and also the plant rhabdovirus, potato yellow dwarf virus (Ahmed *et al*, 1964).

Polyamines have been found in the isometric particles of several plant viruses. They have been identified and their concentration estimated chromatographically (Beer and Kosuge, 1968).

9.1.4 Radioactive tracers

Many different methods are used to introduce radioactively labelled compounds into particular virus or host molecules for study, but few of these methods have been compared. Matthews (1970) found that ^{32}P-labelled orthophosphate and ^{35}S-labelled sulphate were taken up rapidly and efficiently by plants through their roots immediately after they were washed free of soil. This method was more effective than either floating leaf discs on, or putting the cut petioles of excised leaves into, solutions of the radioisotope.

Labelled amino acids or nucleotides may be incorporated by the method used by Zaitlin, Spencer and Whitfeld (1968). The leaf to be labelled is excised and sliced, at right angles to the midrib and under water, into 2 mm wide strips leaving intact the midrib and leaf margin to support the strips. The sliced leaf is washed thoroughly, put in a moist chamber, and the radioisotope spotted on to it directly. Alternatively the strips are cut from the midrib and leaf margin and vacuum infiltrated with the radioisotope solution (Minson and Darby, 1973b).

One of the principal difficulties in labelling molecules while they are being synthesized in intact plant tissue, is that the tissue usually already contains a 'reservoir', or 'pool', of the precursor that is added in radioactive form, and it is difficult to deplete or influence the pool, and thus it has been impossible to do 'pulse-chase' experiments using plant tissue. However the precursor pools are very much smaller in isolated leaf cells or protoplasts, and 'pulse-chase' experiments are possible.

For some purposes the virus or host molecules may be labelled *in vitro*. For example, proteins or nucleic acids may be generally labelled with ^{131}I (Webster, Laver and Fazekas de St Groth, 1962; Commerford, 1971), and there are many ingenious methods for labelling particular parts of proteins and nucleic acids.

Information on the different isotopes available, different counting techniques and the interpretation of counts is given in references in section 2 of the appendix.

9.2 Enzyme methods

Enzymes are used as tools in the biochemical analysis and study of viruses. For example, specific nucleases are used to distinguish between RNA and DNA in preparations of virus particles or in virus-infected tissues, and are also used to fragment nucleic acids in the study of their base sequences (§5.2).

Viruses change the amounts and types of enzymes in infected organisms; some host enzymes are stimulated or new enzymes that are partially or wholly coded by the virus genome are produced. Host enzymes that are stimulated or altered include the polyphenoloxidases and nucleoside mono- and diphosphate phosphotransferases (§3.4). So far the only enzymes isolated that are perhaps virus coded are RNA polymerases (or transcriptases), and they have been found in the particles of a few viruses and in infected plants. RNA polymerases transcribe RNA and produce a copy with the complementary sequence (§5.2). They are usually assayed by their ability to build radioactive acid-soluble precursors of RNA into polynucleotides that can be precipitated by trichloracetic acid. Each enzyme preparation to be tested for polymerase activity is put in a complex assay medium, which consists of a reducing buffer, about pH 8, containing the four nucleoside triphosphates and, sometimes, pyruvate kinase and phosphoenolpyruvate to generate high energy phosphates. Some of the molecules of one of the nucleoside triphosphates are radioactively labelled, usually with ^{32}P, in the innermost (α) phosphate; labelled uridine triphosphate is often used but can give spurious results as it may be terminally added to any transfer RNA present, and it is perhaps better to use labelled guanosine triphosphate (May, Gilliland and Symons, 1970). After the enzyme and RNA template (if required) have been incubated in the assay medium, the RNA is precipitated by acid and its radioactivity measured. Various tests are then made to check that any precipitated radioactivity has been specifically incorporated into new RNA. For example the addition of ribonuclease or the omission of any one of the four nucleoside triphosphates should stop RNA polymerase functioning, whereas adding deoxyribonuclease or actinomycin D should have no effect. Furthermore the labelled RNA product should have a sequence that is complementary to the RNA template, so that when the incubated assay medium is deproteinized, double-stranded RNA should be found; double-stranded RNA resists 5 μg/ml of pancreatic ribonuclease at 37°C for 30 min in double strength SSC (*s*tandard *s*aline *c*itrate; 0.15 M sodium chloride, 0.015 M sodium citrate, pH 7) and can be precipitated by trichloracetic acid, but it is hydrolysed as fast as single-stranded RNA in 0.05 SSC and is then not precipitated by acid. Alternatively the radioactive product in the incubated assay medium should compete with the template RNA in melting and annealing experiments (Black and Knight, 1970).

RNA polymerases have been found in the

particles of clover wound tumor and lettuce necrotic yellows viruses (Black and Knight, 1970; Francki and Randles, 1972). The polymerase in the particles of the latter was not active until the particles had been treated with 1% of the detergent Nonidet P-40, which presumably removed their outer lipid-containing layer, and, like most other virus polymerases, it required magnesium ions for activity.

The RNA polymerases in extracts of barley leaves infected with brome mosaic virus and of cucumber leaves infected with cucumber mosaic virus seem bound to membranes; they were found in the pellets obtained by centrifuging sap extracts at 20 000**g** for 10 min (May and Symons, 1971; Semal and Hamilton, 1968; Semal and Kummert, 1970). They were unaffected by added RNA and so seemed to be bound to their own template RNA. However when 2% of the detergent Triton X-100 was added to the pellet from barley leaves infected with brome mosaic virus, the enzyme became soluble, less stable and more active, and for full activity required added template RNA, particularly that of brome mosaic virus (Hadidi and Fraenkel-Conrat, 1973).

The best way of being sure which protein is specified by a particular nucleic acid is to translate the nucleic acid into protein, using cell-free preparations made from, for example, wheat embryos (Marcus, Luginbill and Feeley, 1968; Marcus, 1970) or mammalian cells (Ball, Minson and Shih, 1973). These have been used to translate satellite virus RNA (Klein *et al*, 1972), and the separate parts of the genomes of brome mosaic (Shih and Kaesberg, 1973) and tobacco rattle viruses (Ball, Minson and Shih, 1973).

9.3 Optical methods

Many optical techniques are useful for virus research because suspensions of virus particles or their components differ optically from the suspending fluid, and much can be learnt from these differences.

9.3.1 *Light scattering*

To the unaided eye concentrated preparations of virus particles mostly appear opalescent or turbid because the virus particles in them scatter light (Fig. 6.7). The amount of light scattered depends not only on the number of particles in the preparation, but is also exponentially related to the size of the particles and to the wavelength of the light

being scattered; scattering only becomes particularly noticeable when the size of the particles is more than 10% of the wavelength of the scattered light. Hence, whereas a preparation containing say 2 mg of tomato bushy stunt virus particles per ml (10^4 nm^3/particle) appears turbid, a preparation containing the same weight of satellite virus particles (2.5×10^3 nm^3/particle) will appear water clear. The amount of light absorbed by a virus preparation is increased by scattering, and this should be corrected for when comparing the optical densities of different virus preparations (§9.3.4). This correction is frequently not made because spherical particles about 25 nm diameter scatter little radiation at the interesting wavelengths around 260 to 280 nm; however, scattering can greatly affect measurements on preparations of larger particles. When a preparation of particles is scattering but not absorbing light, the logarithm of the optical density of the preparation is inversely and linearly related to the logarithm of the wavelength. Hence to correct for scattering, the optical density of a virus preparation is measured in light of wavelengths that are not absorbed, say 350 nm to 650 nm, and then extrapolated arithmetically to estimate the contribution of scattering to the optical density at wavelengths of interest (Englander and Epstein, 1957; Noordam, 1973).

The amount of light scattered at different angles to the incident beam can give an estimate of the molecular weight of the scattering particle irrespective of its shape, size or conformation (Bender, 1952; Zimm, 1948). Though measurements of this type have been used to measure the molecular weights of phage particles and of nucleic acids, they have only occasionally been used in studies of plant viruses (Reichmann, 1959).

9.3.2 *Effect on polarized light*

Another obvious optical property of preparations of the particles of some plant viruses is that they shimmer when stirred. This phenomenon is called streaming birefringence or anisotropy of flow. It showed Takahashi and Rawlins (1932) that sap of TMV-infected tobacco leaves contained anisometric particles (either discs or rods), and this of course was confirmed when X-ray diffraction and electron microscopy revealed that the particles of TMV are rods. The streaming birefringence of preparations of elongated virus particles is most obvious in polarized light and is seen, for example, when a preparation is shaken between two sheets of 'Polaroid' arranged with their planes of polarization at 90°. The uneven movement momentarily aligns the particles in some parts of the suspension; the particles are dichroic and noticeably rotate the plane of polarization of light when enough of them are aligned in close proximity. Preparations containing more than about 10% by weight of TMV particles are spontaneously birefringent, because the particles are so closely packed in the suspension that they align spontaneously to form paracrystals (§§5.5.2 and 6.8).

Flow birefringence may be measured quantitatively to give an estimate of the shape and size of elongated particles. However this method has rarely been used with plant-virus particles (Reichmann, 1958) although much information on the secondary and tertiary structure of virus molecules (Chapter 5) has been obtained by measuring how much they rotate plane polarized light of different wavelengths; these techniques are known as optical rotatory dispersion (ORD) when ultra-violet light is used and circular dichroism (CD) when infra-red light is used.

9.3.3 *Refractive index*

Virus preparations have a greater refractive index than the solvent they contain. The specific refractive increment in water (i.e., the difference in refractive index between the solvent and a 1% w/v solution) of proteins, of both types of nucleic acid, and hence also of nucleoproteins, is close to 0.00180. Thus a sensitive refractometer will give a quick and accurate estimate of the total weight of nucleoprotein in a virus preparation (Englander and Epstein, 1957), although calibration is necessary if the virus particles contain lipid.

The effect of proteins, virus particles etc. on refractive index is the basis of the schlieren and interference optical systems that are used to record the movement of these macromolecules in many of the physical analytical tools, such as analytical centrifuges, and electrophoresis and diffusion apparatuses. The schlieren system is an optical arrangement in which boundaries between solutions of different refractive index change the path of a slit of light; the greater the rate of change of refractive index at a boundary the greater the deviation, so that boundaries of refractive index appear as peaks above a base line

(Fig. 9.2). The area under each peak is proportional to the refractive index change, and hence to the concentration change, at the boundary. The interference optical system divides a light beam by a half silvered mirror, passes one part through the macromolecule suspension and the other (reference) beam through solvent. The two beams are then recombined so that they give a series of

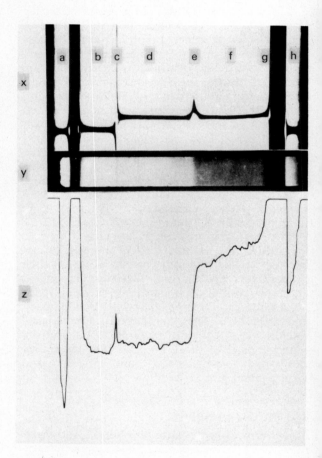

Fig. 9.2 Centrifugal analysis of a purified preparation of red clover vein mosaic virus (a carlavirus); particles sedimenting from left to right. **x,** Schlieren diagram and **y,** photograph of the same cell taken in ultra-violet light, after centrifuging for 12 min at 24 630 rev/min; **z,** microdensitometer record of **y.** (a) and (h) are the inner (centripetal) and outer (centrifugal) apertures of the reference cell; between these is the analytical cell which contains air (b) separated by a meniscus (c) from the virus preparation. The virus particles have already sedimented from the top half of the cell as shown by the boundary (e) between the solvent (d) and the suspension of particles (f), which are accumulating at the bottom of the cell (g). (Courtesy A. Varma.)

interference fringes that reflect differences in the refractive indices along the paths of the two beams.

9.3.4 *Light absorption*

Proteins and nucleic acids absorb ultra-violet radiation of wavelengths between 220 nm and 300 nm because of the cyclic compounds they contain (i.e. the bases and aromatic amino acids). Measurements of the optical density of virus preparations at several wavelengths in blue to ultra-violet light can give information on the concentration of virus particles in a preparation, the nucleic acid content of the particles and the conformation of the nucleic acid. Such measurements will be dominated by the nucleic acid in the virus particles, because nucleic acids absorb twenty or more times as much ultra-violet light as proteins (Fig. 9.3).

The *optical density* of a preparation is measured in a spectrophotometer, and is defined as the ratio of the logarithms of the intensities of incident and transmitted light; in practice the optical density of a virus preparation is measured by comparing the intensity of light transmitted by the virus preparation with that transmitted by the suspending fluid used in the virus preparation. The effect of light scattering must be removed (§9.3.1) before comparing measurements of optical density. The optical density is often used to calculate the *extinction* coefficient ($E_{1\,cm}^{0.1\%}$), which is the optical density of a 1 cm layer of a 1 mg/ml preparation.

The optical density of a preparation in light of

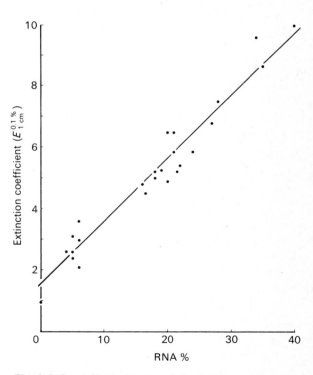

Fig. 9.3 Ultra-violet absorption spectra of suspensions of **a,** RNA from turnip yellow mosaic virus (TYMV) particles; **b,** the nucleoprotein particles of TYMV (33% RNA); **c,** cucumber mosaic virus particles (19% RNA); **d,** tobacco mosaic virus particles (5% RNA); **e,** the protein particles (top component) of TYMV. All the suspensions contained the same weight of material except **a,** which contained half as much as the others. (Redrawn from Kaper, 1968.)

Fig. 9.4 Graph illustrating the relation between the extinction coefficient at 260 nm ($E260_{1\,cm}^{0.1\%}$) of purified preparations of virus particles, and the RNA content of those particles. The line shows the linear regression which best fits the twenty-five records and has the relation

$$(E260_{1\,cm}^{0.1\%}) = 1.531 + 0.205\,(RNA\%)$$

(Data from the C.M.I./A.A.B. Descriptions of Plant Viruses.)

different wavelengths is called the *absorption spectrum*. Proteins characteristically have a maximum in their absorption spectrum at around 280 nm and a minimum near 250 nm (Fig. 9.3e); their extinction coefficients at 260 nm are around 1.0 (Fig. 9.4). Nucleic acids absorb more radiation, their absorption spectra have maxima around 260 nm and minima near 230 nm (Fig. 9.3a) and a maximum/minimum ratio of about 2; their extinction coefficients at 260 nm are around 20 to 25. The precise ultra-violet absorption of a nucleic acid depends on its composition because, although adenine and uracil absorb maximally at 260 nm, guanine and cytosine have absorption maxima near 250 nm and 280 nm respectively.

The average molecular weight of nucleotides is approximately 330, and their average extinction coefficient at 260 nm is about 33. The ultra-violet absorption of bases decreases when they are covalently linked in polynucleotides, because they are then constrained and their resonance properties are altered. A randomly orientated nucleic acid absorbs 10% less than its constituent nucleotides, and the absorption decreases a further 30% when it becomes fully base-paired with a nucleic acid of complementary sequence to form a double-stranded nucleic acid. Because of this decrease in light absorption, or hypochromicity, double-stranded RNA and DNA have extinction coefficients at 260 nm of around 20. When double-stranded nucleic acids are heated above a certain well-defined 'melting' temperature (T_m) the strands separate and become randomly orientated, the so-called helix-to-coil transition, and the optical density increases by about 50% (Fig. 5.6). The T_m depends on the guanine plus cytosine content of the nucleic acid, is increased by increasing the salt concentration in the solvent, and is often in the range 85 to 95°C for DNA and 5°C greater for RNA. T_m may be used to estimate the base composition of both double-stranded RNA (Kallenbach, 1968) and DNA (Marmur and Doty, 1962; Mandel and Marmur, 1968). When a double-stranded nucleic acid is heated so that the strands separate, and then cooled slowly, the strands base-pair again and the optical density decreases; this process is called reannealing. Single-stranded RNA in helically constructed particles, such as those of TMV, is not base-paired and has an extinction coefficient at 260 nm of near 30 Isolated single-stranded RNAs or the single-stranded RNAs in most icosahedrally constructed virus particles often have one quarter or more of their bases paired with other regions of

the same RNA molecule and so have an extinction coefficient around 25, and when heated become hyperchromic over a broad range of temperatures (Fig. 5.6).

The amount of radiation absorbed by a virus particle is the sum of the amounts absorbed by its constituent parts, and thus the approximate extinction coefficient of the particles of a virus can readily be predicted from their composition. However, the extinction coefficient can be determined more accurately by experiment (Fig. 9.4). When the extinction coefficient of virus particles is known, measurements of optical density of purified preparations give quick and sensitive estimates of particle concentration. However, impurities in the preparations can greatly distort such estimates. Another useful figure is the ratio

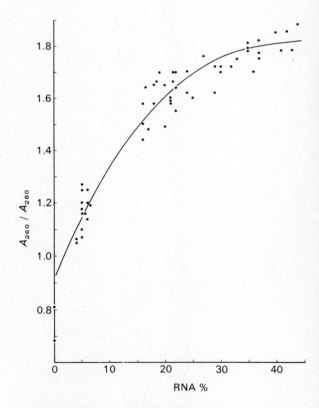

Fig. 9.5 Graph illustrating the relation between the A_{260}/A_{280} ratio of purified preparations of virus particles and the RNA content of those particles. The line shows the quadratic regression which best fits the sixty-two records, some multiple, with the relation.

$$A_{260}/A_{280} = 0.9320 + 0.0454\ (RNA\%) - 0.0006\ (RNA\%)^2$$

(Data from Paul (1959), Sehgal *et al* (1970) and the C.M.I./A.A.B. Descriptions of Plant Viruses.)

of the absorbances of a virus preparation in light of wavelengths of 260 nm and 280 nm (Fig. 9.5). This can be used to give an estimate of the proportion of nucleic acid in nucleoprotein preparations, and is therefore of value not only for characterizing virus particles, but also for assessing the purity of preparations of well-studied viruses. Interpretation of measurements of the A_{260}/A_{280} ratio and the extinction coefficient will, of course, be complicated with virus preparations that contain particles of different types, particularly when the types differ in composition.

Various types of biochemical apparatus use not only refractometric systems to study the behaviour of macromolecules but also systems that measure the absorption of ultra-violet light (Fig. 9.2). An absorption system is of particular value for studies of strongly absorbing materials like nucleic acids or nucleoproteins because its sensitivity is several times greater than that of most refractometric systems.

Absorption of ultra-violet light may also be measured in a micro-spectrophotometer, an apparatus which combines an ultra-violet light microscope with a spectrophotometer, and enables the absorbance of parts of a single cell to be measured. Zech (quoted by Mundry, 1963) used this apparatus to measure the changes in nucleic acid concentrations in different parts of tobacco leaf hair cells at different times after infection with TMV.

9.3.5 *Fluorescence*

Guanine and its derivatives fluoresce blue-violet in ultra-violet light, especially at low pH, and this simplifies their identification in the chromatographic analysis of hydrolysates of nucleic acids, but the fluorescence of nucleic acids is insufficient to be of use in histological studies. However, nucleic acids absorb the fluorescent dye, acridine orange (Schümmelfeder, 1958), and using this dye and a light microscope adapted for fluorescence techniques, the position and type of nucleic acid can be determined in, for example, cell preparations. Double-stranded nucleic acids treated with acridine orange do not absorb much of the dye, and fluoresce pale yellowish green. By contrast single-stranded nucleic acids absorb more of the dye and fluoresce flame-red (Mayor and Hill, 1961; Gomatos *et al*, 1962). The dye may not penetrate thick tissue fully, but treatment with proteolytic enzymes (e.g. 'Pronase') can overcome this problem (Harrison, Stefanac and

Roberts, 1970), and if combined with treatments by specific nucleases the method can be used to determine the type and strandedness of nucleic acid in tissues, virus preparations (Bradley, 1966), and even in specific precipitates in Ouchterlony serological tests (Cowan and Graves, 1968).

The use of antibodies coupled with fluorescent tracers to study the distribution of antigens in, for example, tissue sections is discussed in §8.5.9.

9.4 Centrifugation

Most of the particles of a virus are of one or a few types (§6.7.3), and all the particles of one type generally have the same mass, shape and density. Particles in a gravitational or centrifugal field move and diffuse at a rate which depends on their mass, shape and volume, and on the suspending medium. The densities, diffusion constants and sedimentation coefficients of virus components are distinctive properties that can be used for assay and characterization, and knowledge of them can be exploited during the purification of virus particles (Chapter 6); Fig. 6.6 shows the sedimentation coefficients and densities of plant-virus particles and of the commoner host components found in plant sap (see also Fig. 6.2).

9.4.1 *Basic information*

Figure 9.6 shows the four principal types of centrifuge rotor. The most frequently used is the fixed angle rotor, in which the suspension is held in tubes placed in holes drilled into the rotor at an angle of 18–40° to the axis of rotation. Also frequently used is the swing out rotor (§6.4.1), and increasing use is being made of the zonal rotor for purifying large amounts of material. Analytical rotors are used in centrifuges fitted with optical systems that allow the sedimenting components to be observed and their movement measured; the suspensions are held in cells with windows so that, as the rotor spins, the cells pass through a light beam that is parallel to the axis of rotation, and is part of a refractometric or absorption optical recording system (§9.3 and Fig. 9.2).

In low speed centrifuges the rotors spin at up to 10 000 rev/min or more and produce centrifugal fields of up to 15 000*g* or more. These centrifuges can have fixed angle or swing out rotors that hold from 10 ml to 6 l or more of suspension. They are used to clarify extracts containing virus particles and to collect chemically precipitated virus particles or other materials (Chapter 6).

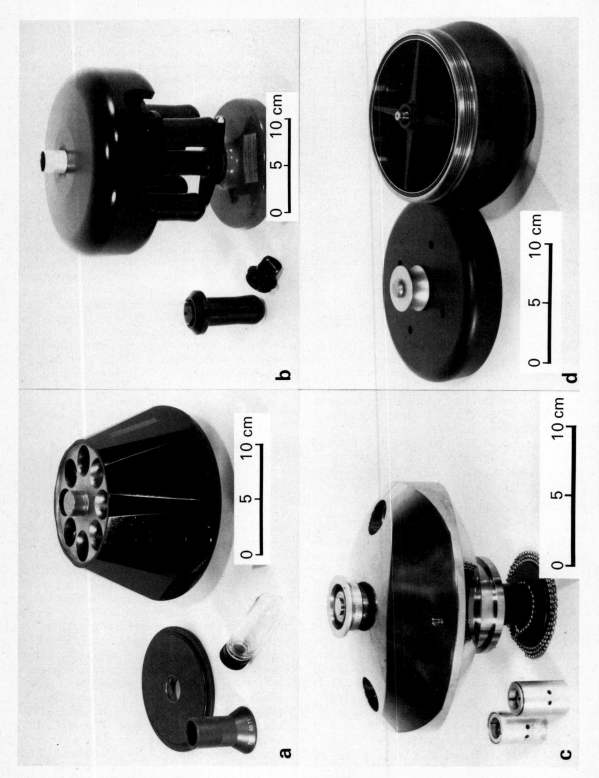

Fig. 9.6 Four types of centrifuge rotor commonly used in virus research. **a,** Fixed angle rotor with lid detached and a single tube; **b,** swing-out rotor with one bucket detached and cap removed; **c,** analytical rotor with two of the cells that contain the suspension being analysed; **d,** zonal rotor with lid removed to show the space in which the gradient is held. (Courtesy B. H. Parr.)

In high-speed (= ultra-) centrifuges, and in analytical centrifuges, the rotor spins in a refrigerated evacuated chamber at up to 75 000 rev/min or more and produces centrifugal fields up to 500 000*g* (Fig. 9.6). Thus not only virus particles but also proteins and nucleic acids can be sedimented (Fig. 6.2). High-speed centrifuges are used to purify, assay and characterize virus particles and their components, and the analytical centrifuge is also used to determine the purity of virus preparations.

9.4.2   *Centrifugation methods*

(a) *Differential centrifugation*   This is the procedure of using brief periods of small-*g* centrifugation alternating with long periods of large-*g* centrifugation to separate mixtures. It is one of the commonest tactics employed for purifying virus particles and is discussed in §6.4.1.

(b) *Boundary sedimentation*   This is the method usually employed when estimating the sedimentation rate of particles in an analytical centrifuge (Fig. 9.2). When a suspension of particles is centrifuged, the particles move away from the air/liquid meniscus (*c* in Fig. 9.2) so that each component forms a boundary, the movement of which can be measured to provide an estimate of the sedimentation rate of that component. The change of refractive index or absorbance at each boundary is related to the concentration of that component in the system, and can be estimated after the optical system has been calibrated. This method of assay cannot be used to determine the concentration of two or more components with similar sedimentation rates and present in large amounts in the same cell, because the centrifugal anomaly called the Johnston-Ogston effect (Schachman, 1959) will distort the estimates.

When virus particles are present in such small amounts that they can only be detected by infectivity tests, their sedimentation rate can still be estimated using a separation cell in an analytical centrifuge (Schachman, 1959). A simpler method is described by Anderson (1968) and Polson and van Regenmortel (1961).

(c) *Rate zonal centrifugation*   This technique is used for purifying virus particles (§6.4.1), for fractionating virus preparations, and also for estimating the sedimentation rates and densities of virus particles.

For this technique, centrifuge tubes are filled with solvent containing an inert dense solute such as sucrose, glycerol or Ficoll (polysucrose) that gives a gradient of density throughout the tube (§6.4.1). The virus preparation is floated, or layered, onto the gradient and then centrifuged. Particles of different types sediment into the gradient at different rates to form distinct bands (Fig. 6.7). The rate at which a particle is sedimenting at a particular point in the gradient depends on its mass and shape, on the centrifugal field, on the difference between its density and that of the gradient at that point, and on the viscosity of the gradient at that point. Some of these factors change as a particle sediments through a gradient, and thus it is difficult to use gradients to calculate sedimentation coefficients directly. However they may be used indirectly by comparing, in the same gradient, the movement of virus particles of unknown sedimentation coefficient with the movement of particles of similar type but of known sedimentation coefficient (Brakke, 1958; Harrison and Nixon, 1959a).

To overcome some of the complexities of interpretation of rate-zonal centrifugation studies, two special types of gradient can be used. One of these is called an isokinetic gradient (McCarty, Stafford and Brown, 1968; Noll, 1967), and is designed to keep the forces acting on the sedimenting particles constant throughout the gradient, so that they move at a constant speed related to their sedimentation coefficient. The other type of gradient is the linear-log gradient (Brakke and van Pelt, 1970a) in which the forces vary logarithmically, so that the logarithm of the distance travelled by a particle is a linear function of the logarithm of its sedimentation coefficient or the period of centrifugation. Both types of gradient have to be designed specifically for each rotor, for a particular temperature, for a particular solute and for particles of a particular density. These gradients have been used to compare the sedimentation rates and densities both of virus particles and of their nucleic acids (Brakke and van Pelt, 1970b).

(d) *Isopycnic banding and sedimentation equilibrium centrifugation*   If virus particles are sedimented in a gradient, part of which is more dense than the particles, then they will eventually sediment to that part of the gradient which has the same density as the particles, and will float there; this is called *isopycnic banding*. However, if centrifugation is continued long enough then centrifugal sedimentation both of macromolecules and of solutes will be balanced by diffusion and all the molecules in the gradient will be in equilibrium; the gradient is then said to have reached *sedimentation equilibrium*.

Isopycnic banding is used for purifying and separating virus components using gradients of sucrose or Ficoll (for lipid-containing virus particles) or dense salts, such as caesium chloride or sulphate, potassium tartrate or rubidium bromide (for nucleoprotein particles or nucleic acids). This technique can also give approximate estimates of the densities of virus particles or other components of preparations.

Sedimentation equilibrium is used to fractionate and estimate the densities of virus particles or nucleic acids in solutions of caesium chloride (for DNA) or caesium sulphate (for RNA). The salts are mixed with nucleic acid or virus particle preparations and the mixture then centrifuged for one day or more at high speed in an analytical rotor or, for preparative work, in a fixed angle rotor; the solute is sedimented and forms a gradient, and the nucleic acid molecules or virus particles sediment centrifugally and float centripetally to band at the level of equal density. The density of a component in such a gradient may be estimated either by sampling the gradient and determining the solute concentration in the sample refractometrically, or by adding materials of known density to the mixture before centrifugation, or by calculations based on the rotor speed, solute concentration etc.

The density of a macromolecule depends greatly on the solution in which it is suspended; for example, TMV-RNA has an estimated density of 1.49 in sucrose solution but about 1.64 in caesium chloride solution. These differences presumably reflect differences in interaction between RNA and solute.

9.4.3 *Uses and interpretation of centrifugation data*

The physical principles underlying centrifugation have been often reviewed (see section 2 of the appendix). The behaviour of particles while being centrifuged depends both on properties of the particles such as their volume, shape and density, and also on properties of the suspending medium such as its viscosity and density, and the effect the medium has on properties of the particles such as their charge, solvation and binding of ions. During centrifugation, particles are moved both by centrifugal force and by diffusion. By measuring the movement of particles transported by these forces it is possible under the correct experimental conditions to compare and assay any of the properties listed above.

Each type of particle sediments at a characteristic speed in a centrifugal field. The sedimentation coefficient (s) is defined as:

$$s = V/\omega^2 r \text{ cm/sec dyne}$$

where V is the rate of sedimentation in cm/sec and the centrifugal field is defined by the radial distance r and the rotor's angular velocity ω in radians/sec ($= 377 \times$ rev/min). This equation may be rewritten as

$$s = \frac{1}{\omega^2 r} \cdot \frac{dr}{dt} = \frac{1}{\omega^2} \cdot \frac{d \ln r}{dt} \qquad (9.1)$$

thus one must measure ω, dr and dt to estimate s. The last term of equation 9.1 is usually determined as the slope of the graph of $\ln r$ versus t, even though this is slightly curved and hence the mean slope (tangent to the curve) is calculated. This slope can be calculated manually, or by computer (Trautman and Hamilton, 1972), or more simply by Markham's ingenious protractor method (Markham, 1962). The values of s are rather small and are usually expressed in svedberg units or S ($1S = 1 \times 10^{-13}$ cm/sec dyne). Frequently the dependence of sedimentation coefficient on concentration of solute is determined, and the extrapolated value at zero concentration in water at 20°C ($= s^0_{20,\text{w}}$) is calculated. The range of S values of virus particles is shown in Fig. 6.6.

Equation 9.1 can be rewritten in terms of the sedimenting particle:

$$s = \frac{M(1 - \overline{V}\rho)}{Nf} \qquad (9.2)$$

where M is the molecular weight of the particle, $\overline{V}$ is the partial specific volume of the particle, ρ is the density of the suspending medium, N is Avogadro's number and f is the translational frictional coefficient. This formula may be used to estimate the molecular weight of a particle of known sedimentation coefficient when $\overline{V}$ and f are known too.

$\overline{V}$ is the reciprocal of the density of the particle and may be determined experimentally, or calculated from the composition of the particle, assuming, for example, that the partial specific volume of nucleic acids is 0.45 to 0.55, of proteins 0.71 to 0.75 and of lipoproteins 0.9 to 1.1. Alternatively $\overline{V}$ may be calculated from the amino acid and nucleic acid composition of protein or nucleo-protein of particles (Schachman, 1957). Different methods of estimating $\overline{V}$ usually give different results (Sehgal *et al*, 1970); however, $\overline{V}$ is closely related to the nucleic acid content of the particle and, in nucleoprotein particles containing single-stranded RNA, is in the range 0.66 to 0.76 (Fig. 9.7).

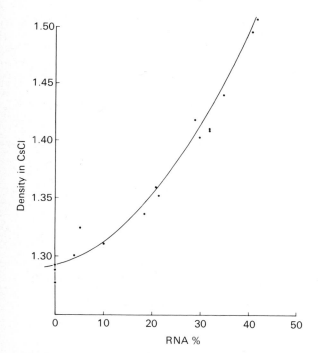

Fig. 9.7 Graph illustrating the relation between the density in CsCl solution of different virus particles, and their RNA content. The line shows the quadratic regression which best fits eighteen records (two multiple) with the relation

   Density $= 1.2922 + 0.0011$ (RNA%) $+ 0.0001$ (RNA%)^2

(Data from Sehgal *et al*, 1970.)

The determination of f is commonly avoided by determining another property of the particle that depends on f. Usually the diffusion coefficient (D) is measured (§9.6) and f calculated from the relation

$$f = kT/D \qquad (9.3)$$

where k is the Boltzmann constant and T is the absolute temperature.

Combining equations 9.2 and 9.3 gives the Svedberg equation

$$M = RTs/D(1 - \overline{V}\rho)$$

where R is the gas constant.

Methods of using data from centrifugation experiments to estimate directly molecular weights of virus components are reviewed by Trautman and Hamilton (1972). In the classical equilibrium method a short column of the virus component preparation is centrifuged at low speed until the particles form a gradient under the opposing influences of diffusion and centrifugal force. Alternatively the preparation may be centrifuged faster and the molecular weight estimated by the meniscus depletion method (Yphantis, 1964); this method has been used to estimate the sizes of proteins from particles of different plant viruses (Paul and Buchta, 1971). Yet another method is to estimate molecular weights from the distribution of particles in bands formed by sedimentation equilibrium in gradients; at equilibrium the distribution of particles in an isodensity band is essentially Gaussian with a standard deviation that is directly correlated to the molecular weight of the particles in the band.

The molecular weights of virus particles or nucleic acids are not directly related to their sedimentation coefficients. Indeed dimer and trimer aggregates of spherical particles sediment at about 1.4 and 1.7 times the rate of monomers. But the molecular weights of particles of the same structure but different size (e.g. helically constructed particles of different length) can be compared from their dimensions, sedimentation coefficients and partial specific volume (Harrison and Klug, 1966).

The sedimentation coefficient of the isometric nucleoprotein particles of a virus relative to that of its RNA-free protein shells depends on the amount of RNA in the nucleoprotein; and this may be estimated using Reichmann's (1965) formula:

$$x = \frac{100((S_N/S_P) - 1)}{((S_N/S_P) + 1)}$$

where $x = \%$RNA, and S_N and S_P the sedimentation coefficients of the nucleoprotein and protein particles respectively. Note that this relationship holds only when the protein shells in the two types of particle are hydrodynamically identical, as in the particles of many plant viruses but not in those of the picornaviruses of mammals.

Estimates of the sizes of nucleic acids are frequently obtained by comparing their sedimentation rates with those of nucleic acids whose sizes have also been estimated in other ways (e.g. light scattering). Spirin (1963) found that many single-stranded RNAs have sedimentation coefficients (s) that are related to their molecular weights (M) by the formula $M = ks^a$, where k is around 1550 and a about 2.1. Estimates obtained in this way are inaccurate because the molecular weights of few RNAs are known accurately and estimates of k and a obtained by different people vary greatly. Boedtker (1968) and Brakke and van Pelt (1970b) sought to refine this method by treating the RNA before centrifugation with formaldehyde to abolish its secondary structure.

Double-stranded RNAs sediment at about two-thirds the rate of their component single-stranded RNA molecules, and their size may be estimated by comparison with the segments of the genome of reovirus (Shatkin, Sipe and Loh, 1968).

The relation between sedimentation coefficient and molecular weight is more complicated for DNA, because virus DNA can occur in many different configurations; in particles of some viruses it is single-stranded, in others is double, in some circular and in others linear. The most carefully determined relation between sedimentation coefficients and molecular weights is that for linear double-stranded DNA of more than 10^6 daltons weight (Freifelder, 1970) and fits the equation

$$s^0_{20,w} = 2.8 + 0.00834 M^{0.479}$$

Some virus DNA genomes are closed circular double-stranded molecules, such as those of cauliflower mosaic virus and the papovaviruses of mammals, including polyoma virus, which has a genome that illustrates how conformation determines the sedimentation behaviour of a molecule. Most of the DNA molecules isolated from polyoma virus particles are coiled coils ($\equiv$ superhelices) of double-stranded DNA of molecular weight 3×10^6 and sedimentation coefficient $20S$. When one of the strands is broken the two strands can rotate, the superhelix uncoils and then sediments at $16S$, but if both strands are broken in the same place the resulting linear molecule sediments at $14.5S$, though the single strands obtained from the linear molecule sediment at $16S$. If the intact superhelix is denatured at high pH

it forms tightly entwined paired single-stranded loops, and these sediment at $53S$.

Superhelical DNA molecules may be separated from linear DNAs with the same sedimentation rate or density by treating the mixture with a dye such as ethidium bromide or actinomycin D. These dyes intercalate between the stacked bases in the helix and change the hydrodynamic behaviour of the DNA molecules.

The density of the particles of a virus is a distinctive character of that virus, and depends most on the composition of the particles. Figure 9.7 shows the relation between the density of various plant virus particles and their RNA content. Most proteins have densities around 1.28 to 1.30, and lipoproteins 0.9 to 1.1. The density of nucleic acids depends on their composition; single-stranded RNAs have densities in caesium sulphate gradients of about 1.62 to 1.65, and double-stranded RNAs have densities around 1.58 to 1.61. DNAs are less dense (about 1.4 in caesium sulphate), and in caesium chloride gradients where they are less hydrated than in caesium sulphate, double-stranded DNAs have densities which are related to their guanine and cytosine (G + C) content (Doty, 1961) by the formula

$$\rho = 1.660 + 0.098 \, (G + C\%)$$

and thus the density of a double-stranded DNA gives an estimate of its composition.

9.5 Electrophoresis

9.5.1 *Virus particles*

The proteins that constitute the outer shell of virus particles have, at their surface, amino acids some of which bear ionizable groups; at low pH the basic groups of lysine, histidine and arginine are charged, and at high pH aspartic and glutamic acids are charged. For the particles of each virus there is one particular pH at which the same number of basic and acidic groups are ionized so that the particles have no net charge; this pH is called the isoelectric point. At all other pH values the particles will carry a net positive or negative charge and will therefore migrate in an electric field. The rate of migration at a particular pH value, or the electrophoretic mobility, is defined as

$$\mu = Q/6\pi\eta a \qquad (9.4)$$

where Q is the net charge on the particles, which have an effective radius of a, and η is the viscosity of the medium.

The particles of different strains of a virus often have different electrophoretic mobilities, and hence this property is rarely used to characterize and distinguish the particles of different viruses, but has been used in studies on strains or mutants of one virus, such as TMV (Chapter 12).

Electrophoresis is also used for purifying virus particles, and is particularly suitable for viruses that have particles of different sizes and/or densities but constructed of the same proteins and therefore with the same electrophoretic mobility. For viruses with elongated particles of different lengths, both Q and a in equation 9.4 will be similarly related to length; however, at pH 8.6, the long and short particles of tobacco rattle virus have slightly different electrophoretic mobilities (Cooper and Mayo, 1972). For isometric virus particles of the same size but different nucleic acid contents both Q and a will be similar for all particle types.

Quantitative estimates of electrophoretic mobility are usually made in buffer in an electrophoresis cell such as that designed by Tiselius. This is in essence a U-tube, in the lower part and one limb of which the virus preparation is put. An electric field is applied through the cell, and the movement in each limb of the boundary between the preparation and the solvent is followed using a refractometric optical system. The same apparatus may also be used for preparative work but, for various reasons, this is more conveniently done by zone electrophoresis in sucrose gradients, contained either in a U-tube (van Regenmortel, 1972) or in a tube held vertically and blocked at its bottom end with polyacrylamide gel.

Electrophoresis of virus particles in polyacrylamide gels of different strengths is a useful way of detecting particles in different size categories in preparations of viruses such as pea enation mosaic (Hull and Lane, 1973), and of estimating their effective radii from the relative sieving effects.

Virus particles and other antigens can also be fractionated electrophoretically in agar or agarose gels, so that, after electrophoresis, the position of the antigens may be determined serologically (§8.5.7). However immunoelectrophoresis, as this technique is called, is difficult to use as a method to estimate electrophoretic mobilities because of the sieving effect of the gel and the electroendosmotic movement of buffer in the gel.

9.5.2 *Virus proteins and nucleic acids*

Electrophoresis is of great value for studying the components of virus particles. The most popular methods involve electrophoresis in polyacrylamide gels (Fig. 5.7), in which the sieving effect ensures that the mobility of a particular molecule is inversely proportional to the logarithm of its size. Thus a mixture of similar molecules (e.g. proteins) of known and unknown sizes are separated electrophoretically, and their positions (and hence their rates of migration) detected by staining, ultra-violet absorption or radioactivity. The sizes and amounts of the 'unknown' molecules in the mixture can then be estimated (Weber and Osborn, 1969).

Proteins differ in charge as well as size, and to obliterate the charge differences the proteins are commonly treated with the anionic detergent sodium dodecyl sulphate, which gives their surfaces a uniform negative charge, or with cetyltrimethylammonium bromide, which gives them a uniform positive charge. However a few proteins, even after treatment with sodium dodecyl sulphate, do not move in polyacrylamide gels at a speed solely determined by their sizes. The sizes of such 'anomalous' proteins can be estimated nonetheless by measuring the sieving effect of different gel concentrations (Hedrick and Smith, 1968; Koenig, 1972; Marjanen and Ryrie, 1974).

Electrophoresis in polyacrylamide or agarose gels is also used to compare the sizes of nucleic acid molecules, particularly RNA molecules, and can be as reproducible and as accurate as methods based on sedimentation coefficients (Kaper and Waterworth, 1973; Daneholt *et al*, 1969). Infective RNA can be extracted from the gels, so that electrophoresis is also useful as a means of preparing samples of RNA molecules with different sizes and biological activities. For comparing or separating RNA molecules of 0.3×10^6 to 3×10^6 daltons it is necessary to use gels containing less than 3% polyacrylamide (Fig. 6.13) but such gels are rather unstable and 1% or more agarose can be added to aid handling; more concentrated gels are used for smaller RNAs. All completely base-paired double-stranded RNA molecules have the same secondary structure, and so their sizes may be compared directly in gels, using, for example, the relatively well-studied RNAs from reovirus particles as standards (Reddy and Black, 1973a). By contrast different single-stranded RNA molecules have different amounts of secondary structure (see §5.2), and this affects their mobility in gels in the same way as it affects their sedimentation rates. Similarly DNAs of different configuration have different electrophoretic mobilities (Aaij and Borst, 1972). One way of removing secondary structure is to heat the nucleic acid with formaldehyde (Boedtker, 1968; 1971), but this also destroys infectivity, and so if it is important to retain infectivity it is best to use formamide (Staynov *et al*, 1972) or to heat without added chemicals (Reignderf *et al*, 1973).

9.6 Diffusion

Diffusion coefficients of virus particles may be measured directly in an apparatus which forms an interface between a suspension of particles, and the solvent in which they are suspended; the suspension and the solvent are put in separate parts of a split cell and the parts then slid together. Diffusion at the interface is then measured using a refractometric optical system (Schachman and Williams, 1959). Alternatively the interface may be made in an analytical centrifuge at high *g*, then the centrifuge may be slowed and diffusion measured at low *g* (§9.4.3).

Diffusion coefficients of particles that are antigenic may be indirectly and less accurately measured in gel-diffusion serological tests, where their diffusion coefficient is compared with that of IgG antibodies (Polson, 1961). Using a special apparatus this method can give accurate results, but it can also give crude but useful results even with the Ouchterlony plate method and readily allows distinction of the band of precipitate containing intact virus particles from that containing their protein subunits (Tremaine and Willison, 1961; see §8.5.7).

Another ingenious method for measuring diffusion coefficients depends on measuring the amount that the frequency of monochromatic light (obtained from a laser) is spread when it interacts with randomly diffusing particles; the frequency of reflected light is spread by Döppler interaction with the moving particles (Harvey, 1973).

When a mixture of particles is washed through a column of agar or agarose gel beads, the particles are sorted according to their ability to diffuse into the gel phase and thus to their diffusion coefficients. Although this technique is useful for fractionating or purifying virus preparations (§6.4.2) or for fractionating mixtures of virus proteins (Davison, 1968), it has not proved useful for quantitative measurements on virus particles (Polson, 1961).

9.7 Electron microscopy

9.7.1 *Introduction*

The electron microscope is perhaps the most useful single tool available to virologists nowadays; unfortunately it is also often the most costly. Some excellent reviews on electron microscopy are listed in section 2 of the appendix.

The reason why it is necessary to use electrons to 'see' virus particles was foreshadowed by Ernst Abbé, who showed in 1873 that the resolution of a light microscope was related to the aperture of its objective lens and to the wavelength of the light being used, but that even when the aperture was maximal the resolution was limited to about half the wavelength of the light used. Thus among virus particles only those of poxviruses, such as vaccinia, are big enough (brick-shaped and 300×200 nm) to be seen in a light microscope. Much greater resolutions are theoretically obtainable using beams of electrons instead of visible light, and modern transmission electron microscopes can resolve 0.2 to 0.4 nm, which is ample for many biological macromolecules.

In electron microscopes, electrons are produced in various ways, accelerated by large voltages, and focused by complex electrostatic or electromagnetic lenses. The pathway of the electron beam in the microscope must be at a very high vacuum, and hence volatile compounds such as water evaporate in the microscope. Biological materials that are actively metabolizing, and so need water, therefore cannot be studied directly in the conventional transmission electron microscope. In such instruments, electrons cannot usually penetrate biological specimens more than 200 nm thick, using accelerating voltages of up to 100 kV; these electron microscopes are therefore best used to study surfaces or very thin sections of materials, or films of dried suspensions of macromolecules. Suspensions used to prepare dried films must contain more than about 10^9 particles/ ml for useful numbers of particles to be seen in each field of view of the microscope.

A specimen is 'seen' in an electron microscope if it scatters, diffracts or changes the phase of electrons passing through it on their way to a fluorescent screen, a photographic emulsion, or an electron sensor. Different biological materials do not affect electrons very differently, and hence must be treated in various ways to increase their electron affecting properties; they can, for example, be treated with stains, or be shadowcast with vapour containing an element of large atomic weight. However, the resolution in electron micrographs of many biological specimens prepared in these ways is limited in practice to about 1.5 nm.

9.7.2 *Types of electron microscopy*

Transmission electron microscopy is the type of microscopy most commonly used in studies of viruses; the electron beam passes through specimens treated in various ways and is recorded. Scanning electron microscopy is used occasionally and will also be described.

(a) *Shadow casting* The commonest method used to increase the contrast of virus specimens in the 1950s was shadow casting (Williams and Wyckoff, 1944). In this technique the specimen, often a preparation of virus particles dried onto a supporting collodion film, is put into an evacuated chamber, and a small amount of a heavy metal such as platinum + iridium, or gold, or uranium is vapourized so that its vapour is deposited on the supporting film from an angle. When the specimen is examined in an electron beam at right angles to the supporting film, the specimen appears to be lit from the direction of shadow casting (Fig. 16.4), and the shape and surface of the specimen is revealed with a resolution limited by the size of the particles of metal vapour. This technique is less often used for virus particles nowadays, but a variant of it (Kleinschmidt and Zahn, 1959) is used with nucleic acid molecules; the molecules are spread and trapped in a monomolecular layer of protein formed at a water/air surface, the layer is transferred to an electron microscope grid and, while rotating, is shadow cast at an extremely oblique angle so that the topography and length of the nucleic acid molecules can be seen (Fig. 5.2).

Replicas of surfaces are also prepared by shadow casting (see section c below).

(b) *Negative staining (negative contrast)* Some of the limitations of shadow casting are overcome by negative staining, a technique that was first introduced by Hall (1955) and developed by Brenner and Horne (1959). In this method the specimen, such as a virus preparation or even sap from a diseased plant, is mixed with a solution of an electron-dense negative stain and then mounted on a supporting film on an electron microscope grid, often simply by spraying, or by placing a drop on a grid and quickly removing it. Bacitracin

can be added to purified virus preparations to make them spread in a very thin layer on the support film, which is often carbon, or a carbon-coated film of nitrocellulose (e.g. 'Collodion'). In the electron microscope the particles are seen in negative contrast (Fig. 2.1), and not only is the outline of each particle seen, but also all the interstices into which stain has penetrated. The most widely used negative stains are solutions of sodium phosphotungstate, uranyl acetate, uranyl formate, ammonium molybdate or sodium silicotungstate, but there are many others. A negative stain suitable for the particles of one virus may be quite unsuitable for others, and the best is found by trial and error (Fig. 5.14), however there has been much success with sodium phosphotungstate, and the helically constructed particles of some plant viruses show most structural detail in uranyl formate (Varma *et al*, 1968).

A virus preparation may be affected by negative staining. For example, ribosomes and the particles of some viruses, such as alfalfa mosaic, are disrupted by sodium phosphotungstate unless treated with formaldehyde, and the isometric particles of some other viruses, though not totally disrupted, lose their nucleic acid core so that only the protein shell is seen in the electron microscope.

Heavy metal salts may stain preparations both positively and negatively; for example uranyl acetate at pH 4 stains positively the nucleic acid core of some isometric virus particles.

(c) *Surface replicas* To study the surfaces of thick or easily distorted surfaces by transmission electron microscopy, thin replicas can be made of these surfaces. Preparation of replicas has been very usefully combined with freeze etching, as was first shown by Steere (1957) in work on crystals of virus particles. In this technique the wet specimen is frozen very rapidly in glycerol at around −150°C. The specimen is then fractured to expose internal surfaces, and is coated with a carbon/platinum film either immediately or after etching the surface by allowing water to evaporate from it. The specimen is then dissolved by heating in sodium hydroxide solution and the replica shadowed and examined in an electron microscope. Much has been learnt of the structure of cell membranes using this method (Fig. 11.7).

(d) *Ultrathin sectioning* The ultrastructure of virus-infected and healthy cells is most often studied by 'fixing' the cells in small pieces of tissue with a suitable chemical, cutting very thin sections from them and examining these by transmission electron microscopy (Chapter 3). The details of the methods that have been used successfully with different materials vary greatly, but have been succinctly summarized by Milne (1972). The tissue is usually fixed in a solution of an aldehyde such as glutaraldehyde, often followed by osmium tetroxide. Then it is dehydrated, and impregnated with and mounted in a hard plastic, such as epoxy resin, Araldite, or various kinds of methacrylate. The tissue is then cut to produce sections 50 to 100 nm thick, using a glass or diamond knife. The sections may then be stained, often with uranium and lead salts, and examined in the electron microscope. With some embedding media, the sections may be treated with specific enzymes or stains to aid recognition of particular compounds, or, if the cells had been fed radioactive compounds before fixation, they may be covered with radiation sensitive film to locate the radioactive compounds (autoradiography). Antigens may be located in sections using ferritin-labelled antibodies (§8.5.9).

(e) *Scanning electron microscopy* Scanning electron microscopes have been available commercially since the mid 1960s, and make it possible to study with relative ease the rigid surfaces of biological materials. Most scanning microscopes have a resolution of about 50 nm and are most useful for studying structures somewhat larger than most virus particles, though very high resolution machines are being made that can study individual heavy atoms in very thin specimens.

For conventional scanning electron microscopy the material is lightly shadow cast with an electron-conducting metal such as gold. The microscope scans the specimen in a regular fashion with a finely focused spot of electrons. The electrons scattered by the specimen are collected and their intensity used to modulate the signal supplied to a television screen that is scanning in phase with the microscope. In this way pictures of the electron scattering surface, even of thick specimens, can be constructed (Fig. 13.2). Moreover the specimen can be rotated and viewed from different angles. By using different collectors it is also possible to distinguish between elastically and inelastically scattered electrons, or even to measure the induced X-rays or the electron-induced fluorescence of fluorescent stains with which the specimen has been treated.

9.7.3 *Quantitative estimates*

The electron microscope can be used to measure the size and number of particular objects, such as virus particles, and also to determine their structure if, as in virus particles, there is a regularly repeating pattern.

Many early estimates of the sizes of virus particles were inaccurate, but since Brandes and Wetter (1959) showed the value of accurate measurements, methods have greatly improved. Each electron microscope has a built-in method for calibrating its magnification, and this may be checked using suitable standards and used to measure the sizes of objects seen, for example, in sections of cells. However built-in calibration systems may be less accurate than mounting objects of known size mixed with the objects of unknown size, so that the two can be examined and photographed together for comparative measurements. In tests to measure the length, diameter or substructure of virus particles, a suitable standard is provided by adding either the particles of TMV, which are 295 ± 6 nm long and have a basic helix of pitch 2.3 nm, or fixed

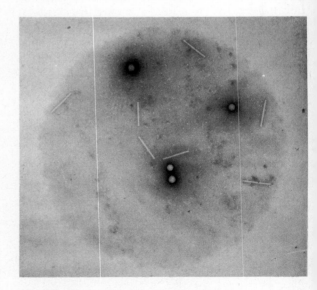

Fig. 9.8 Spray droplet used to assay by electron microscopy the concentration of virus particles in a preparation. The extent of the droplet, which contains seven TMV particles (about 300 nm long) and four latex spheres, is indicated by the phosphotungstate. (Courtesy R. D. Woods.)

catalase crystals which have a lattice spacing of 8.6 ± 0.22 nm (Wrigley, 1968).

It is difficult to count the numbers of virus particles in preparations directly in the electron microscope. The best method is probably that devised by Williams and Backus (1949) and improved by Nixon and Fisher (1958). In this technique a virus preparation is mixed with a suspension of polystyrene latex particles of known concentration. The mixture is sprayed on to an electron microscope grid held in a cascade impactor so as to select small droplets (Fig. 9.8). Several individual droplets are photographed or examined, and the total number of virus particles and latex spheres is compared. The greatest source of uncertainty in this technique comes from the difficulties in measuring the size and concentration of latex spheres, although this does not detract from using the method comparatively.

9.7.4 *Analysis of images*

The outer layer or layers of most virus particles are assembled from subunits arranged in a regular pattern (§5.5). Thus when attempting to determine the arrangement of these subunits by electron microscopy, it is advantageous to use analytical techniques that enhance repeating patterns in the images and, if possible, diminish randomly arranged information. There are various methods for doing this. For instance the photographic superposition method of Galton (1878) has been used to superpose images linearly (Markham *et al*, 1964; Warren and Hicks, 1971) and rotationally (Markham, Frey and Hills, 1963) to study, for example, rod-shaped and isometric virus particles respectively. The superposition method has proved most unreliable and requires, for example, knowledge of the shape of individual subunits *before* the results can be assessed unequivocally. Much more useful and reliable results have been obtained by the Fourier transform method, first used by Klug and Berger (1964) and reviewed by Markham (1968). In this method the electron micrograph image is used as a diffraction grating in a laser diffractometer, and the optical diffraction pattern or transform that is obtained is analysed (Fig. 9.9; Varma *et al*, 1968). A variant of the Fourier method enables images of icosahedrally constructed particles, such as those of turnip yellow mosaic virus, to be superposed and reconstructed (Fig. 9.10; Mellema and Amos, 1972). An alternative method, not yet used with virus particles, but of particular value

for those who find the concept of reciprocal space a problem, is the convolution or Patterson function technique (Elliott, Lowy and Squire, 1968).

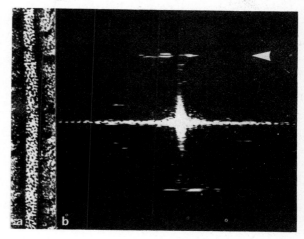

Fig. 9.9 a, Electron micrograph of part of a particle of white clover mosaic virus, a potexvirus, negatively stained with uranyl formate. **b,** The optical diffraction pattern of **a**. In the diffraction pattern, the noticeable asymmetric layer line (arrowed) corresponds to the basic helical repetition of the particle (pitch 3.39 nm), and the faint line, one-third of the distance to that layer line from the origin, shows that there is also some structural repetition every three turns of the basic helix. (Courtesy A. Varma.)

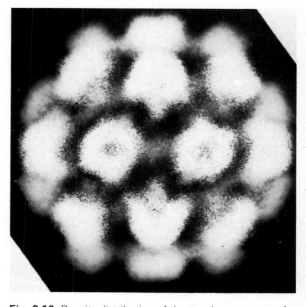

Fig. 9.10 Density distribution of the top three quarters of a model of a particle of turnip yellow mosaic virus reconstructed from the combined Fourier transforms of electron micrographs of four particles. (Courtesy J. E. Mellema; Mellema and Amos, 1972.)

An optical transform of an optical transform is the original image. If the first transform of the image of a virus particle is suitably filtered, and then used to prepare a second transform, it is possible to reconstruct the image of one of the surfaces of the particle (Klug and de Rosier, 1966). An alternative and quite different method for reconstructing the image is the algebraic reconstruction technique (ART) of Gordon, Bender and Herman (1970); discussion of the relative values of the Fourier and ART methods of image reconstruction has been lively (Crowther and Klug, 1971; Bellman *et al*, 1971).

9.8 X-ray diffraction studies

The analysis of the structure of virus particles by X-ray diffraction is a very specialized branch of virology that will not be discussed in this book. For information read the reviews by Holmes and Blow (1965) and Finch and Holmes (1967), and note that the principles involved in the analysis of X-ray diffraction patterns apply also to the analysis of optical diffraction patterns obtained from electron micrographs (§9.7.4).

9.9 Sensitivity of assay methods

It is impossible to give the absolute or relative sensitivity of all the assay methods that have been described in this and the preceding two chapters because many factors greatly affect the sensitivity of different tests.

Biological tests for infectivity are the most sensitive and specific and can detect fewer than 10^6 virus particles per ml. Electron microscopy can only detect about 10^8 particles per ml, and chemical and serological tests are even less sensitive and specific; they can detect 10^9 particles per ml, or perhaps fewer when radioactive labelling is used, whereas physical tests such as optical assay methods are usually least sensitive and will only detect 10^{10} or more particles per ml. Usually five to ten times as much virus is required for making quantitative measurements by a particular method than for merely using the method to detect the presence of virus.

Chapter 10

The effects of inactivators on virus particles

In previous chapters we have described many of the properties of virus preparations. In this chapter we describe the ways in which the infectivity of virus particles can be inactivated. We mainly describe the properties of purified virus preparations, but we also mention three relatively crude tests which are commonly used to characterize a virus by indicating the concentration and stability of its infective particles in sap. These tests determine the so-called dilution end-point, thermal inactivation point and ageing *in vitro* end-point, and they should be done in standard ways (Bos, Hagedorn and Quantz, 1960; Noordam, 1973).

10.1 Routine tests of properties in sap

10.1.1 *Dilution end-point*

This is the point at which sap from infected plants fails to infect mechanically inoculated test plants when diluted increasingly, and is often quoted as the limits between which infectivity is lost. These will obviously depend on virus concentration, but also on the diluent (usually water or 0.01 M phosphate buffer, pH 7.0), the condition and species of source plant, the condition of the assay plants and on whether an abrasive is used to facilitate infection, all of which should be stated. It is rarely worth using dilution steps separated by less than a factor of five. The dilution end-points of different viruses range from 10^{-1} to about 10^{-7}.

10.1.2 *Thermal inactivation point*

In this test, samples of sap in thin-walled glass tubes are placed for 10 minutes in a water bath adjusted to a series of temperatures, usually separated by intervals of 5°C, and then inoculated to test plants. The test is affected by the factors that influence the determination of dilution end-point, and the results are usually quoted as the limits between which infectivity is lost. It has been suggested that this test could be given added precision by determining the half life of infectivity at a range of temperatures. This could be done by heating sap at a given temperature for different periods, plotting the logarithm of the number of lesions produced against time of treatment and determining the half-life from a statistically fitted straight line. However, in practice it may not be worth adopting this method, which also has its limitations, and instead recognize that the results for different viruses given by the original method are only roughly comparable. Thermal inactivation points range from 45°C (tomato spotted wilt virus) to 95°C (strains of TMV), but are mostly between 55° and 70°C; viruses with complex, lipid-containing particles have relatively low thermal inactivation points. Different strains of a virus may have different thermal inactivation points; for example, they are between 80 and 85°C for strain B of tobacco necrosis virus, 85 and 90°C for the serologically closely related strain A, and 90 and 95° for the more distantly related strains D and E (Babos and Kassanis, 1963). Extracts containing the chrysanthemum stunt viroid can be boiled without infectivity being abolished (Hollings and Stone, 1973).

10.1.3 *Ageing* in vitro *end-point*

This test is done by storing infective sap in stoppered tubes at about 20°C, and assaying the infectivity of samples after various intervals, usually chosen to fall on a geometric series. The time at which the sap becomes non-infective is the ageing *in vitro* end-point. The results are affected by the same factors as the other two tests. No attempt is made to separate the different kinds of inactivation occurring in sap, but the viruses that inactivate quickly include several that are inactivated by oxidase systems. The end points range from an hour (Tulare apple mosaic virus) to more than a year (TMV).

10.2 Inactivation of virus particles *in vitro*

Many physical and chemical agents inactivate the infectivity of viruses, although the mechanisms involved are in many instances not well understood. Early studies with plant viruses showed that the infectivity of virus particles could be destroyed without destroying immunogenicity (Bawden, 1935) and led to the development of inactivated vaccines of viruses of man and other animals. However, few of these kinds of inactivation have had any practical application in the control of plant-virus diseases, although the study of inactivation has proved useful in other ways. Understanding inactivating systems, particularly those affecting viruses that have unstable particles, can help in devising methods of preventing inactivation during handling, virus purification, etc. Also, studies of inactivation have been made in attempts to learn something about the structure and composition of virus particles, and this work has given some information that could not easily be obtained in other ways.

In this chapter, we shall use 'inactivation' to mean inactivation of infectivity, and because the integrity of the virus nucleic acid is of prime importance, the virus-inactivating agents will be considered with this in mind. Inactivators either render virus nucleic acid non-infective, or they act on virus protein (or perhaps on some other component of virus) either to remove its protective effect or alternatively to prevent the nucleic acid interacting with infectible sites. The inactivators can be divided roughly into physical, physico-chemical and chemical-biochemical types, and in many instances inactivate exponentially, so that:

$$P = e^{-kt}$$

where P is the proportion of activity surviving the treatment after time t, and k is a rate constant depending on the dose of inactivator, the virus under test and the experimental conditions, such as temperature, ionic environment, etc.

10.2.1 *Physical agents*

(a) *Mechanical treatments* Ultrasonic vibration is one of the few mechanical treatments known to inactivate virus infectivity. The rapid changes from high to very low pressure that occur when sonic waves pass through a liquid will, when enough energy is applied, cause sub-microscopic bubbles to form and then disappear rhythmically. This phenomenon of cavitation generates shearing forces large enough to break the particles of elongated viruses (Fig. 10.1) and to inactivate infectivity exponentially. Serological specificity is little affected by the doses needed to inactivate infectivity and, indeed, ultrasonic vibration can be used to make preparations that are suitable for serological tests by the gel-diffusion method, from virus particles too long to diffuse through agar gel (Tomlinson and Walkey, 1967).

Elongated virus particles may also be broken when viruses are sedimented by ultracentrifugation, and resuspended, though the amount of breakage usually seems small. Different viruses no doubt differ greatly in fragility. Virus particles are also degraded when they are exposed to great pressures.

(b) *Ionizing radiation* Ionizing radiations include X-, γ- and β-rays, together with heavy charged particles such as α-particles, protons and deuterons. X- and γ-rays have the greatest penetrating ability, β-rays (emitted for example by ^{32}P) have less, and the other radiations penetrate only very slightly.

For inactivation of plant viruses, X-rays are the best studied (Ginoza, 1968). They seem to inactivate virus particles in two ways; firstly by causing ionizations in the virus particles, and secondly by producing free radicals and other materials in the suspending medium that inactivate indirectly. Indirect inactivation can be avoided by drying the material to be irradiated, or by adding to the medium substances such as proteins which compete with virus for the free radicals. In dilute suspensions of TMV, inactivation seems to be mainly of the indirect type and is much decreased by adding gelatin to the medium (Lea, 1962). With concentrated TMV (2 mg/ml) such precautions are unnecessary, because indirect inactivation is apparently negligible and, although the appearance of the virus particles is not affected, inactivation can be accounted for by randomly sited ionizations in the RNA strand within them. Inactivation is exponential and there is evidence that about half the primary ionizations in the RNA result in breaks in the strand. With TMV and three other viruses, the protein in virus particles seems to play no part in inactivation, because the calculated target sizes (radiosensitive volumes) of virus particles are about the same as the volumes of their RNAs, and isolated TMV-RNA has approximately the same radiosensitive volume for inactivation of infectivity as intact TMV particles (Epstein, 1953; Ginoza, 1968). The serological reactivity of TMV protein is

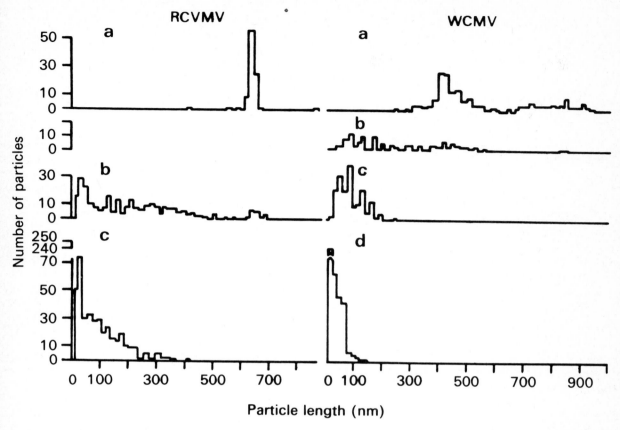

Fig. 10.1 Effect of ultrasonic treatment for **b,** 0.5; **c,** 2.5 and **d,** 17.5 min on the length distribution of particles of red clover vein mosaic (RCVMV) and white clover mosaic (WCMV) viruses. **a,** Untreated samples. (After Varma, Gibbs and Woods, 1970.)

affected only by much larger X-ray doses than are needed to abolish infectivity.

(c) *Non-ionizing radiation* Inactivation of infectivity by ultra-violet radiation has been studied in some detail, usually using low-pressure mercury-vapour lamps equipped with filters to remove radiation of shorter wavelengths, and producing radiation very largely of 254 nm. Inactivation of infectivity of TMV is exponential; however, the rates of inactivation of the nucleoprotein particles of most, and of the extracted nucleic acids of all, plant viruses tested are smaller when the plants used for assaying residual infectivity are kept in visible light than in darkness after inoculation (Bawden and Kleczkowski, 1955). The term photoreactivation is used to describe this ability of some irradiated virus particles or nucleic acid molecules to multiply only in plants given a period of light. However, neither particles of TMV nor of tobacco rattle virus, which are similarly constructed tubes, are photo-reactivable, in contrast to their extracted nucleic acids (Bawden and Kleczkowski, 1959; Fig. 10.2). Evidently the combination between RNA and protein in the intact virus particles is of such a kind that photoreactivable damage is prevented.

With tobacco necrosis virus, the relative efficacy for inactivation by ultra-violet radiation of different wavelengths (action spectrum) is the same for intact virus particles and for extracted virus RNA, and is the same as the absorption spectrum of the RNA (Kassanis and Kleczkowski, 1965). However, the story for the type strain of TMV is different. Here, the action spectrum for inactivation of free virus RNA is correlated with its ultra-violet absorption, but the inactivation spectrum for intact virus particles is quite different from the absorption spectrum (Fig. 10.3). This suggests that the virus protein plays a role, and again emphasizes the intimacy of association of RNA and protein in TMV particles. Because

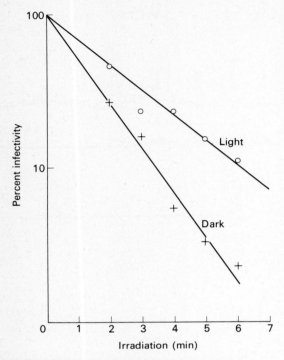

Fig. 10.2 Effect of ultra-violet light (248 nm) at 11 to 14°C on infectivity of TMV-RNA. The assay plants (*Nicotiana tabacum* cv. Xanthi-nc) were inoculated under red lights and then kept either in daylight (upper line), or darkness for the first 6 h (lower line). The quantum yields were: with photoreactivation 1.7×10^{-3}; without photoreactivation 3.0×10^{-3}. (Rushizky, Knight and McLaren, 1960.)

virus particles scatter some of the incident ultra-violet light instead of absorbing it, the absorbance of a virus preparation as measured with a spectrophotometer does not give a true indication of the amount of radiation absorbed, and this is increasingly so as virus concentration increases. Hence, in Fig. 10.3 the action and absorption spectra are also plotted after allowing for light scattering, which is inversely proportional to the logarithm of the wavelength (§9.3.1). The protein in virus particles is also responsible for the difference in susceptibility to inactivation of TMV strains (Siegel, Wildman and Ginoza, 1956), and in potato virus X it gives some protection against both photoreactivable and non-photoreactivable damage (Govier and Kleczkowski, 1970).

The quantum yield for inactivation of infectivity (without photoreactivation) of nucleoprotein particles or extracted RNA of tobacco necrosis virus is about 6.5×10^{-4}. This means that about 1500 quanta of energy are absorbed per inactivation. With TMV-RNA the comparable quantum yield is about 9×10^{-4}, but with intact TMV particles it is only about 4×10^{-5} at 254 nm. Intact TMV particles are therefore much less readily inactivated, per absorbed quantum, than TMV-RNA.

Inactivation of infectivity of these plant viruses by ultra-violet radiation is usually not caused by breakage of the RNA strand, but the absorbed energy is thought to break various essential chemical bonds. Photoreactivable damage seems to involve the production of pyrimidine dimers (bonds between adjacent pyrimidine, probably uracil, residues in the RNA strand) (Carpenter and Kleczkowski, 1969; Tao, Small and Gordon, 1969). Doses of ultra-violet radiation larger than those needed to inactivate infectivity can induce covalent bonding between two RNA molecules in a virus particle (Mayo *et al*, 1973), and break RNA strands. About thirty times larger doses are needed to halve the ability of TMV to combine

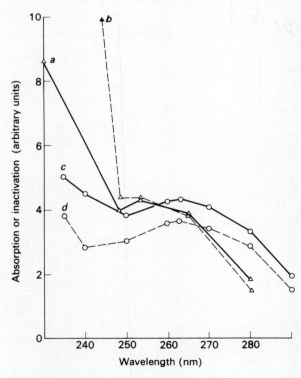

Fig. 10.3 Ultra-violet absorption and inactivation spectra for TMV particles. *a*, Inactivation spectrum; *b*, inactivation spectrum corrected for light scatter; *c*, absorption spectrum; *d*, absorption spectrum corrected for light scatter. Note the difference in shape between *b* and *d* (Courtesy A. D. McLaren.)

and precipitate with antibody than to inactivate infectivity. Here the inactivation is roughly exponential until about a quarter of the original serological reactivity remains; at this point most of the virus particles have already disintegrated into small pieces. Further loss of reactivity and disintegration are increasingly rapid (Kleczkowski, 1962).

With the apparent exception of cucumber mosaic virus (Orlob, 1967), visible light is not known to inactivate the infectivity of plant viruses directly. However, light will inactivate virus particles when some dyes, such as methylene blue, acridine orange or acriflavine, are in the suspending medium. The action is probably on the virus RNA, which also is photosensitized by acridine orange (Sastry and Gordon, 1966).

10.2.2 *Physicochemical factors*

(a) *Heat*. We have already discussed thermal inactivation points in sap (§10.1.2), but particles of different viruses not only have different thermal inactivation points but also different patterns of inactivation. This is well illustrated by comparing TMV and tobacco necrosis virus, for these two have similar thermal inactivation points. Purified TMV particles seem to lose infectivity exponentially on heating, but below 70°C inactivation is slow. Above 70°C, the rate of inactivation increases rapidly, i.e. the temperature coefficient of inactivation is relatively high. Also, infectivity is lost only slightly faster than the virus

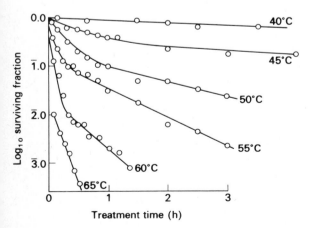

Fig. 10.4 Inactivation of infectivity of particles of tobacco necrosis virus, strain B, in distilled water at 40 to 65°C. The inactivation curves for 45 to 65°C suggest a two-component system. (Babos and Kassanis, 1963.)

protein coagulates, causing the particles to lose their normal shape and serological reactivity. By contrast, tobacco necrosis virus is appreciably inactivated, even at 50°C (Fig. 10.4), has a smaller temperature coefficient of inactivation, and loses 99% of its infectivity before serological reactivity is impaired. A further complication is that purified preparations of several strains of tobacco necrosis virus are not inactivated in an exponential manner. They behave as though the preparations contain two kinds of particle having different inactivation rates, with the relative proportions of the two depending on the temperature (Fig. 10.4). This suggests that one type is produced from the other by heating.

RNA which has been extracted from particles of tobacco necrosis virus is inactivated at exactly the same rate as intact virus particles, so the protein seems to play no role in the loss of infectivity. Extracted TMV-RNA, however, is inactivated exponentially but more rapidly than TMV particles, and it has a smaller temperature coefficient of inactivation (Ginoza *et al*, 1964). Note the parallel between the inactivation of particles of these two viruses by heat and by ultraviolet light; with tobacco necrosis virus both inactivations are caused by effects on the virus RNA, but with TMV the virus protein protects the nucleic acid from inactivation. This kind of protection may be favoured by the tight helical construction of TMV particles, and it is worth noting that viruses with elongated particles have, in general, relatively high temperature coefficients of inactivation. By comparison, many viruses with small isometric particles have smaller temperature coefficients of inactivation, although carnation mottle virus is an exception, for it has isometric particles about 30 nm in diameter, and yet a pattern of inactivation more like that of TMV particles.

It seems probable that thermal inactivation is caused by a single hydrolytic scission of a phosphodiester bond in the virus RNA, giving breaks randomly distributed along the RNA strand (Ginoza, 1968). In helical particles, hydrolysis would not be likely to occur until the protective protein begins to be denatured.

(b) *Freezing and thawing, and desiccation* Most of the particles of many viruses retain their infectivity when stored frozen in preparations or in the leaf but the particles of many of the less stable viruses are inactivated. However, with increasing numbers of successive freezings and thawings even the more stable viruses are inactivated. The conditions of freezing can also be important; thus tobacco necrosis virus is relatively stable in frozen sap or purified preparations but not in frozen leaves (Bawden and Pirie, 1950). The particles of most plant viruses tested, including those of some of the less stable ones, such as tomato spotted wilt, can be preserved by freeze-drying infective preparations (Hollings and Stone, 1970; §6.7.2), and many can be preserved by drying leaf fragments over silica gel at −15°C. Desiccation, however, can also be an effective inactivating treatment.

(c) *pH value* The effect of pH value on thermal inactivation may be large (Lauffer and Price, 1940). For instance, at 71°C purified TMV particles are denatured 3000 times more rapidly at pH 7.05 than at pH 5.8. The stabilities of different plant viruses at room temperature are somewhat differently affected by pH value, and must be determined for each virus, but a high proportion are most stable at pH 7. For instance, particles of TMV tend to disintegrate below their isoelectric point (pH 3.2) or above pH 9.5, as also do those of tobacco rattle virus below pH 4.5 or above pH 10.5. Particles of tomato spotted wilt virus have a narrower range of greatest stability, pH 6–9.5 (Best, 1966), and pH 9 is said to be best for storage of cucumber mosaic virus.

The tendency of virus particles to dissociate into their components at extremes of pH is sometimes used to prepare virus RNA and protein (§5.6), and inactivation of infectivity is facilitated when the virus RNA becomes accessible to various inactivating agents in the medium, such as ribonuclease. However, virus particles need not dissociate completely for their RNA to become accessible to ribonuclease. Particles of bromoviruses swell when put in media above pH 7 without magnesium, and their RNA becomes accessible (§5.7.2).

10.2.3 *Chemical and biochemical agents*

Not all chemical modifications of virus particles abolish infectivity. For example, the sulphydryl groups of TMV protein can be oxidized by treatment with iodine, or the C-terminal threonine residue of TMV protein can be removed by treatment with carboxypeptidase, without abolishing infectivity, although serological specificity can be altered (Harris and Knight, 1955). Less is known about possible modifications in

virus nucleic acid that do not inactivate, but fluorouracil can substitute for about half the uracil residues in TMV-RNA without apparently affecting infectivity (Gordon and Staehelin, 1959) and there is indirect evidence that converting one nucleotide into another by deamination can either give a virus variant or inactivate (Chapter 12). Although we will not discuss the chemical inactivation of viruses in detail, some of the more important kinds of inactivation can be considered under the following headings, which overlap to some extent.

(a) *Fission agents* We have already noted (§5.6) several agents that dissociate virus particles into their protein and nucleic acid constituents; such materials as detergents (e.g. sodium dodecyl sulphate), phenol, urea, acetic acid and alkalis. Their action is predominantly on hydrogen bonding. These fission agents need not necessarily be inactivators in the strict sense. The decrease that they produce in infectivity may be simply because the nucleic acid of most viruses initiates infection less efficiently on its own than when coated with virus protein. Indeed, re-coating TMV-RNA with TMV protein restores much of the infectivity of the original intact virus particles (Fraenkel-Conrat and Williams, 1955). Inactivation may however follow the dissociation of virus particles because the preparations contain traces of potent inactivating agents such as ribonuclease which can then gain access to virus nucleic acid. Some of these fission agents, such as phenol, denature virus protein whereas others such as acetic acid do not.

Some fission agents, for example detergents and alkali, usually attack TMV particles from a specific end. Controlled degradation (e.g. 15–60 seconds treatment with Duponol C at 85°C) can be used to expose short 3′-terminal lengths of RNA (Fig. 10.5), which can then be variously

Fig. 10.5 Particles from a TMV preparation treated with 1% sodium dodecyl sulphate. A TMV particle is partially stripped of protein, leaving a long exposed RNA strand. The postulated location of the local lesion gene is indicated by the arrow. (Kado and Knight, 1966.)

treated, or removed with ribonuclease (Kado and Knight, 1966). In principle it is therefore possible to differentiate the portions of the nucleic acid strand occurring at different positions along the TMV particle (§§5.6.2 and 12.3.2).

The particles of some viruses, such as the bromoviruses and cucumber mosaic virus, dissociate in strong solutions (1–2 M) of salts such as NaCl or $CaCl_2$, and in general the stability of virus particles is affected by the kinds and concentrations of ions in the suspending fluid. The conditions that favour stability or inactivation depend on the virus.

(b) *Organic solvents* Some of these too may denature virus protein, probably exposing virus nucleic acid to inactivators or preventing it interacting with infectible sites. Of these materials, some disrupt the lipid-containing particles of many animal viruses, possibly by some kind of solvent action, and lipid-containing plant viruses may behave similarly. It is worth noting that lettuce necrotic yellows virus, which has large complex particles, is inactivated both by diethyl ether and by chloroform (Harrison and Crowley, 1965).

(c) *Nitrous acid and formaldehyde* These probably react mainly with amino-groups, and the action of nitrous acid is discussed in detail in Chapter 12; it is both an inactivator and a mutagen. Formaldehyde probably reacts according to the equations:

$$(1) \quad R \cdot NH_2 + HCHO \rightarrow R \cdot NH \cdot CH_2OH$$
$$(2) \quad 2R \cdot NH_2 + HCHO \rightarrow R \cdot NH \cdot CH_2 \cdot NH \cdot R + H_2O$$

Both nitrous acid and formaldehyde (Staehelin, 1958) react more rapidly with free TMV-RNA than with RNA in TMV particles (Dobrov, Kust and Tikchonenko, 1972), suggesting that TMV protein confers some protection.

(d) *Oxidizing systems* Some viruses, such as cucumber mosaic, prune dwarf, tobacco streak, tomato spotted wilt and Tulare apple mosaic are readily inactivated by oxidizing systems, whereas many others, such as TMV and tobacco necrosis virus, are not. In crude plant extracts, inactivation seems mainly to result from the reaction of the virus particles with *o*-quinones (Hampton and Fulton, 1961), which are formed by the oxidation of substances such as chlorogenic acid by polyphenoloxidase. Inactivation can be much les-

sened by excluding oxygen, or by adding either reducing agents such as thioglycollic acid or substances such as sodium diethyl dithiocarbamate that inactivate the enzyme by chelating the copper it contains and react with the *o*-quinones (Best, 1939; Pierpoint and Harrison, 1963). Small concentrations of *o*-quinone react with and are thought to tan the protein in particles of readily inactivated viruses, and virus RNA is probably not then freed from virus protein when the particles enter cells. These small concentrations do not inactivate the virus RNA. With other viruses larger concentrations of *o*-quinone are necessary for inactivation to occur; these both degrade the virus particles and inactivate the RNA (Mink and Saksena, 1971).

(e) *Tannins* The leaves of many plants, especially woody species, contain tannins which seem to inactivate virus particles by combining with and precipitating them. The amount of inactivation depends on the ratio of tannin to virus particles in the mixture (Table 10.1) and does not depend on the species of assay host used. The reaction between virus particles and tannin depends on the virus used. Thus tobacco rattle virus is inactivated by concentrations of tannin too small to affect other viruses, such as tobacco necrosis; also, whereas the combination between tannic acid and

Table 10.1 Concentrations of tannic acid and *Phytolacca* protein needed to decrease infectivity to a given extent

A Effect of tannic acid[1]

Concentration of tobacco rattle virus (arbitrary units)	Concentration of tannic acid needed to decrease infectivity by 95 % (mg/l)	Ratio virus/tannic acid
250	47.2	5.3
50	8.9	5.6
10	1.5	6.7

B Effect of protein from *Phytolacca esculenta*[2]

Concentration of TMV (mg/l)	Concentration of protein needed to abolish infectivity (mg/l)	Ratio virus/protein
720	24	30.0
90	12	7.5
10	6	1.7
1.5	3	0.5

[1] Data from Cadman (1959).
[2] Data from Kassanis and Kleczkowski (1948).

particles of tobacco necrosis virus can be reversed by raising the pH value of the medium to 8, the combination between tannic acid and particles of tobacco rattle virus seems irreversible (Cadman, 1959). Tannins therefore inactivate some viruses and inhibit infection by others.

(f) *Enzymes* A remarkable property of the particles of many plant viruses is their resistance to proteolytic enzymes. TMV particles, for example, are not hydrolysed by trypsin, chymotrypsin, pepsin or papain, although both trypsin and papain combine with them and inhibit infectivity. The inhibition of infectivity by trypsin and the combination between virus and enzyme can be reversed by dilution (Stanley, 1934). By contrast, disaggregated subunits of TMV protein are hydrolysed by chymotrypsin. However, intact particles of potato virus X are hydrolysed and inactivated by papain (with KCN but not without), pepsin and trypsin, though the inactivation of infectivity may be caused by ribonuclease impurity acting on virus RNA exposed by the action of the proteolytic enzymes. Limited proteolysis may break the polypeptide chain but have no obvious effect on particle morphology, infectivity or serological activity (Mayo and Cooper, 1973), and it may alter the electrophoretic mobility of virus particles but not their infectivity (Niblett and Semancik, 1969; Geelen, Rezelman and van Kammen, 1973).

In general, pancreatic ribonuclease does not inactivate intact virus particles, presumably because it cannot gain access to the virus RNA, but the infectivity of particles of cucumber mosaic virus (Francki, 1968), and of the bromoviruses at pH 7 or above (§10.2.2c) is inactivated.

(g) *Antibodies* As described in Chapter 8 (§8.5.10), antibodies combine with and precipitate the particles of plant viruses, and may neutralize infectivity. In some instances the combination and neutralization of infectivity can be reversed. The mechanism of neutralization of infectivity is not well understood, but precipitation is not necessary; presumably the virus RNA is prevented from interacting with infectible sites.

In addition to the kinds of chemical inactivation already mentioned there are other methods such as alkylation and treatment with compounds that react with − SH groups or contain heavy metals; these are of more specialized interest.

10.3 Inhibitors of infection

It is important to distinguish virus inactivators from inhibitors of infection. When in the inoculum, many substances inhibit infection; these include some plant proteins and other components of sap, enzymes, normal serum proteins, polysaccharides, various compounds of small molecular weight, etc. (Bawden, 1954). The amount of inhibition they cause depends greatly on the assay-host species (Table 4.2) and little on the virus, and no special ratio of inhibitor to virus is needed for a given degree of inhibition (Table 10.1; compare figures for *Phytolacca* protein and tannic acid). Some inhibitors combine with viruses, but this may have little relevance to inhibition for they also prevent infection occurring in conditions in which there is no combination, and other inhibitors seem not to combine with virus particles. Also, infectivity of a noninfective mixture of virus and inhibitor often can be restored by dilution. The mechanism of action of these inhibitors of infection is not known for sure and is probably not the same in all instances. However *Phytolacca* protein seems to inhibit translation of messenger RNA on wheat and cowpea ribosomes, but not on *Phytolacca* ribosomes, and this may be the basis of its inhibitory activity (Owens, Bruening and Shepherd, 1973). All these facts point to the conclusion that inhibitors prevent infection by effects on the host plant; they seem to have no permanent effect on the viruses. Inactivators by contrast abolish infectivity permanently.

Whether a material is called an inhibitor or an inactivator is to some extent arbitrary. For example, some so-called inactivators, which like antibodies and tannins combine with virus particles, probably do not abolish the potential infectivity of the virus RNA. Moreover, several of the chemical inactivators discussed above are not inactivators in the strict sense. The fission agents, organic solvents and proteolytic enzymes may merely make virus nucleic acid accessible to inactivators from which it is protected in the intact virus particle. By comparison with all these, substances such as formaldehyde and nitrous acid are true inactivators, causing irreversible loss of infectivity by reacting with virus nucleic acid.

Chapter 11

Behaviour of viruses in plants

11.1 Introduction

In their mode of replication, viruses differ greatly from cellular microbes which multiply by the division of mature cells, often in a binary fashion. Hence bacterial cells, for example, are intact at all stages of division, and viable and recognizable as bacteria. By contrast, soon after virus particles enter host cells they partially or wholly disassemble into their constituent macromolecules. Then the virus genome, or messenger RNA transcribed from it, is translated by the host's translation apparatus into new virus proteins, the nucleic acid genome is replicated, and the progeny nucleic acid together with molecules of one or more of the new virus proteins assemble to form progeny virus particles. Thus intact virus particles are not present during part of the replication cycle of a virus; the virus genome is incorporated into the host's protoplasm and is totally dependent on it.

Until recently plant virus replication had been mostly studied in the mechanically inoculated leaves of intact plants, though various other tissues or tissue preparations have been used, including excised leaves, leaf discs, epidermal strips (Dijkstra, 1966) and excised hypocotyls (McCarthy, Jarvis and Thomas, 1970). These studies are limited because at best only 0.1% of the cells in the tissue are infected at the start of the experiment, and the virus spreads from these to other cells so that most of the virus growth observed is in cells infected secondarily at different and unknown times. Another restriction is that 10^5 or more particles must be inoculated to the tissue to obtain a single infection and, because a good deal of the residual inoculum can not be removed, all the interesting early events of infection are studied in the presence of an enormous excess of residual inoculum.

The infectivity of samples of sap taken from leaves at intervals after inoculation follows the trend shown in Fig. 11.1. When the leaves are inoculated, virus particles penetrate susceptible cells, infect them, and replicate. But this takes time, and during this period, often called the 'lag'

or 'eclipse' phase residual inoculum is inactivated and hence the infectivity of samples often decreases. Soon progeny virus accumulates in the infected cells faster than the inoculum is being inactivated, and the infectivity of successive samples increases. The time needed for newly formed virus nucleic acid to become detectable is shorter than that for intact virus particles (Fig. 11.2), presumably because the first is a precursor of the second. Within a few hours the virus spreads from the initially infected cells to surrounding cells and replicates in them, so the infectivity of samples increases exponentially. If the host cells are not killed, infectivity may

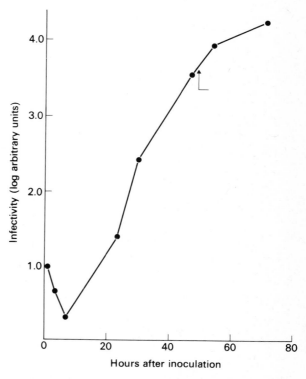

Fig. 11.1 Infectivity of sap of French bean leaves at intervals after inoculation with tobacco necrosis virus. The arrow indicates the time when the lesions appeared. (Redrawn from Harrison, 1956b.)

continue to increase for 2 to 3 weeks and then, as for example with TMV in tobacco, may maintain a plateau of virus concentration until the leaf dies, or, as with alfalfa mosaic virus in tobacco (Fig. 11.3), may decrease. The speed with which these events proceed depends on many factors, and in particular on environmental conditions such as ambient temperature (Fig. 11.3); it may take from 2 days to 3 weeks for infectivity to reach a maximum.

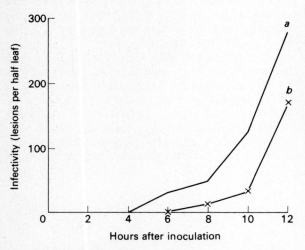

Fig. 11.2 Synthesis of infective RNA and virus particles in French bean leaves inoculated with tobacco necrosis virus. *a*, Infectivity of leaf extracts made using bentonite (virus RNA+virus particles). *b*, Infectivity of water extracts (virus particles only). Note that infective RNA is detectable before infective particles. (Redrawn from Kassanis and Welkie, 1963.)

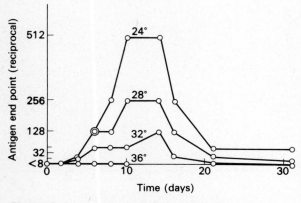

Fig. 11.3 Concentration of alfalfa mosaic virus in sap extracted from tobacco leaves at different times after inoculation. Plants were kept at the indicated temperature after inoculation. Virus concentration was estimated serologically.

The precise details of the biochemistry of virus replication are much more fully known for many bacterial and vertebrate viruses than for any plant virus, mainly because the cells of plants and of plant virus vectors are so much less readily manipulated than those of bacteria and mammals. However, ways have now been found for growing some plant viruses in plant protoplasts (Aoki and Takebe, 1969; Takebe and Otsuki, 1969), cultured callus cells (Murakishi *et al*, 1970), separated leaf cells (Jackson *et al*, 1972) or cells of vectors (Black, 1969). In these systems, infection is more efficient and can be synchronous, and more than half the cells may be infected. Thus TMV can reach a concentration of 10^6 particles per protoplast after incubation for 24 hours at 25°C (Takebe and Otsuki, 1969). There are also methods for translating plant virus RNAs, using ribosomes from wheat embryos or other tissues (Marcus, Luginbill and Feeley, 1968; Marcus, 1970; Davies and Kaesberg, 1974). Hence the relative lack of knowledge of the replication cycles of plant viruses will soon be overcome. Nonetheless, what is already known suggests that plant virus replication does not differ in important biochemical details from that of viruses of other hosts, and few surprises are likely.

11.2 Early stages of infection

The first stage of infection is the entry of virus into the cell, and this happens in nature through breaks in the cell walls made by vectors or by abrasion (Chapter 13). Indeed, no infection occurs when inoculum is gently sprayed onto unwounded leaves. Manual inoculation involves the production of wounds by rubbing the leaf surface; most of the infectible wounds so made become refractory to infection within an hour, probably because the entry points are quickly resealed. Evidence has been advanced that ectodesmata, which are identified in the cell wall by their reactivity with stains, may be involved in the infection of manually inoculated leaves (Brants, 1964; Thomas and Fulton, 1968) but the evidence is not conclusive and there seems to have been no ultrastructural identification of ectodesmata in tissue sections. Many factors influence the success of manual inoculation (Chapter 4), but how they operate is largely unknown.

When intact virus particles enter a cell they disassemble (= 'uncoat') and uncoating may in part control host range. Thus when infective brome mosaic virus RNA was coated with TMV protein, the particles were not infective for barley leaves, which are susceptible to infection by intact particles and by RNA of brome mosaic virus (Atabekov, 1971). Ability to uncoat depends upon the structure of the virus coat protein, and can be greatly changed by a single amino-acid replacement (Bancroft *et al*, 1971). Evidence of uncoating is found by comparing the sequence and timing of events after inoculating French bean leaves with either intact tobacco necrosis virus particles or with the infective RNA extracted from them. In leaves kept at 25°C, newly-formed virus becomes detectable 2 to 4 hours sooner after inoculation with RNA, and the lesions caused by the nucleic acid appear, on average, 4 hours sooner than those caused by the intact particles. It is also possible to stop infection by irradiating the leaves with ultra-violet light immediately after inoculation, but whereas resistance to ultra-violet inactivation starts to increase almost at once in leaves inoculated with nucleic acid, no such increase occurs in leaves inoculated with intact virus particles until 3 hours later (Kassanis, 1960; Bawden and Harrison, 1955). All these observations suggest that uncoating takes, in the example discussed above, about 3 hours. Similar results have been obtained with TMV and other viruses.

Other experiments show that the lack of delay following inoculation with nucleic acid is obligatory, presumably because naked nucleic acid is unstable. By contrast, after inoculation with intact particles long delays are possible in conditions that are unfavourable to replication, such as high temperatures (Bawden and Sinha, 1961).

Another type of evidence for an early stage in infection has come from photoreactivation experiments (§10.2.1). In these experiments preparations of potato virus X were partially inactivated by ultra-violet irradiation, their infectivity compared with suitably diluted but unirradiated inocula, and the inoculated plants kept either in the light or in the dark after inoculation. Light restored some of the infectivity destroyed by ultra-violet irradiation (the phenomenon called photoreactivation), but the virus was not photoreactivable until 15 mins to 2 hours after inoculation, and did not remain photoreactivable for more than 3 hours (Bawden and Kleczkowski, 1955). These results possibly indicate that the virus nucleic acid is not photoreactivable until uncoated.

There is some information on the way in which inoculum particles are uncoated. Cocking and Pojnar (1969) reported that soon after inoculation, TMV particles in tomato callus cells although of the usual length, were narrower, suggesting perhaps that protein had been 'digested' away from the entire surface of the particle. Shaw (1973) found that 8 minutes after inoculating tobacco leaves with TMV, 25% of the virus protein sedimented very much more slowly than intact virus particles; it was not shown whether the removed protein still consisted of intact protein subunits or had been degraded to peptides. The protein was removed from the end of the inoculated particles that contained the 3' terminus of the RNA as fast as from the other end. By contrast, after inoculation with turnip yellow mosaic virus, the RNA in up to 70% of particles is soon liberated and protein coats are produced, with the amount of uncoating increasing with the ionic strength of the inoculum (Kurtz-Fritsch and Hirth, 1972).

However none of these experiments distinguished between the particles that actually infected and the residual inoculum that did not infect. Less equivocal, it is claimed, are the experiments of Machida and Kiho (1970) and Kiho (1970; 1972a, b) in which tobacco leaves were inoculated with radioactively labelled TMV and, at times afterwards, were fractionated to determine the amounts and types of structures containing the inoculum. These experiments showed that inoculated TMV particles are partly uncoated from their 3' ends in the cytoplasm and incorporated into membrane-bound polysomes. Similar polysomes are found in TMV-inoculated pea and soybean plants, even though these species are insusceptible to TMV (Kiho, Machida and Oshima, 1972), and also in susceptible plants inoculated with ultra-violet irradiated TMV (Hayashi *et al*, 1969). In susceptible plants few inoculum RNA molecules become completely free of coat protein, and these are found in a particulate fraction that consists mostly of host cell nuclei; the amount of uncoated RNA is correlated with the number of infections produced. When TMV-RNA instead of particles is used as inoculum, polysomes are formed but are not membrane-bound. Thus it seems that inoculum particles become at least partly uncoated in inoculated leaves and may be translated, irrespective of whether they infect the plant.

It is unlikely that inoculated virus particles start uncoating where they first encounter the surface of the cell, because viruses whose RNAs are much more sensitive to ultra-violet inactivation than intact particles (e.g. TMV) do not, at any time during infection, increase in sensitivity to inactivation by ultra-violet light (Siegel and Wildman, 1956). Thus particles probably uncoat in parts of the cell that are shielded from external ultra-violet light.

The simpler kinds of virus particle, constructed of only two or three different macromolecules, are probably totally disassembled during the biochemically active part of the life cycles of the viruses. However the more complex particles of viruses such as clover wound tumor and lettuce necrotic yellows, are probably only partly disassembled for the active part of their life cycle; infecting particles merely lose their outer layers, and the genome and some of the proteins remain together as a particle core that becomes the centre of virus replication. For example, particles of reovirus (which are similar to those of clover wound tumor and maize rough dwarf viruses) are uncoated in cells by lysosomal hydrolases to form 'sub-viral' particles that are involved in replication of the virus (Silverstein *et al*, 1972; Astell *et al*, 1972). The sub-viral particles still contain the virus-specific transcriptase and are able to synthesize messenger RNA *in vitro* (Gillies, Bullivant and Bellamy, 1972). Similarly, the cores (nucleocapsids) of the bacilliform particles of the rhabdoviruses, including lettuce necrotic yellows virus, seem involved in replication (Bishop and Roy, 1972; Randles and Francki, 1972; Francki and Randles, 1973).

11.3 The biochemistry of virus replication

Many lines of evidence show that virus nucleic acids are replicated and are translated into proteins by separate processes, which are to some extent independent, may occur in different parts of the host cell, and may be separated experimentally using, for example, mutant strains of TMV (Kassanis and Bastow, 1971a), deletion variants of tobacco rattle virus (Sänger, 1968), or chemicals that inhibit one or other process (Francki and Matthews, 1962b; Beachy and Murakishi, 1973). Thus in this section we will discuss first the great diversity of ways in which viruses replicate and transcribe their nucleic acids and then discuss the seemingly more uniform way in which plant virus nucleic acids are translated into proteins.

11.3.1 *Virus nucleic acid synthesis*

There is increasing evidence that viruses with similar types of genome (i.e. the information-carrying nucleic acid in infective particles) usually replicate in the same way irrespective of which type of organism they infect. Thus, although the details of the replication of most plant viruses are poorly understood, they will probably closely resemble those of other viruses with similar genomes.

The genome of some viruses, including many that infect bacteria and animals is DNA; the DNA of some of these viruses is single-stranded, of others double, of some circular and of others linear. The genetic information of all types of virus DNA is probably expressed via messenger RNA in the same way as that of the DNA of pro- and eukaryotes. By contrast, the genome of other viruses is RNA; the RNA of some, including most of those that infect plants, is single-stranded, but of others double-stranded. All RNA genomes seem to be linear but some are divided into one or more segments. The genetic information in some of these virus RNAs is expressed directly, in others it is expressed via complementary messenger RNA (Chapter 5), and in one interesting group of viruses that infect vertebrates it is perhaps expressed via DNA and messenger RNA (Chapter 16).

The ways in which the genomes of viruses are expressed in infected cells have received much attention (reviewed by Sugiyama *et al*, 1972; Gibbs and Skehel, 1973) and as a result most viruses can be placed in one or other of four main groups.

(a) *Viruses whose particles contain single-stranded infective RNA* Much is known about this type of virus for it includes the RNA-containing bacteriophages (ribophages) R17, MS2 and Qβ, the animal picornaviruses including polio and encephalomyocarditis (EMC), and most well-studied plant viruses. The RNA can be liberated from the particles of all these viruses, purified free of coat protein and shown to be one or more pieces of infective single-stranded polynucleotide. The fact that these RNAs are infective is now thought to imply that they serve as messenger RNAs (mRNAs) when they first initiate infection, and are translated directly to produce proteins some of which are required for replication of the virus RNA.

Of this group, most is known about the ribophages (Chapter 16). These viruses have isometric particles about 25 nm in diameter, which contain the RNA genome of about 10^6 daltons weight. In susceptible bacterial cells this genome is translated into three proteins; a RNA replicase subunit, a coat protein and an A (or 'maturation') protein, of which the latter two, together with the RNA, are the sole components of infective virus particles. Once the replicase subunit is synthesized it combines with three host proteins to form a virus-specific replicase; transcription of the virus RNA leading to its replication can then start (Hindley, 1973).

First the replicase attaches to the 3′ end of the virus RNA (Chapter 5), and moves along it to the 5′ end, while synthesizing a complementary RNA molecule as an antiparallel strand, that is, in a 5′ to 3′ direction. When the complementary molecule is complete, the replicase attaches to its 3′ end and in turn copies it to produce a polynucleotide identical in sequence to the initial virus RNA (Stavis and August, 1970). After the initial synthesis of complementary RNA, the replicase preferentially transcribes the complementary RNA, thus ensuring that much more genome RNA is produced than complementary RNA.

There has been much argument about whether the progeny genome RNA is transcribed from free complementary single-stranded RNA or from a double-stranded RNA composed of the parental genome RNA and the complementary RNA. The evidence most often cited to support the second idea is that RNAase-resistant RNA (i.e. double-stranded RNA) is readily isolated from *in vitro* polymerase reaction mixtures (§9.2) and from virus infected cells; this RNA is usually a mixture of completely and partially double-stranded RNA called replicative form RNA (RF) and replicative intermediate RNA (RI) respectively (Amman, Delius and Hofschneider, 1964; Franklin, 1966). However there is now much doubt that these double-stranded RNAs are directly involved in virus RNA replication (Weissmann, Feix and Slor, 1968), because neither of the types of double-stranded RNA will act as templates for RNA synthesis using purified replicase of the ribophage $Q\beta$; furthermore little RNAase-resistant RNA is obtained from infected tissues or polymerase reaction mixtures unless these are deproteinized using detergent or phenol, and thus it seems that much of double-stranded RNA is an artifact and not present in infected cells. Therefore the template for producing the progeny genome-RNA probably is free complementary RNA not extensively base-paired

with the progeny RNA, and to support this hypothesis is the fact that the complementary RNA functions as a template for RNA synthesis using replicase *in vitro*.

The replicase of the ribophage $Q\beta$ is the best-known RNA-dependent RNA replicase. It has been purified in various ways, but during purification its specificity is modified; when impure it will transcribe $Q\beta$ RNA and also synthetic polynucleotides, but when pure it will not transcribe $Q\beta$ RNA unless various specific host factors are added (Stavis and August, 1970). The purest functional $Q\beta$ replicase consists of four different polypeptides with molecular weights of 70 000, 65 000, 45 000 and 35 000. The second largest of these is the product of the virus replicase gene, and the other three are proteins found in uninfected bacteria.

The replication of animal picornaviruses, such as poliovirus and EMC, is similar to that of the ribophages, though the way their proteins are produced is quite different (§11.3.2). The single-component single-stranded RNA genome of poliovirus is replicated via free complementary RNA, which is mostly not base-paired to genome RNA. The replicase of EMC consists of five polypeptides, four of which are almost the same size as those of $Q\beta$ replicase (Rosenberg *et al*, 1972).

Many plant viruses have single-stranded RNA genomes that are infective when removed from their particles. Several of these RNAs can act as templates in protein-synthesizing systems and yield proteins including recognizable coat proteins (§11.3.2) indicating that the coat-protein gene in these genomes is messenger, not the complementary RNA. Furthermore virus-specific replicases and double-stranded RNAs have been isolated from extracts of plants infected with some of these viruses (Ralph, 1969). Thus their RNAs probably are replicated in the same way as those of the ribophages and picornaviruses.

Many plant viruses have, like TMV, a genome consisting of one piece of single-stranded RNA. Others have a similar genome but in two or more pieces (Jaspars, 1974) that are found in separate particles (§6.7.3). These multipartite genomes are of great interest because they enable pseudo-recombinants to be produced and genetic analysis to be made (§§12.2.3 and 12.3.3). They are of two main types. In the first, exemplified by tobacco rattle virus, one of the two pieces of genome RNA can cause infection, albeit of an aberrant type, whereas in the other type all the pieces of the genome are needed for infection to occur. In this second type, the genome may be in two pieces (e.g. comoviruses and nepoviruses) or at least three pieces (e.g. alfalfa mosaic virus, bromoviruses, cucumoviruses). The reason for this difference between the two main types is not established, but whereas the larger piece of genome RNA of tobacco rattle virus evidently provides enough information to elicit a functional virus RNA replicase in cells, apparently none of the genome pieces of the other viruses can do this.

Extracts of a plant infected with one of these single-stranded RNA viruses may contain several sorts of virus-induced RNA. For example, deproteinized extracts of cells recently infected with TMV contain the genome, a larger-than-genome single-stranded RNA, two types of double-stranded RNA (RF and RI) and a small single-stranded RNA (Nilsson-Tilgren, 1970; Jackson, Mitchell and Siegel, 1971; Babos, 1971; Jackson *et al*, 1972; Siegel, Zaitlin and Duda, 1973). Radioactive labelling experiments show that both the double-stranded RNAs are to some extent involved in virus replication, though, as with the ribophages, the double-strandedness of the RNAs may result from the method of extraction, and in the living cell they may be much less base-paired. It seems that the structures in the cell from which RI is produced are directly concerned in TMV-RNA replication because RI is most plentiful early in infection (Kielland-Brandt and Nilsson-Tilgren, 1973) and when a 'pulse' of radioactive nucleotides is given to TMV-infected cells it is incorporated quickly into the RI but can be 'chased' out by adding further non-radioactive nucleotides (Jackson *et al*, 1972). By contrast most of the precursors of RF are perhaps by-products of TMV-RNA replication because RF accumulates late in infection and when it is radioactively labelled *in vivo* most of the radioactivity cannot be 'chased' out by unlabelled nucleotides.

The small single-stranded RNA molecules found in TMV-infected cells are fragments of the TMV genome. They seem not to contain the 5′ terminal sequence of the genome because they are not incorporated into TMV particles either *in vivo* or *in vitro*. In infected cells they are found in polysomes and thus are presumably messenger RNA, perhaps the coat protein gene. Similar RNAs have been found in plants infected with several other viruses (Romero, 1973). These small RNAs may be analogous to the smallest RNAs found in the particles of the bromoviruses (Shih,

Lane and Kaesberg, 1972) and alfalfa mosaic virus. With each of these viruses, the smallest RNA is not strictly part of the genome because although it contains the coat protein gene, that gene is also found in one of the RNA species of the 'true' genome; however, at least with alfalfa mosaic virus, either the coat protein or the smallest RNA is required for infectivity (Bol and van Vloten-Doting, 1973).

TMV-infected plants contain a replicase, which is membrane-bound (Ralph and Wojcik, 1969), but can be solubilized by detergents (Zaitlin, Duda and Petti, 1973). The bound enzyme has RNA template attached, whereas the solubilized enzyme has not, and thus is stimulated by added RNA though it shows no specificity for TMV-RNA. The solubilized enzyme sediments like a protein of molecular weight 160 000. Replicases with analogous properties have been isolated from plants infected with cucumber mosaic virus or with tobacco ringspot virus (May and Symons, 1971; Peden, May and Symons, 1972); neither replicase was all membrane-bound and their molecular weights were estimated by sedimentation to be 123 000 and 120 000 to 180 000 respectively.

The most fully studied plant virus replicase is that found in barley infected with brome mosaic virus (Semal and Hamilton, 1968; Kummert and Semal, 1970; Hadidi and Fraenkel-Conrat, 1973; Hadidi, Hariharasubramanian and Fraenkel-Conrat, 1973). This enzyme is membrane-bound but solubilized by detergent, and also it sediments at a rate which suggests a molecular weight of 160 000. When solubilized, it requires added RNA template and shows some specificity for the RNAs of brome mosaic virus and the related cowpea chlorotic mottle virus; also, of the RNAs from brome mosaic virus particles, the smallest (0.3×10^6 daltons) was the most active as template.

(b) *Viruses whose particles contain single-stranded non-infective RNA* Viruses with this type of genome include the rhabdoviruses, and also several groups of vertebrate viruses including the orthomyxoviruses, paramyxoviruses and leukoviruses (Chapter 16). Best known of the plant rhabdoviruses are lettuce necrotic yellows and potato yellow dwarf viruses, which, like other rhabdoviruses, have large bacilliform particles that contain one strand of single-stranded RNA with a molecular weight of about 4×10^6 (Francki and Randles, 1973; Reeder, Knudson and Macleod, 1972). The genome of lettuce necrotic yellows virus is not infective when deproteinized (Harrison and Crowley, 1965), but if the outer lipid-containing layer only of the particle is removed by detergent, the resulting nucleocapsid is infective (Randles and Francki, 1972) and contains an activated replicase (Francki and Randles, 1973). This replicase produces single-stranded RNAs using the genome as template. These RNAs are much smaller than the genome and are liberated from the nucleocapsid even though complementary in sequence to the genome RNA; they are probably messenger RNAs. The plant rhabdoviruses probably replicate in the same way as bovine vesicular stomatitis virus whose particles contain non-infective RNA that is complementary to messenger RNA; it probably has no ribosome attachment sites and even if translated would yield a nonsense product.

(c) *Viruses whose particles contain double-stranded RNA* The viruses in this group (see Chapters 2 and 16) that have received most attention are the reoviruses of animals (Joklik, Skehel and Zweerink, 1970), and plant viruses such as clover wound tumor and rice dwarf. The genome of these viruses is double-stranded and in ten to twelve parts, each of which is $0.8–2.6 \times 10^6$ daltons in weight, with a total molecular weight per particle of about 15×10^6 (Reddy and Black, 1973a). In infected cells, or in a suitable *in vitro* system, each part of the reovirus genome is transcribed into a single-stranded messenger RNA of equivalent size by a replicase which is a component of the core of the virus particle. Thus protein-free preparations of the RNAs of these viruses are not infective. Each of the reovirus messenger RNAs is translated into a specific protein, the size of which is related to the size of the messenger RNA. Thus each part of the genome is unique and the transcribed messenger RNAs are monocistronic.

Messenger RNAs of reoviruses are transcribed by a mechanism which is unlike that of viruses with single-stranded RNA genomes, but is analogous to that of viruses with DNA genomes. The messenger RNAs are transcribed from the double-stranded templates, but the strands of the template remain base-paired and are conserved intact. Presumably some of the messenger RNAs produced are converted by a replicase into double-stranded progeny genomes.

(d) *Viruses whose particles contain DNA* A great variety of viruses of animals and bacteria have

DNA genomes but so far only one such group of plant viruses, the caulimoviruses, is known. Like the papovaviruses of animals, the caulimoviruses have isometric particles which contain a double-stranded circular DNA genome of about 5×10^6 molecular weight which when isolated from the particles is infective (Shepherd, Bruening and Wakeman, 1970; Russell *et al*, 1971). Caulimovirus particles accumulate in the host's cytoplasm in dense inclusion bodies (Kamei, Rubio-Huertos and Matsui, 1969), but whether replication occurs in these inclusions is uncertain (Favali, Bassi and Conti, 1973).

The replication of cauliflower mosaic virus, like that of all other viruses with a DNA genome and also a few with RNA genomes, is inhibited by the antibiotic actinomycin D (Tezuka, Taniguchi and Matsui, 1971), which inhibits DNA-dependent RNA polymerase enzymes.

11.3.2 *Virus protein synthesis*

There is no evidence that virus proteins are not produced in the usual manner using the host's ribosome system to translate virus-specific RNA, which in some viruses is the virus genome itself and in others is messenger RNA transcribed from the genome. There is no evidence that viruses alter the host ribosomes, but some of the large bacterial viruses do produce additional transfer RNAs (Wilson and Kells, 1972; Wilson and Abelson, 1972).

Studies on the synthesis of bacterial and animal virus proteins have, for reasons discussed earlier in this chapter, made more progress than those on plant-virus proteins, and so in this section we will discuss first some of the results obtained with bacterial and animal viruses.

The single-stranded RNA genome of the ribophages codes for three proteins. During infection the amount and time of appearance of these proteins differs; the replicase is mostly synthesized early in infection, the maturation protein is synthesized at a small but constant rate throughout infection, whereas the coat protein is made throughout infection but at an exponentially increasing rate and in very much greater amounts than either of the other two (Nathans *et al*, 1969). The most likely explanation for these differences is that translation of the three genes specifying these proteins is specifically and independently controlled and that each has separate sites for initiating and terminating translation (i.e. the virus genome is a polycistronic mRNA). The three genes in the ribophage genome are in the order: 5′terminus-maturation protein-coat protein-replicase subunit-3′ terminus. The way in which independent translation of the genes is controlled, and the solution to the dilemma that the genome is translated by ribosomes moving from its 5′ to 3′ end, and yet is transcribed by a replicase moving in the opposite direction is turning out to be a fascinating story though it is not yet complete. It seems that most of the molecules involved have more than one function. For example, with phage Qβ the coat protein molecules also repress translation of the replicase gene, so that as coat protein accumulates during infection, a decreasing amount of replicase is produced. Similarly, the replicase not only transcribes the mRNA but also competes with ribosomes for virus mRNA. When replicase is added to complexes of the virus mRNA and ribosomes, the replicase attaches to both the 3′ end of the RNA and also to the site where ribosomes attach to the coat protein gene; thus no more ribosomes can attach there, or at the replicase gene, which becomes unavailable when no ribosome is attached at the coat protein gene. Ribosomes already attached complete translation before transcription starts, and hence ribosomes and replicase do not try to 'read' the mRNA at the same time and 'run into' each other. These and other mechanisms ensure that progeny molecules of the ribophages are produced at the right stage of infection and in the right amounts to produce the maximum number of progeny particles (Hindley, 1973).

The production of the proteins of animal picornaviruses, such as poliovirus, differs in many important ways from that of the ribophages. Like the ribophages, poliovirus has a genome composed of one piece of single-stranded RNA, which can be translated into proteins both *in vivo* and *in vitro*. However, unlike the ribophages, poliovirus RNA is translated into one large polypeptide chain which is equivalent in size to the entire virus genome (Öberg and Shatkin, 1972). Subsequently, by processes which are not understood, this large molecule is rapidly divided to yield the functional virus proteins; the same polypeptides are synthesized in equimolar amounts at all times during infection – a seemingly wasteful process because only about half the genome specifies proteins found in the particles. This process is perhaps mandatory for viruses of mammals that have an infective genome consisting of a single molecule of RNA, for it seems that

mammalian cells are only able to translate monocistronic mRNAs.

One major difference between the protein-synthesizing systems of plants and bacteria is that bacteria apparently contain only one type of ribosome whereas plants have different types in their cytoplasm, chloroplasts and mitochondria (Wittmann, 1970). However the replication of all plant viruses tested is inhibited by the antibiotic cycloheximide, and this implies that these viruses use the host's cytoplasmic ribosomes. Cycloheximide inhibits replication of TMV, tobacco necrosis, tobacco rattle, eggplant mosaic and broad bean mottle viruses (Zaitlin, Spencer and Whitfeld, 1968; Harrison and Crockatt, 1971; McCarthy *et al*, 1972; Gibbs and MacDonald, 1974), whereas all are unaffected by chloramphenicol, which inhibits translation by chloroplast and mitochondrial ribosomes as well as those of bacteria. When cycloheximide is applied to plant tissues infected with TMV or tobacco necrosis virus late in infection, it stops synthesis of virus protein and the accumulation of virus particles, but it does not stop the synthesis of virus RNA (McCarthy *et al*, 1972; Beachy and Murakishi, 1973); however, when applied immediately after infection no virus RNA is synthesized either, and infection is arrested. These results suggest that a protein (presumably a replicase) is produced by translation early in infection and is essential for virus RNA replication, but once produced is able to replicate the RNA even when production of coat protein, and hence virus particles, is inhibited.

It seems that few proteins are produced by ribosome translation systems in a final active form, but that the primary products of translation are modified in various ways. Thus the translation of most messenger RNAs is probably initiated at the codon for formyl methionine (AUG), or methionine, but that this amino acid is later removed from the N-terminus of the newly synthesized protein. The translation of plant-virus RNAs seems to follow the same pattern, for no coat proteins are known to have N-terminal methionine, and several are acetylated. Furthermore when satellite virus RNA is translated *in vitro* (see below) the coat protein produced has the N-terminal sequence met-ala-lys-, whereas the coat protein produced *in vivo* has terminal ala-lys- (Klein and Clark, 1973).

The genomes of all plant viruses are large enough to code for two or more proteins of the size of their respective coat proteins, and there have been many attempts to find the non-coat proteins in infected plants. Those that have been found are always present in very much smaller amounts than coat protein, and this suggests that plant virus RNAs are probably polycistronic like those of the ribophages; if, like poliovirus, they are monocistronic and are cleaved into individual proteins after translation, then the non-coat proteins must be much more quickly degraded than the coat proteins. In TMV-infected tobacco protoplasts but not healthy ones, Sakai and Takebe (1972) found TMV coat protein and also a protein of about 140 000 daltons. This protein was not a polymer of coat protein because it was not disrupted by guanidine and, unlike the coat protein, it contained histidine. Similarly Zaitlin and Hariharasubramanian (1972) found that TMV-infected plants contained coat protein and also other virus-induced proteins, of which the most plentiful had a molecular weight of about 155 000. Should these large proteins be virus replicase, however, they may be only partly virus coded (§11.3.1a). Indeed, Hadidi and Fraenkel-Conrat (1973) found that active detergent-solubilized, virus-induced polymerase from brome mosaic virus-infected barley plants had a molecular weight around 160 000, but Hariharasubramanian *et al* (1973) found a protein with a molecular weight of 34 500 when the radioactively-labelled enzyme was dissociated; it is unclear whether the undissociated enzyme is a tetramer of the smaller protein, or the smaller protein is the virus moiety of a sequestrated host enzyme. The other three proteins found by Zaitlin and Hariharasubramanian had molecular weights near 245 000, 195 000 and 37 000, and thus the largest alone could require a messenger the same size as the TMV genome. Indeed, some of these proteins are probably virus-activated host proteins although the TMV genome may conceivably be translated like that of poliovirus into one large polypeptide, and subsequently cleaved into smaller functional proteins.

In tissue infected with tobacco necrosis virus, Jones and Reichmann (1973), found virus-coat protein together with small amounts of virus-induced proteins of 64 000, 42 000, 23 000, 15 000 and 12 000 daltons; the tobacco necrosis genome is large enough to code for all these proteins. Tissue infected with both satellite virus and tobacco necrosis virus contained only one additional protein, which was the same size as satellite virus-coat protein but, interestingly, the amount of tobacco necrosis virus-coat protein, but not the other proteins, was greatly diminished (§11.6.2).

Several viruses induce the production of specific inclusions in infected plants. Some of these inclusions are merely aggregates of virus particles but others are partly or wholly made of other materials (Chapter 3). Of the latter type are the pinwheel inclusions (Fig. 3.22) found in plants infected with potyviruses. The pinwheels of each potyvirus contain one virus-specific and serologically distinctive protein of molecular weight 67 000 to 70 000, whereas the coat proteins of potyviruses have molecular weights of 26 000 to 28 000 (Purcifull, Hiebert and McDonald, 1973; Hiebert and McDonald, 1973). Tobacco etch virus, a potyvirus, also induces intranuclear crystals (Fig. 3.20) that seem composed of two virus-specified proteins, of 48 000 and 54 000 daltons (Knuhtsen, Hiebert and Purcifull, 1974). Together the coat, intranuclear and pinwheel proteins would require about two-thirds of the tobacco etch virus genome to code for them.

One way of finding what proteins are coded for by a messenger RNA is to translate it in a cell-free protein-synthesizing system. Attempts to do this with plant virus RNA have been most successful using RNAs with molecular weights less than 10^6 (Davies and Kaesberg, 1974). The first convincing report was that of van Ravenswaay-Claasen *et al* (1967) who translated the top *a* RNA of alfalfa mosaic virus using a ribosome system from the bacterium *Escherichia coli*, and obtained a protein closely resembling the coat protein of the virus. Tobacco necrosis satellite virus RNA has been translated using the *E. coli* system and also one from wheat embryos (Marcus, Efron and Weeks, 1973). Both systems produce proteins slightly smaller than normal satellite virus coat protein, but which may be hydrolysed by trypsin into an almost identical set of peptides (Klein *et al*, 1972). Both systems require charged formyl-methionine transfer RNA to start translation, and initially methionine is the N-terminal amino acid of the proteins produced, although it is removed promptly in the wheat embryo system to expose alanine, which is the N-terminal amino acid of normal satellite virus coat protein (Klein and Clark, 1973).

There has been less success in attempts to translate larger plant virus RNAs such as the genome of TMV. Efron and Marcus (1973) and Roberts and Paterson (1973) used the wheat embryo system to translate TMV-RNA but obtained proteins with a wide range of sizes, though those of about 18 000 daltons yielded tryptic peptides most of which were identical to peptides from TMV coat protein. Figure 11.4 shows the results of a similar experiment in which antiserum prepared against coat protein was used to precipitate proteins obtained by translating the RNA of the bean form of TMV in the wheat

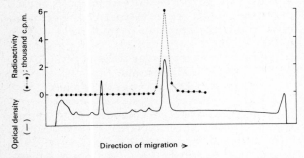

Fig. 11.4 Analysis by polyacrylamide gel electrophoresis of the product of *in vitro* translation of bean strain TMV-RNA using ribosomes, etc. from wheat embryos (Marcus, Efron and Weeks, 1973), and ^{35}S-methionine. After translation, excess unlabelled virus particles were added together with antiserum prepared in rabbits using guanidine-disrupted particles of bean strain TMV as immunogen. The precipitate was heated in sodium dodecyl sulphate and 2-mercaptoethanol, and analysed electrophoretically in polyacrylamide gel. The gel was stained in Coomassie brilliant blue, and then sliced for estimates of radioactivity. Note that there are several stained bands of protein, perhaps serum proteins, but that the radioactivity is confined to the coat-protein band; indistinguishable results were obtained using mixtures of other radioactive amino acids. (Courtesy P. B. Goodwin.)

embryo system; radioactive methionine was incorporated into this protein (presumably at its N-terminus) even though methionine does not normally occur in the coat protein of this virus, and the protein produced is perhaps slightly larger than the added carrier protein.

Interesting results have been obtained by translation of the four species of RNA of brome mosaic virus; these weigh 1.1×10^6, 1.0×10^6, 0.75×10^6 and 0.28×10^6 daltons, the three largest species being essential for infection, and the two smallest each containing the coat-protein gene (Shih, Lane and Kaesberg, 1972). Shih and Kaesberg (1973) found that these RNAs, particularly the smallest, were very effective messengers in the wheat embryo system. The smallest RNA seemed to be a monocistronic messenger RNA for the coat protein, because the protein it gave and normal coat protein yielded almost identical peptide patterns when treated with trypsin or with cyanogen bromide. The 0.75×10^6 dalton RNA gave a small amount of coat protein together with a larger amount of a slightly larger protein,

and the latter yielded quite different peptides from the coat protein. The two largest RNAs gave mixtures of several larger proteins. Mixtures of RNAs that included the smallest RNA yielded coat protein almost exclusively, and it was evident that the smallest RNA competitively inhibited translation of the other RNAs.

11.4 Cellular sites of virus replication, and of assembly and accumulation of virus particles

It is difficult to find out where the different stages of the biochemical life cycles of viruses occur in cells because the different techniques that can be used all have disadvantages. For example, microscopic observations of infected cells merely show where there are changes and do not distinguish between the direct and indirect effects of virus infection. Similarly, autoradiography shows most clearly where compounds are accumulating and may not show where ephemeral stages in a life cycle occur, and the difficulties of experiments in which cell extracts are fractionated are quite obvious. However, despite these drawbacks, information from several types of experiments indicates that different plant viruses use different parts of host cells for replication sites.

In a series of elegant experiments Zech (1963) followed changes in the absorption of ultra-violet light by different parts of TMV-infected hair cells, and found that within an hour of infection there is a great increase in absorption of 265 nm light by the nucleus, and later by the cytoplasm, though absorption of 280 nm light by the cytoplasm also increases. Using phase contrast microscopy, Bald (1966) found that material is periodically emitted from the nucleolus of TMV-infected cells and passes through the nucleus into the cytoplasm; this material stained like RNA. Furthermore Smith and Schlegel (1965) found that when tritiated uridine was supplied to TMV-infected tobacco cells, it accumulated first in the nucleolus, but later moved to virus inclusions in the cytoplasm. These three types of experiments, together with those of Machida and Kiho (1970) (§11.2) suggest that the nucleus, and especially the nucleolus, is involved in TMV-RNA replication. However, this conflicts with the conclusions of Ralph, Bullivant and Wojcik (1971), who found that the double-stranded virus RNA in TMV-infected tobacco cells is associated with cytoplasmic membranes, not nuclei.

De Zoeten and Schlegel (1967) found that clover yellow mosaic virus (a potexvirus), like

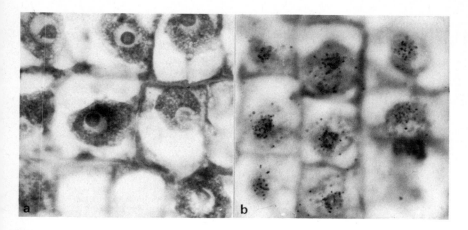

Fig. 11.5 Autoradiogram of cells from the root tips of *Vicia faba* treated with actinomycin D (to inhibit DNA-dependent RNA synthesis in the nuclei) for 5 hours and then supplied with ^{3}H-uridine together with actinomycin D for a further 30–60 min. **a,** Virus free cells. **b,** Cells infected with clover yellow mosaic virus (a potexvirus). Note the incorporation of radioactivity into the nuclei, and particularly the nucleoli, of the virus-infected cells. (Courtesy D. E. Schegel; De Zoeten and Schlegel, 1967.)

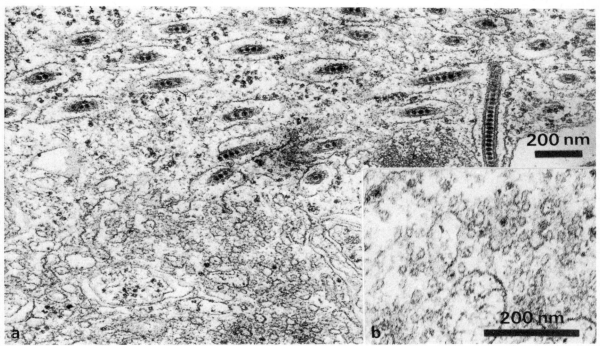

Fig. 11.6 Part of an inclusion body from a leaf palisade cell of *Chenopodium amaranticolor* infected with strawberry latent ringspot virus. The upper part contains tubules cut in transverse and longitudinal section. The tubules are filled with a single row of nucleic-acid-containing virus particles. The lower parts of **a,** and **b,** contain membraneous structures and granular material, thought to be empty virus protein coats. (After Roberts and Harrison, 1970.)

TMV, caused tritiated uridine to accumulate first in the nucleolus (Fig. 11.5), whereas broad bean mottle virus (a bromovirus), tobacco ring-spot virus (a nepovirus) and tomato spotted wilt virus, all caused tritiated uridine to accumulate in the cytoplasm rather than any particular cell organelle.

Experiments with cowpea mosaic virus (a comovirus) show that most of the double-stranded RNA in cowpea leaves is located in cytoplasmic inclusion bodies, which contain masses of membrane bounded vesicles (Assink, Swaans and van Kammen, 1973). This suggests that these inclusions are sites of virus-RNA synthesis, as also may be the similar inclusion bodies induced by nepo-viruses. For example, the inclusions found in leaves infected with strawberry latent ringspot virus probably have a role in virus synthesis and assembly because they contain separate areas rich, respectively, in nucleic acid-containing and nucleic acid-free virus-like particles, as well as membraneous vesicles (Fig. 11.6).

In extracts of tissue infected with deletion variants of tobacco rattle virus (§12.2.4), the infective RNA is partly protected from leaf ribonuclease, is associated with easily sedimented material (Cadman, 1962), and ultrastructural evidence (Fig. 3.15) suggests that involvement of mitochondria in synthesis of the virus RNA should not be ruled out (Harrison, Stefanac and Roberts, 1970).

There is good evidence that the chloroplasts are the site of RNA synthesis of turnip yellow mosaic and other tymoviruses (Ushiyama and Matthews, 1970). The chloroplasts of tymovirus-infected plants contain many peripheral vesicles

formed by invaginations of both chloroplast membranes (Figs. 3.14 and 11.7); these accumulate supplied tritiated uridine (Bové, 1972), and both virus replicase and double-stranded RNA are associated with the chloroplast membrane (Bové, Mocquot and Bové, 1972).

As already described (§11.3.2), experiments with cycloheximide and chloramphenicol suggest that the RNAs of several plant viruses are translated by ribosomes in the cytoplasm not the chloroplasts. This has been substantiated in several ways for TMV; all methods first detected TMV particles or protein confined to the cytoplasm 15 to 20 hours after infection (Nagaraj, 1965; Milne, 1966). However, in tissue infected with tobacco etch virus, virus-specified proteins accumulate as crystals in the nucleus (Fig. 3.20) and as pinwheels in the cytoplasm, although their site of synthesis is not known.

The way in which virus particles are assembled

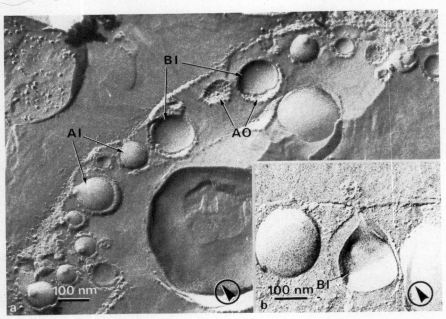

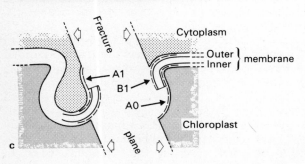

Fig. 11.7 a and **b** Surface of a fracture plane through the edge of a frozen chloroplast isolated from leaves of *Brassica pekinensis* infected with turnip yellow mosaic virus. Note the vesicles that are characteristically found around the periphery of chloroplasts infected with tymoviruses. Each vesicle seems to be an invagination of both chloroplast membranes into the chloroplast. Each chloroplast membrane is two-layered and fractures most readily between these layers; thus fracturing usually exposes one or other of four different faces. **c,** Diagram illustrating the fracture faces and the lettering used in the micrographs. Arrow in **a** and **b** indicates direction of shadowing. (Courtesy T. Hatta; Hatta, Bullivant and Matthews, 1973.)

in vivo is also not known, though it is probable that for the simpler virus particles, like those of TMV, the *in vivo* and *in vitro* processes (§5.7) are similar. Particles of TMV usually accumulate in and are confined to the cytoplasm, but possibly assemble and can accumulate elsewhere; for example, some particles of the U5 strain of TMV are found in chloroplasts (Shalla, 1968), though RNA synthesis of this strain seems not to occur in the chloroplasts (Betto *et al*, 1972), and its coat protein synthesis is inhibited by cycloheximide but not chloramphenicol (Gibbs and MacDonald, 1974).

Viruses with larger genomes and more complex particles also use various sites for the different stages of replication. Several double-stranded RNA viruses produce electron-dense amorphous cytoplasmic inclusions called viroplasms. These occur, for example, both in plants and in leafhoppers infected with wound tumor virus (Shikata and Maramorosch, 1967). The particles of maize rough dwarf virus seem to be assembled in similar viroplasms, and only later disperse through the cytoplasm (Bassi and Favali, 1972; Favali, Bassi and Appiano, 1974; Fig. 11.8). Although one of the first effects of infection with cauliflower mosaic virus is on the nucleus (Favali, Bassi and Conti, 1973), dense cytoplasmic inclusions (Fig. 3.21) are produced later which may be the site of later virus replication (Rubio-Huertos *et al*, 1972).

Virus particles with lipid-containing layers usually acquire these layers as the core of the particle passes through a membrane. For example, the bacilliform particles of the rhabdo-viruses are formed when the helically constructed core of the particle is extruded through a cellular membrane, taking part of the membrane with it. It seems that different plant rhabdoviruses may use different cellular membranes for this process. However the most carefully studied of the plant rhabdoviruses, sowthistle yellow vein and potato yellow dwarf viruses, both in their plant and insect hosts, acquire their outer membrane by budding through the nuclear membrane into the cytoplasm or into the space between the inner and outer nuclear membranes (reviews by Hummeler, 1971; Knudson, 1973). This behaviour contrasts with that of the rhabdoviruses of vertebrates, all of which acquire their outer membranes when passing through the outer cell membrane. The site of synthesis of rhabdovirus RNA is not known, and it is likely that the several proteins found in rhabdovirus particles are not synthesized in nuclei, even though they are first detected there by fluorescent antibodies. Thus rhabdovirus proteins are presumably transported into the nucleus, some assemble there with the genome to form the core (or nucleocapsid), while others replace proteins in the nuclear membrane, and finally the nucleocapsid passes through the membrane acquiring part of it as a coat.

Tomato spotted wilt is another virus that has lipid-containing particles, which it seems are assembled as they bud into membrane-bounded vesicles in the cytoplasm (Milne, 1970), and carrot mottle virus particles may be assembled similarly at the tonoplast membrane (Murant *et al*, 1969, Murant, Roberts and Goold, 1973; Fig. 5.14 h).

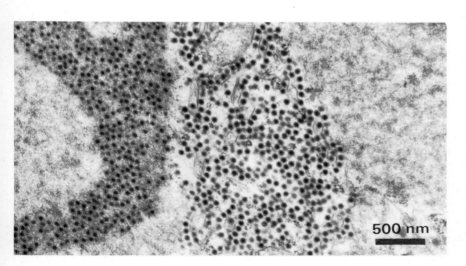

Fig. 11.8 Section of cytoplasm of maize leaf cell infected with maize rough dwarf virus, showing viroplasm containing cores of virus particles (left), and complete virus particles outside the viroplasm (middle). (Courtesy O. Lovislo and M. Conti; after Lovislo and Conti, 1966.)

11.5 Movement and accumulation of viruses in plants

11.5.1 *Movement*

Viruses probably move from cell to cell within plants through plasmodesmata but move from one part of a plant to another through the vascular system.

One way of measuring the rate of cell-to-cell movement is to inoculate plant leaves and then remove their epidermis at different times after infection. For example, Welkie and Pound (1958) found that when the epidermis was stripped from cowpea leaves inoculated with cucumber mosaic virus after 2 hours at 28°C then no lesions developed, but after 3 hours 10% developed and after 7 hours, 100%. Virus moved from the epidermis more quickly with increase of temperature: for example estimates of the time taken at 16, 20, 24 and 28°C for 50% of the lesions to be unaffected by stripping were 14, 8, 6 and 4 hours respectively. However most of this time is probably taken up by virus replication, and there may be little delay between the synthesis of new virus and its movement into adjacent cells. The speed of cell-to-cell spread can also be estimated from the rate of expansion of lesions, and 14 μm/hour was recorded for TMV in inoculated *Nicotiana glutinosa* leaves at 25°C (Rappaport and Wildman, 1957). Another method was used for potato mop-top virus, which produces browning in newly infected cells in response to a particular decrease of temperature, so that the boundary of the virus infected zone can be located after different times simply by changing the temperature. The figures obtained for movement in tobacco leaves were 38 μm/h at 22°C and 16 μm/h at 14°C, and in potato tuber tissue, 10 μm/h at 14°C (Harrison and Jones, 1971a, b).

It is not only intact virus particles (Fig. 11.9) that move from cell to cell. Some defective strains of tobacco necrosis virus and TMV, and deletion variants of tobacco rattle virus, do not form nucleoprotein particles but nonetheless produce lesions that expand at normal speeds in inoculated leaves, although the two last-named viruses do not spread to other parts of the plant in the normal way, through the plant's vascular system, but merely spread from cell to cell along the stem (Siegel, Zaitlin and Sehgal, 1962; Cadman, 1962).

Long-distance spread in the plant is normally much more rapid than cell-to-cell movement, and there is good evidence that many viruses move through the phloem; they move rapidly, will not pass through a steamed section of stem, may not attain detectable amounts in parts through which they have passed, suggesting that they have replicated little if at all in these parts, and their direction of movement can often be correlated with that of carbohydrates (reviewed by Bennett, 1956). Indeed, the particles of some viruses, for example beet yellows, occur predominantly in phloem tissue (Fig. 3.16) and have been found in sieve tubes by electron microscopy.

Viruses can move quickly in stems; for example Bennett (1934) used the leafhopper vector *Circulifer tenellus*, a phloem feeder, to inoculate and sample the phloem of beet leaves, and estimated that beet curly top virus moved 15 cm in 6 minutes. Similarly, after the initial period needed for mechanically inoculated virus to reach the phloem, TMV and potato virus X were found to move 80 cm in tomato stems in 12 hours (Capoor, 1949).

Samuel (1934) followed the spread of TMV in tomato plants by inoculating a terminal leaflet of each. At different times after inoculation the plants were cut up, the parts incubated for the virus in them to replicate, and finally sap from the parts rubbed on indicator plants. The results (Fig. 11.10) show the typical pattern of invasion following mechanical inoculation; slow spread within the leaf followed by rapid spread to other parts of the plant once the virus has entered the phloem.

The direction of movement of virus in the plant

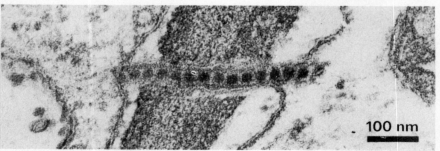

Fig. 11.9 Section showing particles of strawberry latent ringspot virus in a plasmodesma linking two spongy mesophyll cells in a *Chenopodium amaranticolor* leaf. (After Roberts and Harrison, 1970.)

100 nm

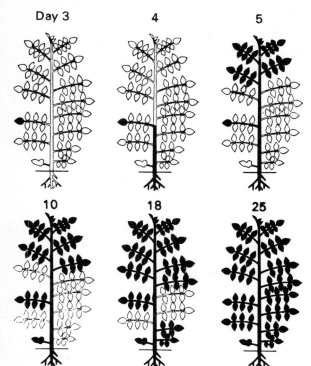

Day 3   **4**   **5**

10   **18**   **25**

Fig. 11.10 Movement of tobacco mosaic virus in tomato plants. Plants were inoculated on one leaflet and then, on the days indicated, the plants were cut into pieces, which were incubated, and sap from them then inoculated to indicator plants. Virus-containing portions shown in black. (Redrawn from Samuel, 1934.)

can be controlled by, for example, defoliating parts of the plant or keeping leaves in darkness so as to direct the movement of metabolites, notably carbohydrates, in the phloem. Perhaps for this reason many viruses spread only slowly from root to shoot, though there may also be other reasons. Little is known about the way viruses pass from the parenchyma to the phloem or from the phloem to the parenchyma, but an inability to enter the phloem may be the reason why some viruses are restricted to the inoculated leaves.

In some circumstances, viruses can move through xylem vessels. Schneider and Worley (1959a, b) found that when a plant of a bean cultivar that was systemically infected with southern bean mosaic virus was grafted to a cultivar that was normally only a local lesion host, the virus could move either up or down the xylem; also, when injected into the xylem of the second cultivar, the virus caused systemic infection.

Viruses rarely enter mature leaves. There is

also evidence that many viruses do not become established in the meristematic regions of root or shoot, and all that have been studied are in low concentration there. The reason for this pattern of distribution is not clear, but the pattern can be disturbed, for when infected tomato roots are treated with ethylene diamine tetraacetate, greatly increased amounts of TMV accumulate in the root tips (Crowley and Hanson, 1960).

11.5.2 *Accumulation*

Different viruses attain very different concentrations in plants. At one extreme, a litre of oat leaf sap may yield 25–100 μg of barley yellow dwarf virus particles, whereas at the other a litre of tobacco leaf sap may yield 2 g or more of TMV particles. The particles of neither of these viruses is particularly unstable, and even allowing for possible differential losses during purification and for the probable localization of barley yellow dwarf virus in specific oat tissues, it still seems likely that the two reach very different concentrations in the cells they infect. The virus concentration attained in tissues depends on both virus and host. Thus TMV reaches five- to ten-fold higher concentrations in tobacco than in tomato. It also depends greatly on the environmental conditions and on how long the plants have been infected. Most viruses reach their highest concentrations in the first leaves to be infected systemically, smaller amounts occurring in leaves produced subsequently. This is particularly well seen in diseases caused by nepoviruses, where early in infection the leaves usually show bright ringspots but later leaves are almost symptomless; this change is correlated with a great decrease in the virus concentration in the leaves (Price, 1936a). Diseases of the mosaic type may go through cycles of more and less severe symptoms in successively formed leaves, again with corresponding changes in virus concentration (Paul and Quantz, 1959). Although the severity of symptoms may be correlated with virus concentration for any one combination of host species and virus, the relative severity of symptoms caused by two different viruses in the same host or by one virus in two different hosts may not indicate relative virus concentration.

Viruses attain their maximum concentration sooner in separated cells, such as protoplasts or leaf or callus cells, than in intact leaf tissue. The maximum TMV concentration in protoplasts is attained 1 day after infection, in separated cells

2 to 4 days after infection, and in tissues after 4 to 10 days; somewhat less virus is obtained from the protoplasts and more from the tissue (Takebe and Otsuki, 1969; Murakishi *et al*, 1971; Otsuki, Shimomura and Takebe, 1972; Jackson *et al*, 1972). Presumably one important difference between the tissue and the protoplasts or separated cells, is that the two last are mostly infected simultaneously, whereas the growth curve in tissues reflects the total concentration of virus in a group of serially infected cells of which only a very small proportion are infected initially. The effect of virus on protoplasts may also differ from its effect on the tissue from which the protoplasts were obtained. For example, at 22°C TMV caused necrosis and accumulated to only a limited extent in leaf discs of tobacco varieties that contained a necrotic gene, whereas protoplasts from similar leaves were not killed, and accumulated normal amounts of virus (Otsuki, Shimomura and Takebe, 1972).

Temperature, nutrition, intensity and duration of light, and availability of water all affect virus concentration. The greatest concentrations often are attained in conditions in which the host plants grow best. Thus Bawden and Kassanis (1950) found that growth of tobacco was increased more by applying both nitrogen and phosphate fertilizers than by applying either alone, and there were parallel effects on concentration of TMV; plants receiving both nitrogen and phosphate yielded forty times as much virus as those receiving neither. However, in contrast to nitrogen and phosphorus, potassium, which caused stunting, often further increased the virus concentration in sap. In poorly fertilized plants, nitrogen seemed to be used preferentially in virus synthesis, for when nitrogen fertilization increased virus concentration, it increased other plant proteins more.

Temperature has particularly large effects on virus concentration (Fig. 11.3) with the optimum for virus accumulation depending on both host and virus. For instance, turnip mosaic and cauliflower mosaic viruses have different optima for accumulation in cabbage, and the concentration of turnip mosaic virus is differently affected by temperature in cabbage and horseradish (Pound and Walker, 1945; Pound, 1949). In some conditions, inactivation of viruses in tissues can be detected, particularly at high temperatures. Thus, when tomato plants infected with tomato bushy stunt virus are placed at 36°C the amount of virus that can be extracted from them progressively decreases. Infectivity, however, decreases more rapidly than antigen titre, suggesting that the particles can be inactivated without losing their protein coats (Kassanis, 1952). Some viruses are known to fall in concentration with time over a wide range of temperatures, and others, for example cowpea chlorotic mottle virus in cowpea (Kuhn, 1965), decrease in specific infectivity (relative infectivity per unit weight virus) with increasing age of infection. Work with tobacco necrosis virus in French bean at a range of temperatures suggests that virus synthesis, and inactivation of infectivity, go on simultaneously, with the virus content of the tissues being the resultant of the two contrasting processes; increase of temperature from 10 to 30°C speeds up both processes, but inactivation speeds up more than synthesis, and becomes dominant at the higher temperatures. Also, the amount of virus inactivated is related to the amount in the tissues, and the rate of inactivation in the leaf is greater than in sap at the same temperature (Table 11.1; Harrison, 1956a).

Kassanis and Bastow (1971a) used normal and defective strains of TMV to study the relative rates of synthesis of virus RNA and coat protein

Table 11.1 Effect of temperature increase on the relative virus content of French bean leaves inoculated with tobacco necrosis virus. (Data from Harrison, 1956a.)

Days after inoculation	Temperature after inoculation			
	22°C	22°C for 2 days then 30°C	22°C for 1 day then 30°C	30°C
1	5.17[1]	5.20	5.22	0.18
2	382	313	3.88	0.36
3	3160	49.5	2.74	2.37

[1] Relative infectivity of sap in arbitrary units.

at different temperatures and at different times after infection. They assessed the amount of virus RNA and protein, both free and in particles, in plants kept at 20°C or at 35°C. They used the type strain, which produces stable particles at both 20°C and 35°C, various other strains which produce stable particles at only one of these temperatures, and one strain (PM2) which produces no intact particles at either temperature. The amount of RNA of strain PM2 in plants attained a maximum, at all temperatures tested, about one week after infection (Fig. 11.11); the lower the temperature the greater the amount of RNA at the maximum, and the longer the RNA persisted. Free virus protein in plants infected with PM2 reached a stable plateau about two weeks after infection. By contrast, type strain TMV produced intact particles and no free RNA or protein at both temperatures; at 35°C, the particles were produced sooner but attained a smaller concentration. The other defective strains behaved like type strain at temperatures where they formed intact particles, and like PM2 at temperatures where either no particles or only defective particles were formed. These experiments show that TMV-RNA accumulation (i.e. synthesis minus degradation) reaches a maximum one week after infection. However, when normal TMV protein is produced, the protein and the progeny RNA assemble to produce stable particles and the RNA is protected from degradation.

The processes involved in the inactivation of infectivity *in vivo* are not at all clear. Treating cucumber leaves with an inhibitor of polyphenoloxidase 2 hours after inoculating them with prune dwarf virus, which is inactivated by quinones, increased the amount of virus extractable from them 3 to 6 days later, whereas treating leaves with a substrate of the enzyme decreased the amount of virus (Hampton and Fulton, 1961). Whether these treatments affected the initiation of infection or whether they affected the course of virus accumulation is not clear. Also, although oxidase systems may have effects on some plant viruses, no evidence could

be obtained that they were involved in the decrease, with increasing age of infection, of the specific infectivity of cowpea chlorotic mottle virus (Kuhn, 1965). Such decreases in specific infectivity may indicate increasing numbers of breakages of the nucleic acid strand within the virus particles or its increasing fragility on extraction, for the proportion of low molecular weight RNA in broad bean mottle virus particles is larger when obtained from long-infected plants than from recently infected ones (Kodama and Bancroft, 1964).

11.6 Interactions of viruses in mixedly infected plants

The behaviour of viruses in a plant infected with more than one virus depends on many factors, such as the relatedness of the viruses, the relative amounts of each virus in the mixture and the environment of the infected plant. In Chapter 12 we discuss the kinds of genetic interaction that produce virus variants and heterologously coated particles, and here we discuss interactions of other sorts.

11.6.1 *Related viruses*

Even the most rigorously cloned virus isolate usually contains a small but finite proportion of

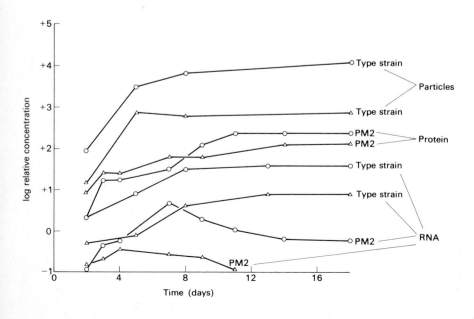

Fig. 11.11 The relative concentration of particles and total extractable RNA (infectivity estimates) of type strain TMV, and of protein (serological estimates) and extractable RNA (infectivity estimates) of the PM2 strain, in tobacco plants at different times after inoculation, and kept either at 20°C (circles) or 35°C (triangles). (Redrawn from Kassanis and Bastow, 1971a.)

variants. These variants have presumably arisen as mutants of the cloned virus, and many survive in harmony with the parent unless they have some selective advantage or disadvantage, or unless conditions change enough to give them that selective differential.

There are presumably similar selection pressures on the different parts of the genome of viruses with multipartite genomes, though these pressures presumably ensure that the different parts of the genome are maintained in proportions that most favour virus growth and transmission. When these proportions are altered experimentally the virus may be disadvantaged; thus there are obvious disadvantages to tobacco rattle virus, if it loses the smaller part of its bipartite genome, for this part contains the coat protein gene. However even an experimentally produced imbalance in the proportion of long and short particles in tobacco rattle virus inocula is reported to decrease the infectivity of the inocula, and the yield of virus from the inoculated leaves (Morris and Semancik, 1973); this effect is specific and is not obtained by mixing long and short particles of two tobacco rattle virus strains that do not produce pseudo-recombinants (§12.3.3).

There are also other kinds of interaction between related viruses that do not result in the production of variants. Some of these phenomena are best seen when two virus strains are mixed for use as inocula, and others when plants already infected with one strain of a virus are inoculated with a second.

(a) *Inoculation of strain mixtures* Sadasivan (1940) found that the number of necrotic lesions produced in tobacco and *Nicotiana sylvestris* by the tomato aucuba mosaic strain of TMV was decreased by adding to the inoculum the type strain of TMV, which does not itself produce necrotic lesions in these plants. By contrast, adding sap containing a strain of potato virus X had no larger an inhibiting effect than healthy sap, although it decreased the number of lesions caused by a second strain of potato virus X, which in turn was not affected by the type strain of TMV. The specificity of this effect led to the idea that once a virus particle has initiated infection, a second particle of the same virus cannot infect at the same site. Whether this is true is questionable, because when plants are inoculated with a mixture of two strains, both of which cause local lesions, both strains can be isolated from some of the lesions; the proportion of mixtures increases with the concentration of virus in the inoculum. Also, Benda (1956) found that when individual leaf hair cells were mechanically inoculated with a mixture of two strains of TMV, both strains could be found in some of the resulting lesions. However, in none of these instances is it certain that the two strains were multiplying in the same cell and not in adjacent cells.

Lesion number can also be decreased by using unrelated viruses, and the virus specificity of the effect is not as great as was at first thought. For instance, Thomson (1960) found that the type strain of TMV could decrease the number of lesions produced by turnip mosaic virus as much as the number produced by the tomato aucuba mosaic strain of TMV; but usually a related virus strain decreases lesion numbers more than unrelated viruses.

There is little evidence how virus strains compete, but ultra-violet-inactivated virus has little effect on the number of lesions produced by control virus, suggesting that some function of infective nucleic acid, not of protein, is involved. Work with the U1 and VM strains of TMV also suggests this. In *Nicotiana glutinosa*, U1 causes large local lesions and VM causes small ones; also, VM is twice as susceptible as U1 to inactivation in the leaf by ultra-violet radiation. Rappaport and Wu (1962), using mixed inocula found that VM decreased the number of lesions produced by U1, and that this effect could be partially removed by ultra-violet irradiating the inoculated leaves; the number of U1 lesions was slightly increased even when irradiation was delayed for a few hours after inoculation, perhaps because, at some sites, infection was delayed.

A mixture of two strains that both infect a plant systemically usually causes systemic symptoms intermediate in severity between those caused by each strain alone. In some instances, such intermediate effects do not persist, for one strain becomes dominant and the symptoms come to resemble those in plants infected by it alone. Indeed, it seems reasonable to expect that in any one set of conditions one strain of a pair will usually have a selective advantage over the other. By contrast, two unrelated viruses usually together cause more severe symptoms than either alone, and in some instances are synergistic. This led Holmes (1956) to suggest that virus isolates could be tested for relationship by inoculating them together and singly, and observing the severity of systemic symptoms. Although the method seems promising, it has not often been used.

(b) *Acquired resistance to super-infection* McKinney (1929), Thung (1931) and Salaman (1933) independently found that plants systemically infected with one strain of a virus did not develop additional symptoms when inoculated with another strain, and this phenomenon is the basis of the *cross-protection* test which can be used for establishing the identity of a virus isolate (§12.4.2c). The phenomenon is particularly well shown by viruses that cause diseases of the ring-spot type (Price, 1936b). Plants infected with any one of these at first show severe systemic symptoms and later produce leaves that look normal although they contain the virus (this is sometimes called acquired tolerance, or recovery; Fig. 3.1). When these normal-looking leaves are inoculated either with the virus they already contain or a closely related one, they give no reaction, even though virus-free leaves of similar age react strongly. By contrast, when inoculated with an unrelated virus they develop local and systemic symptoms (Fig. 12.7). When two distantly serologically related strains of a virus are used, the virus inoculated second may cause mild symptoms in the inoculated leaves and may become systemic; eventually it may become dominant in newly produced leaves (Matthews, 1949; Harrison, 1958).

Over the years an increasing number of exceptions to these general rules have come to light. First, there are isolates, thought on other grounds to be strains, that do not protect plants from one another. For instance, Bennett (1955) found that plants of water pimpernel (*Samolus parviflorus*) systemically infected with less virulent isolates of beet curly top virus could often later be infected by more virulent isolates. However, there is little other information on the relatedness of these isolates. Another example is provided by potato virus Y and its tobacco veinal necrosis strain, neither of which completely protects plants from the other, although they are closely serologically related (Bawden and Kassanis, 1951).

Secondly there are viruses that interfere with one another although on other criteria they seem not to be closely related. The best-known example is that of tobacco etch virus which suppresses potato virus Y when inoculated to young plants systemically infected with potato virus Y (Bawden and Kassanis, 1945). However, the significance of this observation was misunderstood for many years, until it became apparent that both are potyviruses and are distantly serologically related (Bartels, 1963–4).

There are also several examples of plants infected with one virus being less susceptible than healthy plants to infection with an unrelated virus. This is only to be expected, because a virus disease, like many other factors, presumably alters the whole physiological state of the plant and hence its susceptibility to infection. Although these exceptions show that results of cross-protection tests must be interpreted with caution, and that the virus gene or genes responsible for cross protection are distinct from those for the coat protein, nonetheless the tests can be most useful and there are no examples of complete and reciprocal protection between a pair of viruses that on other criteria would be considered unrelated.

When the challenge virus is inoculated by grafting instead of by rubbing leaves with virus preparations, the protection afforded by the first virus is more likely to break down. This may be because graft inoculation ensures a small but continuing supply of challenge virus, whereas after mechanical inoculation of sap, the challenge virus probably has only a few hours in which to become established. Alternatively it may be because different tissues are exposed to virus in the two instances; after graft inoculation, but not after leaf inoculation, the challenge virus might be introduced directly into the phloem and transported to virus-free meristematic tissues in the growing point of the plant.

Protection against related viruses occurs only in the presence of the protecting virus. Thus when shoots are cured of tobacco ringspot virus infection by keeping them at high temperature, they become fully susceptible to reinfection (Benda and Naylor, 1958). There is no good evidence that plants produce substances analogous to the antibodies or interferon produced when viruses infect vertebrates. However, Wallace (1944) found that when beet curly top virus was transmitted to healthy tobacco or tomato plants by grafting, a mild disease developed but when the leafhopper vector was used a severe one resulted. He thought this might indicate that, as well as the virus, the infected scions contained antibody-like substances that were not transmitted by the vector. Other work has given no further evidence of virus-specific antibody-like materials and shows that this phenomenon depends on the site of inoculation, because when the infective leafhoppers were confined to the older parts of the test plants instead of on the shoot tips a mild disease developed (Price, 1943).

The mechanism causing the protection in these tests is not understood. The challenge virus may grow to a limited extent but cause no additional symptoms, as happens when plants infected with a deletion variant of tobacco rattle virus are inoculated with a normal isolate (Cadman and Harrison, 1959). However, in other instances the challenge virus apparently fails to complete the early steps of replication. Now that there is increasing evidence of the specificity of replicases, it is conceivable that the challenge virus nucleic acid is irreversibly and ineffectually bound to the replicase of the virus already in the cell, which may well have a similar but not identical replicase recognition (attachment) site, whereas viruses with either identical or quite different replicases would be able to coexist in the one cell (Gibbs, 1969).

11.6.2 *Unrelated viruses*

When a plant is infected simultaneously with two different viruses, the most usual situation is that the presence of one virus decreases the concentration reached by the other (interference), or that they have little effect on one another. In a few instances, the presence of one virus may considerably increase the concentration attained by the other. Thus potato virus X reaches a greater concentration in tobacco leaves also infected with potato virus Y than in those infected by potato virus X only (Rochow and Ross, 1955). The stimulus provided by potato virus Y is proportionately greater at temperatures above the optimum (20 to 24°C) for accumulation of potato virus X and, interestingly, potato virus Y accumulates optimally at 28°C (Ford and Ross, 1962). Furthermore, Close (1964) found that five viruses that multiplied and caused symptoms at 31°C increased the concentration attained by potato virus X, whereas four that multiplied poorly at 31°C did not stimulate potato virus X. How this stimulus is provided is not understood.

Total dependence of one virus on another is seen with tobacco necrosis virus and its satellite virus (Chapter 2), a situation analogous to that of parasitism among organisms. Satellite virus multiplies only in tissues in which tobacco necrosis virus is multiplying; on its own it has no effect on plants (Kassanis and Nixon, 1961; Kassanis, 1962). The two viruses are, however, serologically unrelated, and their proteins are quite different. Although the genome of satellite virus is just large enough to code for two proteins the size of its coat protein, only the coat protein seems to be translated from the genome, both *in vivo* and *in vitro*; presumably the additional RNA has other functions. Satellite virus probably therefore depends on tobacco necrosis for its replicase and any other enzymes needed for replication. This dependence is specific and not only are other similar viruses (e.g. tomato bushy stunt virus) unable to assist satellite virus, but also different satellite virus isolates depend on different strains of tobacco necrosis virus (Uyemoto, Grogan and Wakeman, 1968; Kassanis and Phillips, 1970). Satellite virus inhibits the replication of tobacco necrosis virus (Kassanis, 1962; Jones and Reichmann, 1973), and the amount of inhibition is correlated with the concentration of satellite virus in the inoculum. Furthermore two different satellite viruses that use the same helper virus specifically interfere with one another in plants infected with all three (Kassanis and White, 1972). This interference occurs in the first few hours of infection, and probably involves competition for a specific early virus enzyme, such as a replicase.

Acquired resistance to infection is the term used for the phenomenon observed when a plant infected with a necrotic lesion forming virus is inoculated with the same or a different virus (Ross, 1961a, b). The virus inoculated second produces fewer and smaller lesions than in virus-free plants of the same age, and the resistance develops both in inoculated leaves and in non-infected tip leaves. In tobacco, systemic resistance is associated with increased peroxidase and catalase activities, and with more rapid lesion formation; it may be caused by faster accumulation of quinones at infection points, leading to the earlier development of a barrier to virus spread (Simons and Ross, 1971a, b).

11.7 Inhibitors of virus replication

Several substances with molecules small enough for them to penetrate unwounded cells either slow down or stop virus replication when they are applied to leaves infected some days previously. Most are analogues of purines or pyrimidines, but some are antibiotics and others are miscellaneous materials. All these substances damage uninfected plants, although some have little macroscopic effect at the concentrations needed to inhibit virus multiplication. Some of the base analogues, such as 2-thiouracil, are incorporated to a small extent into virus nucleic acid (thiouracil instead of uracil), and virus preparations from thiouracil-treated plants have a slightly smaller specific infectivity than those from untreated plants (Francki and Matthews, 1962a). Others, such as 5-fluorouracil, are incorporated to a greater extent without detectably affecting specific infectivity. TMV particles containing this analogue are indeed infective, for they can be shown to have an altered sensitivity to inactivation by ultra-violet radiation as compared with normal TMV particles.

Base analogues seem to inhibit by competition, for the effects of thiouracil are prevented by simultaneously treating with uracil (Commoner and Mercer, 1952) and those of azaguanine are prevented by guanine. Such inhibitors of virus replication may well act as a result of their effects on the metabolism of the host plant. Thus synthesis of tobacco leaf nucleic acid and protein, but not that of amino acids, is impaired by thiouracil; and accumulation of TMV in tobacco is inhibited by thiouracil most strongly in physiological conditions in which TMV replication is most favoured in control plants (Bawden and Kassanis, 1954). Also, the efficiency of the inhibitor depends on the identity both of the host plant and of the virus. For example, accumulation of tobacco necrosis virus is inhibited by thiouracil in tobacco but not in French bean, and thiouracil inhibits the accumulation of TMV in tobacco but has little effect on that of cucumber mosaic virus, whereas azaguanine inhibits cucumber mosaic virus more than TMV (Badami, 1959). Thiouracil has a rather different effect on the multiplication of turnip yellow mosaic virus in leaves of Chinese cabbage (Francki and Matthews, 1962b). Whereas untreated infected plants contain a third to a half as many virus protein shells as virus nucleoprotein particles, thiouracil-treated plants may contain one and a half times to twice as many (Fig. 11.12; to obtain these figures from schlieren diagrams of the kind illustrated, it is necessary to take into account the different specific refractive increments given by the same numbers of turnip yellow mosaic virus protein and nucleoprotein particles, which differ in weight (§9.3.3)). There is little effect on the total amount of virus protein (shells plus nucleoprotein) found in the leaves, and no excess of free virus RNA was found in the treated plants.

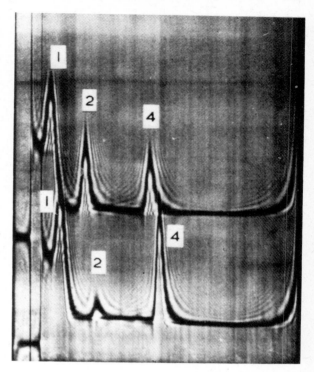

Fig. 11.12 The effect of 2-thiouracil on the production of intact and empty particles of turnip yellow mosaic virus. Schlieren diagram of a centrifugal analysis of sap from plants of *Brassica pekinensis* 22 days after infection. Upper pattern, plants sprayed daily with 2-thiouracil for the 10 days before sampling; lower pattern, plants not sprayed. Peaks in the diagram correspond to; 1. plant protein; 2. virus protein particles; 4. virus nucleoprotein particles. Note change in the ratio 2:4 caused by 2-thiouracil. (Courtesy R. E. F. Matthews; Francki and Matthews, 1962b.)

The effects of antibiotics have already been mentioned (§11.3.2). Cycloheximide inhibits protein synthesis on cytoplasmic ribosomes and decreases the accumulation of particles of all the plant viruses tested. By contrast chloramphenicol, which inhibits protein synthesis on chloroplast ribosomes, is largely ineffective, and actinomycin D, whose main action is to inhibit DNA-dependent RNA synthesis, decreases the accumulation of tymoviruses (Albouy *et al*, 1970), cauliflower mosaic virus (Tezuka, Taniguchi and Matsui, 1971) and a few others.

Finally, there are other compounds that inhibit virus multiplication, such as the fungal product trichothecin ($C_{19}H_{24}O_5$; the isocrotonyl ester of the tetonic alcohol, trichothecolone), whose effect depends on the host plant and not on the virus (Bawden and Freeman, 1952).

The behaviour of inhibitors thus presents many puzzling features, and it seems best to assume that they affect many different processes involved in virus replication, and that the relative importance of these processes and the relative extents to which they are affected will depend on the inhibitor, the host plant, the virus and the environmental conditions.

Chapter 12

Variation, strains and classification

A characteristic property of all living organisms is that when they reproduce, the progeny may differ from the parents. It is the differential selection of the new types or variants that probably leads to evolution. Viruses behave in the same way, and although each virus culture or clone that is maintained experimentally consists mostly of genetically identical virus, it usually includes some variants that have been derived spontaneously by processes that we do not know how to stop.

Jensen (1933) was the first to show that genetic variants are found when plant viruses multiply. He passed an isolate of TMV through a series of ten local lesions in an attempt to obtain a pure culture, and then inoculated the virus back to a systemically susceptible variety of tobacco, some of the tip leaves of which developed small yellow spots in addition to the usual green mosaic. From the different yellow spots Jensen obtained a range of isolates that caused different symptoms from the parent culture but which, because of the selection procedure used, were probably all progeny of one virus particle.

Apart from its importance in virus evolution, the occurrence of variants is of great practical importance to plant pathologists. For instance, variants that induce contrasting types of symptom may damage crops to different extents, and may complicate virus identification; variants that differ in host range may differ in ecology, and may make necessary more complex programmes of breeding for resistance to infection (§15.5.3); and variants with different vector specificities (§13.2.11) will differ in ecology and hence control measures used for one variant may be ineffective for another.

In this chapter we first discuss the experimental production and properties of variants, and the known basis of variation among plant viruses. We then deal with the related problems of how to detect relationships between naturally occurring strains of a virus, and how to assess affinities between, and classify, viruses and virus strains. We use the word *variant* to describe any novel virus isolate but confine the use of the word *strain* to describe naturally occurring variants.

12.1 Types of variants of plant viruses

Variants of plant viruses may be usefully placed in one or other of three categories:

(a) Those with a genome that is from a single parent, but which has been altered by some chemical or physical event. Most commonly the alteration is in one or more bases in the nucleotide sequence of the genome, and the progeny are called *mutants*. Sometimes part of the genome has been lost and the remaining part is viable though lacking certain functions; these are *deletion mutants*.

(b) Those with a genome that is derived from more than one parent, and that have a novel combination of parental genes. When the formation of the new genome involves the breakage and renewal of covalent links in nucleic acids, the process is called recombination and the progeny are called *recombinants*. This type of exchange occurs for example during meiosis in macro-organisms; it has also been found in various viruses with DNA genomes (Fenner, 1970), but there are only two claims for its occurrence in viruses with a single-stranded RNA genome (Cooper, 1969; Pringle *et al*, 1970). However, the exchange of parts of different genomes to produce variant progeny is much more common in those viruses with either single or double-stranded RNA that have genomes divided among two or more pieces of nucleic acid that are analogous to chromosomes. In mixed infections these multipartite genomes may reassort and segregate, like chromosomes, to produce new types of progeny, which we will call *pseudo-recombinants* to distinguish them from the true recombinants with which they are often confused.

(c) Those that have acquired novel properties by *complementation* when, for example, a virus unable to produce one or other of its proteins, or a particular type of protein, is assisted by another virus, or a variant of the first, in a mixedly infected host. One particular type of complementation is called 'heterologous coating' or 'genome masking' (§12.2.5), and occurs when the genome of one virus becomes assembled into particles using the coat protein of another virus and in this way is enabled to survive or to change its vector specificity (Rochow, 1970). The novel properties that progeny acquire by complementation are lost when the complemented progeny are transmitted alone to a new host, but complementation is a way in which defective variants of plant viruses can be maintained experimentally, because successful infections by plant viruses often involve the inoculation of many thousands of virus particles and thus variants and the complementing strain are transmitted together.

12.2 Production and maintenance of variants

12.2.1 *Methods of detecting and separating variants*

When White Burley tobacco leaves infected with TMV and showing a mosaic of light and dark green tissue are examined carefully, a few yellow or whitish spots can often be found (Fig. 12.1). When the virus particles from such spots are used to inoculate other plants, these usually develop symptoms of a new and different type; a variant

Fig. 12.1 Tobacco leaf systemically infected with TMV. Note the two yellow spots, from which mutants can be obtained.

has been found. The variant can then be inoculated to tobacco cv. Xanthi-nc, in which TMV produces necrotic local lesions, and each necrotic local lesion removed and inoculated individually to White Burley tobacco plants, which mostly develop the new kind of symptom. By this method the variant is separated from the parent virus culture. This experience illustrates three important points. The first is that most variants are recognized because their effect on the host plant differs from that of the parent virus; they may or may not also have different *in vitro* properties. The second point is that for a variant to be found, a population of variant virus particles is needed; the variant has to multiply, often in some kind of competition with the parent virus. No variant that is at a great selective disadvantage is likely to be detected, because at least 10^5 virus particles are required to infect a plant, and the variant must therefore multiply to this extent at least before it can be recognized. The third point is that if the variant is genetically stable it can usually be separated from the parent virus by *passage* (i.e. consecutive transmission) through a series of single lesions, which should be well separated from one another and taken from leaves that were carefully rinsed with water after inoculation to remove as much non-infecting residual inoculum as possible.

It is probable that many of the lesions caused by viruses with unipartite genomes result from infection by single particles, and by passage through a series of single lesions this chance is correspondingly increased. Thus passage through local lesions is an important step in obtaining a culture of a variant virus. With viruses for which no local-lesion host is available, the cultures can be obtained by serial passage using inoculum diluted so that less than about 10% of the plants inoculated are infected. Whichever selection method is used, the variant virus can then be bulked by transmitting it to a host in which systemic symptoms develop, or by propagating it in inoculated leaves.

If the isolated variant is a mutant of the parent it is no more immune to mutation than the parent and may produce *revertants* (i.e. variants with the properties of the parent or, as geneticists call it, the *wild-type*). Thus the relative frequency of different mutants in a population of this sort depends on the spontaneous mutation rate and on the relative selection pressures on parent and mutant. In most virus clones one particular genome type usually has greatest selective advantage and may constitute up to 99% of the clone (Siegel, 1965; Sehgal and Krause, 1968), but by altering the conditions (e.g. by changing the temperature or the host plant, or by selective passaging) other mutants can be selected and maintained if the altered selection pressure is continued. By contrast, variants that are recombinants, pseudo-recombinants or deletion mutants differ from their parents in large parts of their genomes and hence, once they are isolated, the probability that they will revert by mutation is negligible.

Most of the variants of plant viruses that have been studied differ from the wild-type strains in their effects on host plants. Obviously symptom expression is not the only genetically determined character of a virus, but to search for other types of variant is often less convenient. In genetical studies of animal and bacterial viruses (Fenner, 1970; Hayes, 1969) much use has been made of temperature-sensitive (*ts*) mutants. These will multiply at one temperature, the *permissive* temperature, but not at another, the *restrictive* temperature, which does not greatly affect multiplication of the wild type. Work of this type is also done with plant viruses. For example, Robinson (1973a) transferred each of many single-lesion isolates of tobacco rattle virus to two *Chenopodium amaranticolor* plants, one of which was kept at 30°C and the other at 20°C; each isolate that produced at least ten times as many lesions at the permissive temperature of 20°C as at the restrictive temperature of 30°C was then selected for further study. Isolates of TMV resistant to high temperature or sensitive to low temperature have also been detected in similar ways (Jokusch, 1968).

Another class of mutants, called *nonsense* mutants, have proved of great value in genetical studies of bacterial viruses (Hayes, 1969), but have not yet been obtained and used in plant virus studies. In nonsense mutants, one or more bases in essential genes are mutated so that codons which originally specified particular amino acids (§5.2 and Table 12.2) have been changed into one or other of the chain-terminating codons (i.e. UAA, UGA or UAG). Thus translation into the proteins specified by the genes is terminated prematurely and the partially formed proteins are useless. The reason why nonsense mutants of bacterial viruses can be detected and studied is that there are mutant strains of bacteria that contain nonsense *suppressor* genes. These bacteria produce unusual transfer RNAs that translate chain terminating codons as if they were coding for particular amino acids, and thereby they suppress premature termination. Nonsense mutants can be particularly valuable because they mostly have much smaller reversion rates than other types of mutants, but strains of plants with suppressor genes have not yet been found.

12.2.2 *Experimentally produced mutants*

There are claims that variants have been produced by treating virus-infected plants with ionizing or non-ionizing radiation, heat, or chemicals such as 5-fluorouracil or nitrogen mustard, but all are unproven because the treatments may have merely selectively favoured pre-existing variants in the virus population rather than produced new ones. Similarly the mutagenic action of ionizing or non-ionizing radiation on plant virus preparations is also unproved (Mundry, 1957; 1960).

By contrast, there is good evidence that mutants are produced when virus preparations are treated *in vitro* with chemicals such as nitrous acid or hydroxylamine (Mundry and Gierer, 1958; Mundry, 1960; Schuster and Wittmann, 1963). However, not all treatments that modify virus particles chemically also affect them genetically. For instance, removing the C-terminal threonine residue of TMV coat protein by treating the virus particles with carboxypeptidase does not affect infectivity or the quality of the virus progeny obtained from treated inocula.

Nitrous acid is well known as a deaminating agent which inactivates plant viruses in an exponential manner. However Mundry and Gierer (1958) found that many of the infective virus particles remaining after partial inactivation with nitrous acid produced progeny which were unlike those produced by the particles in the parent virus culture. They gave several kinds of evidence that treating either purified TMV particles or TMV-RNA with nitrous acid produces mutants. First they found that the parent virus culture gave many chlorotic lesions but very few necrotic ones when inoculated to leaves of tobacco cv. Java, and that nitrous acid treatment could increase the proportion of necrotic lesions from 0.2% to 15%. However, because of interference between the necrotic- and chlorotic-lesion forming variants, undiluted inocula containing a small proportion of particles of the necrotic variant produce fewer necrotic lesions than more dilute inocula. Hence, an increase in the proportion of necrotic lesions can be the result of not only an increase in the proportion of the necrotic-lesion variant, but also the loss of interference caused by inactivation of the chlorotic-lesion strain, and these two possibilities must be distinguished. To do this Mundry and Gierer (1958) and Mundry (1959) diluted the control untreated inoculum so that it caused the same number of lesions as the treated inoculum on a tobacco cultivar which reacts necrotically to all TMV strains. They then compared the numbers of chlorotic and necrotic lesions in Java tobacco to assess the effect of the mutagen. Table 12.1 gives the results of such an experiment. It

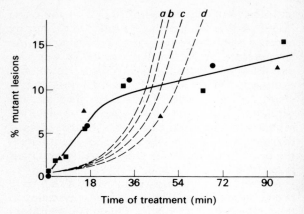

Fig. 12.2 Effect of nitrous acid on type strain TMV: mutation not selection. Abscissa, time of treatment with nitrous acid (18 min for inactivation to 37 %). Ordinate, percentage of mutant lesions among the lesions on Java tobacco. ● and ■, experiments with TMV-RNA; ▲, experiment with TMV nucleoprotein. The broken lines show the theoretical result if the proportion of necrotic lesions had increased merely because the mutants were more resistant to inactivation than type strain TMV. *a*, completely resistant, *b*, 20 times, *c*, 5 times and *d*, 3 times more resistant than type strain TMV. After 18 min treatment, the percentage of mutants was four times greater than could be explained by selection. (Data from Mundry, 1959.)

Table 12.1 Effect of nitrous acid on the number and proportion of necrotic local lesions produced by type strain TMV in Java tobacco. (Data from Mundry, 1959.)

RNA in inoculum (mg/l)	Treatment with 1 M NaNO$_2$ at pH 4.8 (min)	Number of chlorotic lesions/ leaf	Number of necrotic lesions/ leaf	Percent necrotic lesions
19	0	138	0.3	0.21
19	16	92	5.4	5.6
19	32	56	6.6	10.5
1.9	0	42	0.13	0.31

shows that the treated inoculum caused more necrotic lesions in Java tobacco than untreated inoculum, and this was particularly noticeable when the untreated inoculum was diluted so that it had a similar total infectivity to the treated one. Also, the relationship between the proportion of necrotic lesions and the time of treatment with nitrous acid (Fig. 12.2) indicates that the proportion of necrotic-lesion mutants in the treated inoculum did not increase merely because these mutants were more resistant to inactivation than the particles causing chlorotic lesions. Finally it showed that the number of necrotic lesions in Java tobacco was greatest when the inoculum was inactivated so that about 37% of the original infectivity survived (i.e. when each infective unit had received an average of one inactivating dose of nitrous acid). Thus the production of mutants, like inactivation, seems to be a first-order reaction; in other words, nitrous acid produces point mutations, and at a rate about 30 times the spontaneous rate (Hennig and Wittmann, 1972). In these experiments only a particular class of mutant was studied, but when tests were made

for several different classes, mutants were obtained from 85% of the lesions caused by inocula inactivated to 0.5% survival. The proportion of different classes of mutants was independent of the time of treatment, and also did not depend on whether intact TMV or TMV-RNA was treated.

By comparing the chemical and mutagenic effects of nitrous acid, it is estimated that about half the deaminations of bases in TMV-RNA abolish its infectivity for Xanthi-nc tobacco and that many of the other half produce mutants. The bases in free TMV-RNA are deaminated as follows:

Adenine → hypoxanthine, which behaves like guanine when transcribed.
Cytosine → uracil
Guanine → xanthine and possibly other compounds; xanthine behaves like guanine when transcribed.

Adenine and guanine are deaminated 150% as rapidly as cytosine. In intact TMV particles, however, the total rate of deamination is only 20% of that for TMV-RNA in the same conditions, and furthermore only adenine and cytosine are affected, and cytosine is deaminated almost twice as rapidly as adenine (Schuster and Schramm,

Table 12.2 The codon dictionary (the genetic code) for *Escherichia coli*

First position	Second position				Third position
	U	C	A	G	
U	Phe	Ser	Tyr	Cys	U
	Phe	Ser	Tyr	Cys	C
	Leu	Ser	CTC	CTC	A
	Leu	Ser	CTC	Trp	G
C	Leu	Pro	His	Arg	U
	Leu	Pro	His	Arg	C
	Leu	Pro	Gln	Arg	A
	Leu	Pro	Gln	Arg	G
A	Ile	Thr	Asn	Ser	U
	Ile	Thr	Asn	Ser	C
	Ile	Thr	Lys	Arg	A
	Met	Thr	Lys	Arg	G
G	Val	Ala	Asp	Gly	U
	Val	Ala	Asp	Gly	C
	Val	Ala	Glu	Gly	A
	Val	Ala	Glu	Gly	G

This is the codon dictionary for the RNAs that are translated into proteins, the messenger RNAs. It is important to remember that the information in DNA, double-stranded RNA and some single-stranded RNAs is stored in sequences that are complementary to those in messenger RNAs (§5.2).

Each amino acid is specified in the mRNA by a codon which is made up of three nucleotides, each of which may be adenine (A), cytosine (C), guanine (G) or uracil (U). The first member of this triplet is indicated in the left-hand column, the second across the top, and the third but less restricted member is shown in the right-hand column. For example aspartic acid is coded for by GAU or GAC. UAA, UAG and UGA are codons causing termination of the polypeptide chain (CTC). Ala, alanine; Arg, arginine; Asn, asparagine; Asp, aspartic acid; Cys, cysteine; Glu, glutamic acid; Gln, glutamine; Gly, glycine; His, histidine; Ile, isoleucine; Leu, leucine; Lys, lysine; Met, methionine; Phe, phenylalanine; Pro, proline; Ser, serine; Thr, threonine; Trp, tryptophan; Tyr, tyrosine; Val, valine.

1958; Schuster and Wilhelm, 1963).

The effect of deamination is to alter the coded message carried by the virus RNA. When the parts of the message that code for protein are 'read' by the ribosomes, each successive sequence of three nucleotides designates one amino-acid residue, and hence the order of these triplets determines the sequence of amino-acid residues in virus-coded proteins. These proteins include the virus-coat protein or proteins, and probably also others, some of which are likely to be enzymes involved in virus replication. The specific nucleotide triplets coding for each amino acid have been determined (Table 12.2) by studying the effects of synthetic polynucleotides on the incorporation of amino acids into protein using a cell-free system obtained from the bacterium *Escherichia coli*. Note that different triplets can code for the same amino acid; this is what is meant when the code is said to be degenerate. Other evidence suggests that the same code applies to other organisms and also to plant viruses (Gibbs and MacIntyre, 1969). With viruses, the specificity of the nucleotide code leads to two predictions; firstly that deaminations induced by nitrous acid should lead to alterations in the amino-acid composition of virus proteins, and secondly that some of these changes should be irreversible by nitrous acid. Both predictions have been confirmed using TMV mutants with altered coat proteins (§12.3.1).

12.2.3 *Experimentally produced pseudo-recombinants*

An increasing number of plant viruses are being found to have genomes divided among two or more species of RNA (see also Chapters 2 and 11). For example, brome mosaic virus has a tripartite genome (Lane and Kaesberg, 1971) and tobacco rattle (Lister, 1966) and raspberry ringspot (Harrison, Murant and Mayo, 1972) viruses have bipartite genomes. With some viruses, the pieces of the genome are packaged into obviously different particles and can easily be separated by fractionating virus preparations into classes of particles with different biophysical properties; with others, the particles are indistinguishable or all parts of the genome are packaged together, but if the RNA species are of different sizes they can be extracted from the particles and then separated by methods such as polyacrylamide gel electrophoresis. Sometimes an RNA species from one strain of a virus can be exchanged with the corresponding species from another strain to reconstitute a viable virus that we call a pseudo-recombinant and that behaves differently from either parent strain. Each RNA species carries one or more genes that form a linkage group, and thus genes in different RNA species can readily be obtained in novel combinations whereas those in the same species cannot. Pseudo-recombinants can easily be produced in the laboratory and are perhaps important in nature in increasing the range of variability stemming from a given number of mutations.

It seems that viable pseudo-recombinants are produced only when the parent isolates are fairly closely related. For example, viable pseudo-recombinants can be reconstituted from the genomes of serologically distinguishable strains of raspberry ringspot virus but not from those of the two different nepoviruses, raspberry ringspot and tobacco ringspot (Harrison, Murant and Mayo, 1972). Similarly, serologically distinguishable strains of tobacco rattle virus can be used but not the very distantly related Brazilian and Scottish strains, nor tobacco rattle virus and the allied pea early-browning virus (Frost, Harrison and Woods, 1967; Lister, 1968). With brome mosaic and cowpea chlorotic mottle viruses, which are serologically related, a viable genome can be reconstituted from the smallest RNA (RNA-3) of cowpea chlorotic mottle virus and the largest two RNA species of brome mosaic virus but not from RNA-1 and RNA-2 of the cowpea virus and RNA-3 of brome mosaic virus (Bancroft, 1972). Thus although only serologically related viruses have been found to form viable pseudo-recombinants, a serological relationship does not guarantee that the reconstituted genome will be viable. As the degree of serological relatedness decreases, so too does the probability that the reconstituted genome will be viable, and at some point along the scale only a proportion of the possible combinations of RNA species from two viruses may be viable.

12.2.4 *Defective and deletion variants*

Some isolates, of viruses which normally produce characteristic nucleoprotein particles, fail to produce nucleoprotein particles; these have come to be known as defective strains. For example, plants infected with the longer of the two particles of tobacco rattle virus contain the larger virus RNA species of this virus but no nucleoprotein particles; this is because the smaller RNA of the virus contains the gene for the virus coat protein (Sänger, 1968). Similar isolates can readily be obtained experimentally by using diluted inocula, and also from naturally infected plants (Cadman and Harrison, 1959); they seem to result from the chance separation of long and short particles during transmission. There is no evidence that these defective strains can be transmitted by vector nematodes, and they would seem a dead end in the evolutionary sense. Isolates with these properties (Table 12.3) have been called NM isolates, or defective isolates. They lack a specific part of the genome and thus might more properly be described as deficient or deletion variants

Table 12.3 Properties of defective (NM) and normal isolates of tobacco rattle virus. (Data mostly from Cadman, 1962, and Cadman and Harrison, 1959.)

Property	Normal isolates	NM isolates
Infective particles	Long and short nucleo-protein rods	RNA only
Precipitation end-point of sap tested with antiserum against particles of the virus	$<1/20$	No detectable antigen
Dilution end-point of sap	10^{-4} to 10^{-6}	$<10^{-2}$
Stability in sap 10 min at 50°C 3 hr at 20°C 3 days at —10°C	Infectivity unaffected	Infectivity greatly decreased, or abolished
Infectivity after centrifuging sap at 9000 *g* for 10 min	Mostly in supernatant fluid	Mostly in sediment
Infectivity of extracts made from sap using phenol	Much less than of sap	Much more than of sap
Symptoms produced	Often evanescent	Often persistent
Time before systemic symptoms appear	5–8 days	Usually longer
Transmissibility by *Trichodorus* nematodes	Readily transmitted	Probably not transmitted

instead of defective, because there is nothing imperfect about the part of the genome they do possess.

Deletion variants of an allied kind have been obtained by culturing wound tumor virus in vegetatively propagated sweet clover plants for up to 24 years without transmission by its insect vector. Many isolates obtained in this way have lost their vector transmissibility, and their ability to infect cultured cells of the vector. Also, whereas vector-transmitted isolates typically have a genome containing 12 segments of double-stranded RNA, the genomes of the variants have subnormal amounts of one or more segments, or have one or more segments each partly or completely replaced by a smaller piece of double-stranded RNA (Reddy and Black, 1974).

A different class of variants has been obtained by treating TMV with nitrous acid. They produce coat proteins with an abnormal tertiary structure, and which cannot assemble with TMV-RNA to produce the typical tubular nucleoprotein particles. An example is the PM2 strain of TMV which has two amino acid changes in its coat protein, isoleucine for threonine at position 28, and aspartic acid for glutamic acid at position 95 (Wittmann, 1965). However this protein can aggregate to form helical rope-like structures (Siegel, Hills and Markham, 1966).

Defective strains of tobacco necrosis virus perhaps represent a third class of variant. The best studied example (Kassanis and Welkie, 1963) occurred spontaneously and produces no nucleoprotein particles but much non-coated virus RNA. There is no evidence that any coat protein is produced by this defective strain of the virus, which has a unipartite genome.

Viroids (§2.2.4) are naturally occurring pathogens that possibly comprise an even more extreme type of defectiveness, because they are not known to contain genes coding for any polypeptide (Davies, Kaesberg and Diener, 1974). However they are also not known to have arisen from a less defective pathogen and may have some quite different origin.

The genomes of these different pathogens are all more or less unstable *in vitro*, but all except the variants of wound tumor virus can be transmitted from plant to plant by using special methods to extract the virus nucleic acid in solution while protecting it from inactivation. Tissue extracts made using phenol are, as a rule, very infective but extracts made using a buffer of about pH 9, or sodium perchlorate or bentonite clay, are also worth testing (§4.2.3c).

12.2.5 *Complementation variants; heterologously coated particles*

In some kinds of mixed infection, the nucleic acid of one virus or virus strain can become coated with protein of another virus or strain. This phenomenon is a form of complementation and is known as heterologous coating (or genome masking). It is clearly shown in conditions in which one of the viruses is defective, for example, when tobacco leaves are infected with both the TMV mutant Ni 118 and another strain of TMV, such as the legume strain (Atabekov *et al*, 1970; Kassanis and Bastow, 1971b). At 35°C, Ni 118 multiplies to produce free virus RNA, much insoluble coat protein and a few nucleoprotein virus particles, but in leaves also infected with the legume strain

a large amount of virus nucleoprotein with Ni 118 infectivity is made. This Ni 118 infectivity is associated with the variant particles and is neutralized by antiserum to the legume strain but not by antiserum to TMV type strain, which is serologically identical to Ni 118.

Heterologous coating seems not to be confined to serologically related strains, and is reported to occur in oat plants infected with two apparently serologically unrelated barley yellow dwarf virus isolates (Rochow, 1970). In this instance, the two virus isolates have different aphid vectors, but both could be transmitted by one aphid species from the mixedly infected plants or from virus preparations obtained from these plants, but not from mixtures of virus preparations from singly infected plants. Also, the ability to transmit was abolished by treating the preparations with antiserum to the normally-transmitted isolate but not with antiserum to the isolate not normally transmitted. Evidently the non-transmissible isolate had become transmissible because its nucleic acid was coated with the protein of the transmissible isolate.

12.2.6 *Spontaneous or natural variants*

All virus cultures probably contain variants that have been derived spontaneously by processes which presumably mostly involve mutation and selection.

The frequency of variants in a population can only be estimated when one knows how to identify the variants. By inoculating preparations of TMV particles from tobacco to *Nicotiana glutinosa*, and then transmitting separately from each lesion back to tobacco, Kunkel (1940) found that 0.5% to 2% of the lesions gave variants he could detect from their pathological effects. To this substantial proportion must be added the variants that were not recognized in these tests. In practice, spontaneous variants are often recognized because they multiply preferentially in response to some change of environment. Thus when TMV-infected tobacco plants are kept at 35°C, an avirulent variant, the 'masked' strain, is selected because it multiplies preferentially at the raised temperature (Kassanis, 1957a). Similarly, when TMV is inoculated to sea holly, *Eryngium aquaticum* (Johnson, 1947), a strain that systemically infects sea holly is selected and this is avirulent for tobacco. However, some other claims that viruses are attenuated by passage through a particular species may have resulted from the selection of one component of a mixture of unrelated viruses.

In nature most viruses are selected for transmissibility by vectors, and hence it is not surprising to find that if this selection is relaxed, by maintaining a virus in plants for long periods without using vectors, the vector transmissibility of the virus may be lost. This occurs with some aphid-transmitted viruses and has also happened with some laboratory strains of clover wound tumor virus (Black, 1953), which normally is transmitted by the leafhopper *Agallia constricta* (§13.2.11).

Although there is much evidence that selection pressures act on naturally occurring variants, there is no proof that particular kinds of variants are produced in response to particular conditions; it seems that the genetic change must precede selection, not the reverse.

12.3 Genetic analysis

Genetic analysis is the experimental analysis of the functions and organization of the genome. Ideally it involves firstly the biochemical analysis of the structure and function of the proteins produced by individual genes, though usually only the function or effect of these genes can be assessed. Secondly it involves studying the position, and if possible the control of expression, of different genes within the genome.

Compared with the genomes of the smallest organisms, such as mycoplasmas, those of plant viruses are very small. For example, the citrus stubborn disease organism, *Spiroplasma citri*, has a DNA genome of about 10^9 daltons whereas it seems that the genomes of all plant viruses studied to date are less than about 10^7 daltons (Chapter 2) (or double for those with a double-stranded genome). This relative simplicity should make genetic analysis easier, but this has not proved to be so, principally because most plant viruses seem to lack recombination, the prime tool of the classical geneticist. However, mutants induced by treatment with chemicals, notably nitrous acid, have provided much information on the coat protein of TMV, and the gene that produces it, and more recently pseudo-recombinants of viruses with multipartite genomes have enabled analyses to be made that are analogous to the early segregation studies of Mendel with plants.

Thus genetic analysis often involves the prior production of suitable mutants. Most of the isolates used in the first studies were coat protein or symptom variants. Host range mutants, drug resistant mutants and temperature-sensitive mutants with unaltered coat protein (Jockusch, 1968; Robinson, 1973b; Dingjan-Versteegh and van Vloten-Doting, 1974) are likely to receive more attention in the future.

12.3.1 *Coat protein mutants of TMV*

The coat protein of TMV has been used to study chemically-induced mutation in detail. According to Hennig and Wittmann (1972) about a quarter of the 300 TMV mutants examined have altered coat proteins; incidentally, this is good evidence that the coat protein gene is only one of several genes in the TMV genome. Of 57 coat protein mutants induced with nitrous acid, 30 have a single amino-acid replacement in the protein, 12 have 2 and 1 has 3 replacements. Moreover 53 of the 57 replacements can be explained by the conversion of either cytosine to uracil or adenine to guanine (§12.2.2).

These results are readily explained if one assumes that the genetic codes used by TMV and *E. coli* (Table 12.2) are mostly the same. For instance, Fig. 12.3 illustrates that the possible effects of single changes of cytosine to uracil on the nucleotide triplet (codon) for proline (CCC)

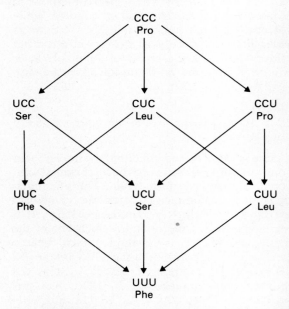

Fig. 12.3 One set of codon changes produced by nitrous acid. Note that the changes are not reversible. Symbols as in Table 12.2.

are to convert it firstly to either serine or leucine and then by a further change to phenylalanine. These changes, but not the reverse changes, have been found to be common among the nitrous acid mutants of TMV (Table 12.4). These results also support the idea that any particular nucleotide residue has a fixed position in one particular codon and that the codons do not overlap one another.

Thus most of the nitrous acid mutants have only one amino-acid replacement, and when there is more than one the replaced residues are not closer in the polypeptide than one would expect for two random events. When the observed replacements are compared with those expected from the known chemical action of nitrous acid and the sequence of amino-acid residues in TMV coat protein, many of the expected changes have been recorded (Wittmann and Wittmann-Liebold, 1966), although there are some notable absences. Perhaps these absences reflect changes that would make the coat protein non-functional, or perhaps the codon dictionary for *E. coli* (Table 12.2) is not identical with that for tobacco plants, a somewhat heretical thought.

In spontaneous mutants, by contrast with nitrous acid induced mutants, only about half the amino-acid replacements (one to three per mutant) can be explained by codon changes that could be produced by nitrous acid, but again all the changes could reflect a single nucleotide change in the codon (Table 12.4). Naturally occurring strains tend to differ more from one another than these mutants. For example the coat protein gene of the vulgare strain of TMV is calculated to have 18% of its sequence different from that of the dahle-

mense strain and 26% different from that of the U2 strain, dahlemense and U2 differ by 30%, and all three strains differ from Holmes' ribgrass strain by 53 to 56% (Hennig and Wittmann, 1972) (§12.4.5).

Alterations in coat protein may also change properties such as electrophoretic mobility, serological specificity (see §§8.2 and 12.4.5; also Fig. 8.1) and heat stability (Table 12.4; Jockusch, 1966), from those of the parental form.

In summary, the studies on the effect of nitrous acid on TMV confirm that nitrous acid can act as a mutagen in the way expected from its chemical effects, and they have produced a major advance in understanding the chemical basis for variation. They also indicate that the codons do not overlap and that the *E. coli* codon dictionary is very similar to and possibly the same as that for tobacco plants.

12.3.2 Gene mapping in TMV

Because TMV seems not to undergo recombination, other methods have been used to try to determine the relative or absolute position of particular genes. For example when TMV particles are treated with the detergent sodium dodecyl sulphate, protein is removed from that end of the particles which contains the 3′ terminus of the RNA. The reaction may be stopped after various amounts of protein have been removed (§5.6.2). The partially stripped virus is then exposed to nitrous acid which produces mutations more rapidly in free than in protein-coated RNA. Thus the position of genes can be

Table 12.4 Some amino-acid replacements in the coat protein of TMV mutants. (Data from Hennig and Wittmann, 1972.)

Mutant	How obtained	Replacement	Probable codon change	Position of replacement in coat protein[1]
Ni 109	Nitrous acid	Glu → Gly	GAA → GGA	97
Ni 118[2]	Nitrous acid	Pro → Leu	CCU → CUU	20
Ni 462	Nitrous acid	Thr → Ile	ACU → AUU	5
		Ser → Leu	UCG → UUG	55
383	Nitrous acid	Asn → Ser	AAU → AGU	25
		Thr → Ala	ACU → GCU	81
		Pro → Leu	CCU → CUU	156
A 14	Spontaneous	Ile → Thr	AUU → ACU	129
B 13	Spontaneous	Asn → Lys	AAU → AAA	33
GK 1	Spontaneous	Asn → Ser	AAU → AGU	73
430	Spontaneous	Pro → Leu	CCU → CUU	20

[1] The 158 amino-acid residues in the coat protein of TMV are numbered from the N-terminus.
[2] The particles of this mutant are relatively heat labile (Jockusch, 1966).

estimated from the amount of protein that must be removed before the mutation rate for that gene increases. Using this method, Kado and Knight (1966) found that the local lesion gene of TMV was about a quarter of the length of the RNA from the 5′ end.

12.3.3 *Pseudo-recombinants of viruses with multipartite genomes*

Brome mosaic and cowpea chlorotic mottle viruses have two of the best studied multipartite genomes. Working with a number of naturally occurring variants, and mutants produced by treatment with nitrous acid, Bancroft and Lane (1973) made pseudo-recombinants between these and the wild-type viruses or between different mutants. To make all possible combinations between the three essential RNA species of the crossed isolates is laborious, but evidence of this type indicated the locations of the determinants for several characters. The smallest essential RNA species (RNA-3) determines the coat protein composition, the type of systemic symptom in cowpea, and some characteristics of the local lesions formed in *Chenopodium hybridum*. Some other lesion characteristics are determined by RNA-2 and the ability to produce lesions in *C. hybridum* at 32°C is determined by RNA-1. Host range seems to depend on more than one gene, because pseudo-recombinants containing RNA-1 and RNA-2 of brome mosaic virus, and RNA-3 of cowpea chlorotic mottle virus, only infect plants that are hosts of both parent viruses (Bancroft, 1972).

Raspberry ringspot virus has also been used for this type of analysis but, in this instance, naturally occurring variants were used, and the virus has a bipartite, not a tripartite, genome. Some characters are determined by RNA-1 and others by RNA-2; a few are controlled by both RNA species (Table 12.5). For instance, RNA-1 and RNA-2 can act in a reinforcing or a counteracting manner in controlling the type of lesion produced in *Chenopodium quinoa*. As a consequence, reciprocal crosses may give pseudo-recombinants some of which are more virulent, and others less virulent, than both parent strains (Fig. 12.4). There is also dual control of systemic symptoms in *Petunia hybrida*; RNA-1 determines whether or not symptoms develop and RNA-2 determines the symptom type, if any. Thus a determinant in RNA-1 is epistatic to one in RNA-2 (Harrison *et al*, 1974).

Table 12.5 Location of genes in the bipartite genome of raspberry ringspot virus. (Data from Harrison *et al*, 1974.)

Character	RNA[1] species that determines the character
Serological specificity	2
Nematode transmissibility	2
Systemic yellowing in *Petunia hybrida*	2
Suppressor of *P. hybrida* symptoms	1
Infection of Lloyd George raspberry	1
Systemic infection of *Phaseolus vulgaris* cv. The Prince	1
Severity of systemic symptoms in *Chenopodium quinoa*	1
Lesion type in *C. quinoa* and *C. amaranticolor*	1+2

[1] RNA-1 has a molecular weight of 2.4×10^6; RNA-2 of 1.4×10^6.

A problem in work of this sort is to decide whether different properties determined by the same RNA species are expressions of the same or of different genes. Thus the characters listed in Table 12.5 might all be controlled by just two genes. One approach is to obtain revertants for one property and then to see whether other properties have also reverted. Another is to characterize the gene products. This is relatively straightforward for virus coat protein, but more difficult for other virus-coded proteins. Cell-free systems in which messenger RNAs are translated into polypeptides may be useful here. For instance, systems from either *E. coli* or wheat embryo translate satellite virus RNA into a single polypeptide which has the same amino-acid sequence as the satellite virus coat protein, though it is slightly smaller (Klein *et al*, 1972). Similarly the wheat embryo system translates the RNAs of brome mosaic virus into various distinctive proteins, one of which is the coat protein (Shih and Kaesberg, 1973), and it also translates TMV-RNA into various polypeptides ranging in size from 10 000 to 140 000 daltons (Roberts and Paterson, 1973).

12.4 Classification of viruses; aims, methods and uses

12.4.1 *The aims*

To the plant pathologist, the classification and grouping of viruses is of great practical importance

Fig. 12.4 Lesions produced in *Chenopodium quinoa* leaves by pseudo-recombinant, and reconstructed parent, isolates derived from raspberry ringspot virus strains E and LG. RNA constitution of the isolates was **a,** RNA-1(E)/RNA-2(E); **b,** RNA-1(E)/RNA-2(LG); **c,** RNA-1(LG)/RNA-2(LG); **d,** RNA-1(LG)/RNA-2(E). Note that the two kinds of pseudo-recombinant have produced lesions that are respectively more necrotic and more chlorotic than either parent. (Harrison *et al* 1974.)

because, if a newly found virus can be shown to be related to a previously described group of viruses, then much can reasonably be predicted about the behaviour of the new isolate by analogy with the better studied members of the group, even to the extent of predicting its vectors and ecology in the field, and hence the ways in which its spread might be limited.

Both classification and identification are parts of the process of data storage and retrieval; data must be acquired, collated, stored and organized before it can be used efficiently for other purposes, such as the identification of new isolates and the prediction of their properties. Classification starts with the investigation and description of the properties of the individuals that are to be compared and grouped; for each virus it is best to describe the properties of one or more carefully cloned isolates. The greater the number of properties of the isolates that are investigated and included in the data used as the basis of the classification, the more useful the classification is likely to be.

The properties of each described isolate are then compared with those of other isolates. There are no firm rules about how to group isolates but it seems that most virologists find two types of taxonomic category useful; isolates that differ in very few properties are grouped as strains of one virus, and viruses with many similar properties are placed in virus groups. Thus these two taxonomic categories are analogous to the species and genera of the Linnaean classifications of higher organisms, and they can be given special names that are useful in communications between people working with the viruses.

There have been many attempts to define the *virus 'species'* and the *virus group*, but most definitions offer little help to those trying to classify viruses as they do not emphasize that expediency is often the principal factor in the creation of these taxa. Thus a practical definition of an individual virus 'species' is that it is a collection of strains whose known properties are so similar that there seems little value in giving separate names. Similarly, a virus group is a collection of viruses that it is especially useful to define by their shared properties because the groupings thus formed help with such problems as identifying newly found viruses and predicting their properties. Indeed it was partly for these reasons that we briefly described the main groups of plant viruses in Chapter 2.

Complete hierarchies of higher categories have also been proposed in attempts to devise general classifications of all viruses (Lwoff and Tournier, 1966; Hansen, 1966). In most of these a limited set of intuitively selected characters is used, producing an artificial hierarchical classification that seems to have little practical value except to satisfy tidy minds; for discussion and references see Gibbs and Harrison (1968) and Gibbs (1969).

12.4.2 Properties for comparing viruses

The distinction between virus strains, 'species' and groups can also be clarified by studying the range of variants that can be obtained from single cloned isolates (§12.2), and that we are therefore sure are the variant progeny of individual parents. It is found that variant progeny usually differ from the parent in one or another of a restricted range of properties, such as the severity of symptoms they cause (Fig. 12.5), the composition of their coat protein, or the other properties listed in part C of Table 12.6. More rarely they differ in properties listed in part B, but never in those listed in part A except when they are defective or deletion variants and have lost particular functions. Thus experience suggests that virus isolates which share all their properties except perhaps a few of those in parts B and C of Table 12.6 are related strains of a single virus, whereas those that differ by several of the properties listed in parts B and C but not A are best considered members of a virus group, and those that differ by several properties in part A are members of different virus groups.

The properties by which virus strains most often differ are perhaps genetically unstable because they are properties that are noticeably altered by changes to one or a very few codons in a gene. Those properties that are more stable genetically perhaps require simultaneous changes of several particular codons to produce a viable variant gene, or are determined by several interacting genes.

Harrison *et al.* (1971) list about fifty properties that are useful for classifying viruses, and Table 12.6 gives some of them. These properties are of two types. One type can be measured in absolute

Table 12.6 Characters that differ between virus isolates

A *Unrelated viruses* usually differ from one another in one or more genetically stable properties such as the:

    Type, strandedness and number of pieces of their nucleic acid genome

    Size, shape, symmetry and gross composition of their particles

    Size and number of polypeptides in their particles

    Hydrodynamic behaviour of their particles

    Ability of their particles to react with one another's antisera

    Ability of their particles to be photoreactivated after ultra-violet irradiation

    Type of symptoms induced in host plants

    Taxa of their natural vectors

    Type of relation with their vectors

B *Viruses in one virus group* share most or all properties in A but may:

    Differ in the amino-acid composition of their coat proteins so that their particles have different electrophoretic mobilities, and are serologically only distantly related or unrelated

    Have different, but related, vector species

    Differ in their host range and symptom severity, though still causing similar types of symptoms in host plants

C *Virus strains* or the *variant progeny of one virus* usually share the majority of their properties but often:

    Differ slightly in the amino-acid composition of their coat proteins, and hence differ slightly in the serological specificity, electrophoretic mobility or stability of their particles

    Differ in vector species specificity or the ease of transmission by vectors

    Cause symptoms of different severity

terms using analytical apparatus (see Chapter 9); such properties as the gross composition and dimensions of the virus particle. Some of the properties of this type are genetically stable and are useful for defining groups of viruses. By contrast there are other properties that are assessed indirectly and comparatively, using the reaction of living organisms to compare viruses with one another. The properties compared in this way include the ability to infect a range of hosts and the symptoms that a virus causes, its vector specificity, its serological specificity and its

Fig. 12.5 Effects of strains of cucumber mosaic virus on tobacco plants. **a,** Necrotic strain; **b,** yellow mosaic strain; **c,** green mosaic strain; **d,** healthy plant.

behaviour in cross-protection tests (§11.6.1). It seems that these comparative tests are very specific and only viruses that share most properties and hence we consider to be related interact in serological and cross-protection tests. However, with the exception of serological behaviour, we have little idea of what virus properties are being compared in these tests, though the properties seem to be genetically unstable and hence are useful both to group and to distinguish between related isolates.

Although we have insufficient space to discuss all the properties by which viruses differ, some of the comparative biological tests warrant special attention. In addition, vector specificity is dealt with in §13.2.11.

(a) *Host ranges* The host range of a virus is, in general outline, a fixed and relatively stable character of that virus, although strains may vary in the severity of the symptoms they produce in a host (Fig. 12.5). Thus it is very important to determine the host range when describing or identifying new virus isolates. However there are few guidelines for people wishing to determine the host range of a new isolate, because this property of each virus seems to be quite uncorrelated with any of its other properties, and even closely allied viruses may have totally different host ranges; but although we do not know what controls the host ranges of viruses some generalizations are nonetheless possible.

Some of the host ranges of individual viruses bear a relation to the taxonomy of their hosts. For example, the known hosts of sugarcane mosaic virus, a potyvirus, are mostly non-festucoid grasses (chloridoid, panicoid and andropogonoid grasses) and few festucoids can be infected. The converse is true for barley yellow dwarf virus, and barley stripe mosaic virus shows no preference for festucoid or non-festucoid grasses (Watson and Gibbs, 1974). Furthermore, most non-festucoid but few festucoid hosts of sugarcane mosaic virus show obvious symptoms of infection, and again the converse is true for barley yellow dwarf virus.

The host ranges of different strains of a virus are usually closely similar, but should one infect more hosts than another, then frequently the host range of the second strain is contained within that of the first; the first strain merely has additional hosts and often gives more severe symptoms than the second on shared hosts (Bald and Tinsley, 1967; Tosic and Ford, 1972). By contrast related viruses may have overlapping or totally different hosts; for example, sugarcane mosaic, potato Y and bean common mosaic viruses are distantly related potyviruses yet share few, if any, hosts. However, it seems that many plant viruses have hosts confined to or concentrated in caryophylloid families (for plant taxonomic groupings see Young and Watson, 1970), such as the Caryophyllaceae, Portulaceae, Polygonaceae and Cruciferae, and tenuinucellate families, such as the Solanaceae, Compositae and Labiatae, but do not infect most crassinucellate families. It is not clear whether these patterns reflect virus properties or the whims of virologists when choosing test plants. However, widely used test plants, such as those listed in §4.2.3b or given by Bos, Hagedorn and Quantz (1960) and Noordam (1973), are the most valuable for virus identification.

(b) *Serological behaviour* In Chapter 8 we described how antisera are produced and used, and here we shall briefly discuss the serological behaviour of the particles of related viruses. It is noteworthy that, with few exceptions, only the particles of viruses that share many different properties (e.g. particle morphology, composition and behaviour, and vector type) and would thus be considered to be closely related, have been found to be serologically related. An exception is that antisera prepared against cocksfoot mild mosaic virus particles, which are isometric, are claimed to react with the particles of TMV and some tymoviruses, and antisera to TMV react with cocksfoot mild mosaic virus particles (Bercks and Querfurth, 1971b) though there are no other reasons for supposing that the viruses are related. The explanation of these unexpected reactions is uncertain but it is possible that there are some similarities in the surface conformations of the particles of these viruses, or that particles of the viruses adsorb a common host plant protein that does not react in standard control tests.

Let us consider the more usual situation in which antisera are available against two related but distinguishable viruses. Virus A is likely to react better with antiserum A than is virus B, and virus B is likely to react the better with antiserum B. The difference in some instances is shown by end-point titrations of the antisera, but where an antiserum has the same titre against two virus isolates, the two can often be differentiated by cross-absorption tests or by spur-formation tests in gel (§8.5.7; Fig. 8.6b and c; Table 8.1). Usually, viruses that are differentiated in cross-absorption tests but not in antiserum end-point titrations are considered to be strains. Related viruses that differ serologically to the extent that the homologous antiserum titre is scarcely altered by absorption with the heterologous virus have been called serotypes, though this word has another meaning when used for animal viruses. When the antigenic difference is associated with important differences in other properties, this may justify separate vernacular names. Viruses showing relationships between these extremes are also usually referred to as strains but no hard and fast rules can be laid down because the proportion of strain specific antibodies in an antiserum depends on the way the antiserum is prepared and on the individual animal that produced it (§8.4.1).

The size and composition of the immunogenic regions of virus particles are not known, but work with TMV mutants has enabled serological differences to be correlated with changes in the amino acid composition of the virus coat protein (Fig. 12.6; §8.2). This has shown that a serological difference between two virus strains may mean only that their coat proteins differ in a single amino-acid residue, whereas other strains may be serologically indistinguishable and yet differ by several amino-acid residues in serologically inactive regions of their coat protein. In general, the serological specificity of most plant virus isolates is a relatively stable property, and there are few reports of major antigenic changes in experimentally cultured viruses. In viruses of vertebrates, by contrast, serological specificity can change quickly, probably because the antibody-producing system of the host is constantly selecting antigenic variants.

(c) *Cross protection; virus-induced resistance to superinfection* This phenomenon (§11.6.1) is seen when plants that are systemically infected with one strain of a virus do not develop additional symptoms after inoculation with a second strain.

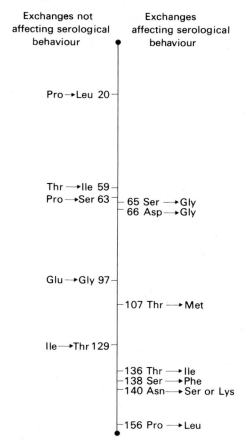

Fig. 12.6 Effects of different amino-acid replacements in the polypeptide chain on the serological behaviour of TMV. These results were obtained by testing a series of mutants of TMV, each differing from the type strain by one or two amino-acid replacements. The numbers indicate the position of the residues replaced relative to the N-terminal end of the polypeptide. Symbols as in Table 12.2. (Data from von Sengbusch, 1965.)

Fig. 12.7 Cross-protection test. Leaf of tobacco plant chronically infected with tomato black ring virus. Left half, challenge-inoculated with tobacco ringspot virus; right half, challenge-inoculated with tomato black ring virus. Only the unrelated virus has produced lesions.

Only related viruses protect against one another, and not all of these. Thus cross-protection tests are usually used to confirm that two viruses are related, but failure to cross protect does not prove that they are unrelated. There is often some correlation between closeness of serological relationship and ability to cross protect (Matthews, 1949); however this correlation is not absolute (Harrison, 1958) and the genetical or biochemical basis of cross protection is not known.

Cross protection is best shown by viruses that cause diseases of the ringspot type, because plants that have been long infected with one of these viruses usually show few or no symptoms on their youngest leaves and hence give unequivocal results when inoculated with the second, or challenge, virus in cross-protection tests (Fig. 12.7). Cross-protection tests should be done reciprocally if possible between each pair of isolates that is to be tested, because distantly related strains may protect non-reciprocally. Indeed, cross-protection tests are less generally applicable for studying relationships than serological tests, because whereas cross-protection tests are mainly effective for identifying closely related strains, serological tests can also detect more distant relationships.

12.4.3 *Classification methods*

The successes and failures of attempts to classify viruses on the basis of observed characters (i.e. *phenetically*) have repeated, on a shorter time-scale, attempts to classify higher organisms, and can be divided into three phases.

In the earliest classifications little was known of the entities being classified, and so only a few, usually obvious or superficial, characters were used to produce classifications. In the 1920s to 1940s several such classifications were devised for plant viruses (Johnson, 1927; Smith, 1937; Holmes, 1939; McKinney, 1944). Most of these were somewhat artificial classifications linked to nomenclatural systems, and were based on the few virus characters then known, namely symptoms, hosts and vectors.

In the second phase, greater knowledge of the viruses being classified allowed taxonomists to search for, and group by, 'important' characters; ones that were basic to the 'essence' or 'nature' of the viruses. The first attempts to make more natural classifications were after Bawden (1941) suggested that plant viruses should be grouped on characters such as the morphology, chemistry and serological specificity of their particles, and on cross-protection behaviour. Similar ideas were applied by Brandes and Wetter (1959), who showed that viruses with filamentous or rod shaped particles could be grouped according to the most common length of their particles, and that viruses grouped in this way shared many other properties such as the taxa of their natural vectors and the stability and serological specificity of their particles.

The third phase of classification has depended on the ideas of Adanson (1763), who classified plants. Adanson realized that much time was required to amass the data and 'experience' used by classical taxonomists to judge the 'importance' of different taxonomic characters, and also recognized that 'important' characters were, in the main, grouped and correlated characters. Therefore he devised an objective way of attaining the same result more quickly. He collected information on *all* observable characters of the organisms being classified, and made a series of separate classifications each based on a single character. He then examined these classifications to find which of them grouped the organisms in the same way; these are the correlated characters (or 'important' characters of classical taxonomists) and are the best for defining the groups of organisms.

Adanson's methods did not become popular until recently because the task of collating and handling sufficient data was tedious without the help of computers. However recent progress has been fast and many classification methods based on Adansonian principles and able to simulate the intuitive methods of classical taxonomy, are now widely used. They were first used for animal viruses by Andrewes and Sneath (1958), but with more success by Bellett (1967a, b), and have also been used for plant viruses (Gibbs, 1969; Tremaine and Argyle, 1970); in most instances they confirm and extend the groupings already produced by other methods.

Virus classification has therefore followed the path predicted by Mayr (1953) when he said 'The history of all classification . . . shows that early attempts at classification are based on superficial similarities and very often on single characters, while all improvements of classification are due to an ever more deeply penetrating analysis and broadening of the basis of classification by including more and more characters.' He also pointed out that 'the soundest classifications are those based on the greatest number of clues', the reason being that the more characters are used, the more likely is information from all parts of the genome to have been sampled, and the more *phyletic* (i.e. reflecting evolution) the classification will be.

12.4.4 *Classification using a computer*

Over the past decade there has been much discussion on the relative merits of classical and computer methods of classification. Because viruses have relatively small genomes, they provide relatively few characters that can be used for classification. Thus the human brain has sufficient capacity both to memorize and to compare these characters, and hence for most virus characters a 'classical' approach requires little experience. It is mainly in the handling of large amounts of numerical data, such as the amino-acid composition or sequence of virus proteins, or the results of comparing many viruses in serological titrations, that computer classification techniques are of great value.

One of the few real differences between classical and computer taxonomists is the way in which they represent the classifications they devise. Most of the former devise *monothetic* classifications, which have rigid and successive divisions, each of which is based on one or a few characters. Thus each taxon has a unique minimal set of characters that is both necessary and sufficient for its definition, and the characters are of graded importance. Organisms (and probably viruses) seem not to evolve in this way. Also, this type of system is apt to misclassify atypical species, the extent of misclassification depending on the assumed importance of the characters in which the species is atypical. Thus a classical taxonomist would indubitably classify together the NM variants of tobacco rattle virus and the viroids, because none of them produces nucleoprotein particles, a character that would for sure be considered of great importance. By contrast, most computer taxonomists prefer *polythetic* classifications, in which groupings at any level of the classification are defined by groups of characters, none of which is considered more important than others, and none of which must be shared by all members of a particular group. Individuals are therefore arranged in groups based on overall resemblance, a seemingly more natural arrangement even though it has the disadvantage that the groups may overlap.

A great variety of computational methods are now available to assist the taxonomist who is overwhelmed with data. The first step in computer taxonomy is in some way to compare the characters of the individuals so as to obtain a numerical estimate of their relative similarity. This is usually called a similarity coefficient, and can be visualized as a distance. It is relatively easy to devise a suitable similarity coefficient from simple numerical data; for example, the similarity of two viruses could be assessed from the base composition of their nucleic acids by calculating the difference between the viruses in the amounts of each base and then taking the mean of the differences, or by calculating a correlation coefficient from the same data. For more complex data, or mixtures of data, more complex methods are required (Lance and Williams, 1966). One particularly complex problem is to express the similarity of two sequences as a similarity coefficient. This can be done in various ways but most give results that are difficult to interpret, though two of the simplest are the nearest neighbour (Gibbs, *et al*, 1971) and diagram (Gibbs and MacIntyre, 1970b) methods. When several individuals are to be classified, similarity coefficients between all pairs of individuals are calculated.

The next step in classification is to examine the similarity coefficients by some method, usually called a cluster analysis, to determine which individuals cluster together, and to determine the relationships of these clusters. This is rarely straightforward, because usually the similarity of, for example, individual A to B when 'seen' from individual C is not the same as when 'seen' from individual D. Thus some computational technique must be used both to find the most representative position of each individual within the group being classified, and to present this information to the taxonomist in a digestible form. There are several ways of doing this, but the two most commonly used illustrate two extreme approaches to the problem. An example of the first method is the agglomerative hierarchical sorting method of Lance and Williams (1966, 1967). The similarity coefficients are examined by the computer to find the 'closest' pair of individuals, their similarity is noted, and a new individual is synthesized with properties that are an average of this pair. The similarities of this new individual to all others is then computed, the similarities are again examined to find the closest pair, and the procedure repeated until all have been combined into one individual. The results are represented as a dendrogram (Fig. 12.8) in which the horizontal scale indicates the level at which different

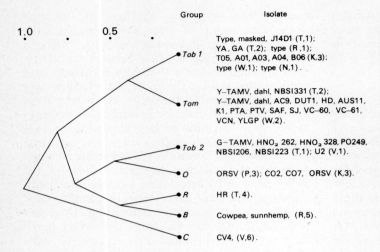

Group | Isolate

Tob 1 — Type, masked, J14D1 (T,1); YA , GA (T,2); type (R ,1); T05, A01, A03, A04, B06 (K,3); type (W,1); type (N,1) .

Tom — Y—TAMV, dahl, NBSI331 (T,2); Y—TAMV, dahl, AC9, DUT1, HD, AUS11, K1, PTA, PTV, SAF, SJ, VC—60, VC—61, VCN, YLGP (W,2).

Tob 2 — G—TAMV, HNO₂ 262, HNO₂ 328, PO249, NBSI206, NBSI223 (T,1); U2 (V,1).

O — ORSV (P,3); CO2, CO7, ORSV (K,3).

R — HR (T, 4).

B — Cowpea, sunnhemp, (R,5).

C — CV4, (V,6).

Fig. 12.8 A classification of tobamoviruses computed from the amino-acid composition of their coat proteins; agglomerative hierarchical classification, 'Gower metric' similarity coefficient/flexible sorting strategy ($\beta = -0.25$). (Lance and Williams, 1967.) Horizontal scale gives the level of dissimilarity at which the clusters fused. Clusters lettered as follows: *Tob 1* =tobacco type strain/orchid cluster; *Tob 2* = tobacco (U2) cluster; *Tom* =tomato (dahlemense) cluster; *O* =orchid cluster; *R* =ribgrass cluster; *B* =bean cluster; *C* =cucumber cluster. Data sources given in parentheses; K. Kado, van Regenmortel and Knight (1968); N. Nozu and Okada (1970); P. Paul *et al* (1965); R. Rees and Short (1965); T. Tsugita (1962); V. van Regenmortel (1967); W. Wang and Knight (1967). Source hosts of the isolates also indicated in parentheses: 1, tobacco; 2, tomato; 3, orchid; 4, plantain; 5, bean; 6, cucumber.

individuals or groups fused (i.e. their similarity). Note that the dendrogram is analogous to the pendant ornament known as a mobile; each cluster can rotate around its point of fusion. Note too that in this agglomerative sorting method the position of each individual in the dendrogram is determined by its closest neighbours and is usually little affected by distantly related individuals.

The other approach to determine the most representative position of each individual in a classification is the principal coordinates method of Gower (1966). In this, the position of each individual is computed statistically by a least squares method from all the relevant similarity coefficients, and the result usually represented in

a two- or three-dimensional pattern (Fig. 12.9); when there are clearly defined clusters of individuals in the classification, then relative positions can be represented in two or three dimensions with very little loss of information. Note that this method differs radically from the agglomerative method in that the position of each individual is averaged statistically from its apparent positions as seen from all other individuals in the classification, not just the nearest.

It is important to realize that the final classification obtained is determined both by the method used to calculate the similarity coefficients and the way in which the similarities are sorted and represented (Lance and Williams, 1967), so that

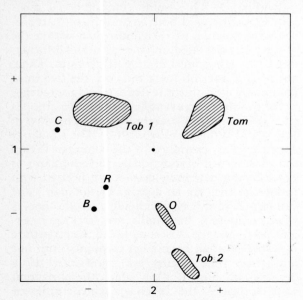

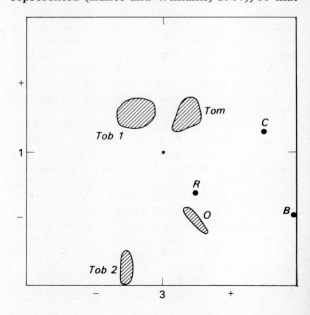

different classifications can be obtained from the same basic data. This illustrates the general point that 'computer taxonomy' is no less subjective than 'classical taxonomy'; the judgement of taxonomists is needed for both.

Additional computer programmes can determine which characters of the individuals contribute most to (i.e. are most closely correlated with) the clusterings or groupings obtained in the classifications, and thus we can identify the correlated or 'important' characters (§12.4.5; Table 12.7).

12.4.5 *Affinities of viruses within a group; the tobamoviruses*

To understand plant virus classification more clearly it is useful to see how the extents of the affinities between serologically related virus isolates can be assessed, and whether such isolates can easily be assigned to different 'species'. We will use the tobamoviruses as an example, because much is known about their biological and pathological behaviour as well as about the composition of their particles.

The main shared characteristics of the tobamoviruses were listed by Harrison *et al* (1971) as:

Straight tubular particles with sedimentation coefficient of about 190*S*, helical symmetry with pitch 2.3 nm, containing about 5% single-stranded RNA of molecular weight 2×10^6, protein subunits of molecular weight about 17 500, infective particles about 300 nm long, thermal inactivation point of 90°C, longevity *in vitro* of years, concentration in sap often over 1 g/l, symptoms mostly mosaics and mottles, spread mechanically, efficient natural vectors not known, members serologically related.

Many members of this group are known. Martyn (1968, 1971) lists about 150 named

Fig. 12.9 Diagrams illustrating a principal co-ordinate analysis (Gower, 1966) of the tobamoviruses computed from the amino-acid composition of their coat proteins using the similarity coefficients computed for Fig. 12.8. The isolates formed close clusters as in Fig. 12.8, and so only the positions of the clusters are indicated and lettered as in Fig. 12.8. The number of amino acids in the coat proteins of virus isolates in these clusters, and the contributions to this analysis of the estimates of the different amino acids, are given in Table 12.7. This classification, expressed as three co-ordinates, accounts for more than two-thirds of the differences between ('information' about) the individuals estimated from the amino-acid composition of their coat proteins.

Table 12.7 Amino acids for distinguishing between the tobamoviruses

a Number of amino acids (range) in the coat proteins of the tobamovirus groups in Figs. 12.8 and 12.9.

	Tob 1	Tom	Tob 2	O	R	B	C
Ala	13–15	11–12	17–18	11	18	12	20–21
Arg	11–12	8–10	8	9–10	11	11	8–10
Asp+ asn	17–19	17–19	22	20	17	17	17–20
Cys	1	1	1	1	1	0	0
Glu+ gln	15–16	18–21	16	15	22	16	10
Gly	5–6	6–7	4–5	7	4	4	5–9
His	0–1	0	0	0	1	1	0–1
Ile	8–9	6–8	8	8–9	8	10	6–7
Leu	12	12–14	11	14	11	15	13–18
Lys	2–3	2	1	1	2	1	3–4
Met	0–1	1	2	3	3	0	0
Phe	8	7–10	8	7	6	6	9–11
Pro	8	8–9	10	9	8	8	6–8
Ser	14–17	14–16	10	12	13	18	24
Thr	16–17	14–17	19	21	14	19	10–12
Trp	3	3	2	3	2	1	1–2
Tyr	4	4–6	6	5–7	7	8	4
Val	13–14	14–18	12	10	10	12	7–14
Total	158	158	158	158	156	159	160

b Correlation coefficients between the co-ordinates in Fig. 12.9 and the amino-acid contents of the coat proteins

Co-ordinate 1		Co-ordinate 2		Co-ordinate 3	
Amino acid	Correlation[1]	Amino acid	Correlation	Amino acid	Correlation
Pro	−0.92	Arg	−0.89	Cys	−0.77
Lys	0.83	Ile	−0.71	Leu	0.76
Ser	0.72	Glu	0.63	Ala	−0.63
Thr	−0.69	Cys	0.54	Tyr	0.62
Tyr	−0.65	His	−0.47	His	0.52
Trp	0.65			Trp	−0.48
Ala	−0.61				
Asp	−0.53				
Val	0.52				
Met	−0.51				

[1] Correlation coefficients significant (p = 0.001) at ±0.46.

isolates mostly obtained from tobacco or tomato, and mostly distinguished from one another by their pathological effects. Some of the best known are listed below.

(a) The *type* (=*vulgare* or *U1*) *strain*, which occurs commonly, principally in tobacco and other members of the Solanaceae in many parts of the world.

(b) *Tomato mosaic strains*, which commonly

cause a yellow or green mosaic of tomatoes (Bewley, 1923), though some strains cause leaf enations (Ainsworth, 1937). However the strains prevalent in tomato showing mosaic seem to change; for example, most isolates obtained in the U.K. in the 1930s produced, in tobacco plants, a systemic mosaic indistinguishable from that caused by type strain TMV, whereas those obtained in the 1960s gave necrotic local lesions (Broadbent, 1962).

(c) *Odontoglossum ringspot virus* (Jensen and Gold, 1951) and other tobamoviruses from *Cymbidium, Cattleya* and other orchid species (Paul *et al*, 1965).

(d) *Holmes' ribgrass strain* (Holmes, 1941). A distinctive strain found in *Plantago* in many parts of the world.

(e) *Bean, or cowpea, strain* (Lister and Thresh, 1955), which is perhaps the same as sannhemp mosaic virus (Capoor, 1950), and is prevalent in beans (*Phaseolus* spp.) in the tropics. These strains typically infect various leguminous species systemically.

(f) *Cucumber green mottle mosaic virus* (CGMMV = *cucumber virus 3*; Ainsworth, 1935), which is perhaps related to bottle gourd mosaic virus and other tobamoviruses of cucurbits, and is world wide in occurrence. Unlike other tobamoviruses, these isolates infect cucurbits systemically and may fail to infect solanaceous species.

(g) *Sammons' opuntia virus* (Sammons and Chessin, 1961), and other tobamoviruses of cacti in North America. This has been studied in less detail than most of the other viruses, but does not infect several solanaceous species.

Many attempts were made to correlate the pathological behaviour of these tobamoviruses with other properties. Several kinds of evidence indicated that, of the better studied isolates, CGMMV is perhaps the most distinct (for references see Price, 1964). There was some evidence that the relatedness of TMV strains was correlated with the amino-acid composition of their coat proteins (Knight, 1947), but this could not be widely tested because of the very great technical problems of analysis during the 1930s and 1940s. However the taxonomic value of the properties of TMV proteins was confirmed by biological and other studies (Siegel and Wildman, 1954) and, when automatic amino-acid analysers were introduced, by estimates of the compositions of the coat proteins of many more TMV strains and mutants (Hennig and Wittmann, 1972).

There is now much quantitative information on the composition of tobamovirus particles, from which classifications can be computed and it is interesting to compare the estimates of relatedness obtained from different types of information.

(a) *Nucleic acids* There have been many estimates of the base composition of tobamovirus RNAs. When some thirty records of the base ratio of eighteen isolates were analysed by computer, only the isolates from cucumber formed a cluster separate from the other isolates. The mean base ratio of the cucumber isolates is G 25.5, A 26.0, C 18.7, U 29.8, and of all the other isolates is G 25.2, A 29.2, C 18.9, U 26.7. There was considerable variability in the data for the type strain, probably reflecting experimental error. It seems that base ratio has considerable evolutionary stability and that more precise estimates, such as those on tomato isolates by Mosch, Huttinga and Rast (1973) will be needed to distinguish between the majority of tobamoviruses.

Estimates of the proportions of different oligonucleotides produced by treating RNAs with specific nucleases probably would distinguish between the non-cucurbit tobamoviruses but too few estimates of oligonucleotides have been made to judge this (Rushizky and Knight, 1960; Rushizky, Sober and Knight, 1962). Similarly, tests of hybridization between single- and double-stranded RNAs (§§5.2 and 11.3.1a) are more specific than base ratio estimates, but only a few comparisons of tobamoviruses have been made by this method (A. Siegel, personal communication), and show, for example, that type strain double-stranded RNA hybridizes fully with RNA from the particles of the type, J14D1 and YA isolates, but not with that from the particles of the U2, dahlemense, ribgrass and cucumber strains.

(b) *Proteins* Much more is known about tobamovirus coat proteins than about their nucleic acids. Analytical methods for proteins are simpler and, for the same effort, are more discriminatory than those for nucleic acids, mainly because proteins are constructed of twenty different kinds of unit instead of just four.

Tsugita (1962) was first to propose a classification of tobamovirus variants and strains based on the composition of their coat proteins. By the classical taxonomic approach he defined four clusters monothetically on such characters as the methionine and histidine content, and N-terminal amino acids of the proteins. This classification correlates with pathological data in that, firstly, only isolates of Tsugita's group A (cluster *Tob 1*, Fig. 12.8) infect *N. sylvestris* systemically, whereas those of his other three groups give local lesions. Secondly, 13 isolates of TMV from tomato from different parts of the world all belong to Tsugita's group B (cluster *Tom*), even though isolates in both cluster *Tob 1* and cluster *Tom* can infect both tobacco and tomato (Wang and Knight, 1967); recent data on a further eighteen tomato isolates (Mosch, Huttinga and Rast, 1973) show that all but one also belong to the *Tom* cluster. Thirdly, tobamoviruses from orchids belong only to either cluster *Tob 1* or to one of two additional clusters defined on the same criteria (van Regenmortel, 1967).

Figure 12.8 is a dendrogram of a classification computed from the composition of tobamovirus coat proteins. Figure 12.9 is a principal co-ordinate classification of the same data indicating the relative positions in three 'dimensions' of the major groups listed in Fig. 12.8. Table 12.7 lists the correlation coefficients between the co-ordinates (dimensions) in Fig. 12.9 and the amino-acid contents of the virus proteins in that classification, and thus indicates which polythetic characters were 'important' in the classification illustrated in Fig. 12.8 and 12.9.

It can be seen that the computed classifications agree closely with Tsugita's and van Regenmortel's proposals, and give an additional cluster containing the bean strain of TMV. Both classifications indicate that the clusters are clearly distinct from one another, and agree with the data on RNA composition in finding that the cucumber strain is the most distinct of all the tobamoviruses analysed.

The analysis of the principal co-ordinates classification (Table 12.7) shows that the amino acids selected by Tsugita to group the viruses seem of lesser importance in defining the computed clusters than other amino acids such as, in order of decreasing 'importance', proline, arginine, lysine, cysteine, serine, leucine, isoleucine and others. Furthermore this analysis illustrates the value of polythetically defined classifications; Tsugita's monothetic classification would, for example, have mis-classified some unusual histidine-containing mutants of the type strain (Rombauts and Fraenkel-Conrat, 1968), whereas the criteria in Table 12.7 would place them in cluster *Tob 1* with their parent.

Recently the sequences of amino acids in an increasing number of proteins have been determined, and this has included, so far, the coat proteins of individual isolates from six of the seven tobamovirus clusters shown in Fig. 12.8. The sequence of these six proteins are all clearly related, though that of the cucumber strain is again least like the others. Figure 5.9 shows the close similarity of the type and dahlemense sequences and, by contrast, the greater dissimilarity of the type and ribgrass isolates. Figure 12.10 represents a classification computed from

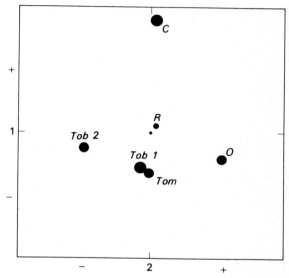

Fig. 12.10 A principal co-ordinate analysis (Gower, 1966) of the similarities of the amino-acid sequences of the coat proteins of six tobamoviruses. 'Runs indices' of similarity were computed from 'diagrams' like those of Gibbs and MacIntyre (1970b), and used for the analysis. Each virus is represented by a letter indicating its cluster in Figs. 12.8 and 12.9. Data from Kurachi *et al* (1972) for a cucurbit strain (*C*), and Dayhoff (1972) for the type (*Tob 1*), dahlemense (*Tom*), U2 (*Tob 2*), ribgrass (*R*) and odontoglossum ringspot (*O*) strains. In these comparisons no distinction was made between aspartic acid and its amine or between glutamic acid and its amine. The position of the individuals in the third co-ordinate is indicated approximately by the size of the dot and the lettering; the three co-ordinates together account for three-quarters of the differences between ('information' about) the individuals obtained from these sequence comparisons.

estimates of the similarities of the known sequences, and this agrees closely with the classifications based on the compositions of coat proteins.

Serological tests have also been used to compare tobamoviruses, but few of these have compared several isolates and antisera at the same time. Most complete are those of van Regen-mortel (1967; 1975), Kado, van Regenmortel and Knight (1968) and Hollings (1974). The general conclusion from these tests is that the isolates belong to the same clusters as are shown in Fig. 12.8, with the *Tom* cluster being closest to the *Tob 1* cluster, and all the other clusters, plus Sammons' opuntia virus, being distinct from one another. Thus analyses of the data on amino-acid composition and sequence, and serological relatedness, all indicate similar patterns of affinity.

(c) *Cross-protection tests*   Most cross-protection tests with tobamoviruses have used the type strain as one of the pair, and hence no general picture of cross protection within this group can be drawn. Cross protection occurs, for example, between different *Tob 1* isolates, and between *Tob 1* and *Tom* isolates, but *C* isolates do not protect unequivocally against *Tob 1* isolates (Fulton, 1950; Rochow, 1956).

In general, therefore, the groupings produced within the tobamoviruses, by analysis of data on their coat proteins, correlate quite well with the natural hosts of the isolates; and the data on base ratio and cross-protection behaviour do not contradict these groupings. The clusters that are most distantly related to the *Tob 1* isolates contain *C* isolates and Sammons' opuntia virus, respectively. The bean, ribgrass, orchid and *Tob 2* isolates also each form distinctive clusters; whether they all merit the rank of 'species' is a matter of opinion but they must be serious contenders. By contrast, *Tom* isolates seem closer to *Tob 1* isolates and could be considered as strains, although it might be worthwhile giving them different names, such as tomato mosaic and tobacco mosaic viruses, because isolates in the two clusters cause economic damage in tomato and tobacco crops, respectively.

Additional clusters of tobamoviruses will undoubtedly be found; for example, the isolate from frangipani (*Plumeria acutifolia*) (Francki, Zaitlin and Grivell, 1971) has particles with an amino-acid composition quite different from those of the other tobamoviruses we have discussed.

The relationship of the tobamoviruses to other viruses is largely a matter for speculation, although Kassanis, Woods and White (1972) noted the similarities between the particles of potato mop-top virus and those of certain defective isolates derived from type TMV, and reported finding a distant serological relationship between the viruses. This possible relationship is of

interest because TMV is not known to have an efficient natural vector, whereas potato mop-top virus is readily transmitted by a plasmodio-phoromycete fungus (§13.2.10).

No other viruses are as well studied as the tobamoviruses and it is uncertain whether similar clusterings of isolates will be found in other virus groups. In some groups, virus species seem easier to differentiate, in others more difficult. For example, in the potyviruses, 'species' seem particularly hard to delineate, both among the isolates that infect legumes, and among those that occur in solanaceous plants; there are many intergrading types among these viruses.

12.4.6   *Delineation of virus groups*

We have explained the reasons for considering many characters of viruses when assessing their affinities. This has been attempted by simple comparisons (Harrison *et al*, 1971) and by analysis using a computer (Gibbs, 1969). The two methods give similar results, namely, the groups of viruses briefly described in Chapter 2.

The kinds of problem encountered in attempting to discern virus groups are illustrated by the comparisons made in Table 12.8. Most of the viruses that have been assigned to one group, the nepoviruses, share many properties and differ in few, whereas cowpea mosaic and turnip yellow mosaic viruses differ in several properties, both from one another and from the nepoviruses. However, there are two viruses, cherry leaf roll and strawberry latent ringspot, that seem less closely related to typical nepoviruses. Should these two be included in the nepovirus group or not? In practice it is probably best to consider them as more peripheral members of the nepovirus group, but the conclusion to be drawn is that all viruses are unlikely to fall neatly into distinct watertight compartments at this or any other level in classifications of the kind that are most useful to virologists.

12.4.7   *Uses of classification; virus identification*

In §12.4.1 and Table 12.6 we showed that the properties of viruses differ in the extent to which they are shared by allied viruses, and that some properties, such as particle morphology and other particle characters, were best for forming virus groups whereas others, such as pathological effects and cross-protection behaviour, were most useful for differentiating between closely related

Table 12.8 Comparison of characters of nepoviruses, cowpea mosaic virus and turnip yellow mosaic virus

Character	Nepoviruses[1]							Other viruses[1]	
	AMV	CLRV	RRV	SLRV	TobRSV	TomRSV	TBRV	CpMV	TYMV
Wide host range	+[2]	+	+	+	+	+	+	0	0
Ringspot symptoms	+	+	+	0	+	+	+	0	0
Seed transmission	+	+	+	+	+	+	+	0	0
Concentration in sap 10–50 mg/l	+	+	+	+	+	+	+	+	0
Thermal inactivation point 55–70°C	+	+	+	+	+	+	+	0	0
Particles about 28 nm diameter	+	+	+	+	+	+	+	+	+
Particle outline angular	+	+	+	+	+	+	+	+	0
RNA genome in two pieces	(+)	+	+	(+)	+	(+)	+	+	0
RNA mol. wt about 1.4 and 2.4×10^6	+	0	+	+	+	0	+	+	0
Coat protein mol. wt 5 to 6×10^4	+	+	+	0	+	–	+	0	0
Nematode vector	+	(+)	+	+	+	+	+	0	0
Persistence in vector > 1 wk	+	–	+	+	+	+	+	+	+

[1] Abbreviations of virus names: AMV = arabis mosaic, CLRV = cherry leaf roll, CpMV = cowpea mosaic, RRV = raspberry ringspot, SLRV = strawberry latent ringspot, TobRSV = tobacco ringspot, TomRSV = tomato ringspot, TBRV = tomato black ring, TYMV = turnip yellow mosaic.
[2] Symbols: + = character possessed, 0 = character not possessed, – = no information, ( ) = information tentative.

viruses. In the early days of plant virology only the latter type of test was available, and hence virus identification consisted of comparing these 'differentiating characters' of the new isolate with those of all previously described viruses or, in other words, it was only possible to ask the question 'Is this new virus the same as, or very closely related to, any previously described virus?' Nowadays, however, with a better understanding of virus properties and virus classification, those fortunate enough to have the use of an electron microscope and, perhaps, an analytical centrifuge, can identify a virus more logically and quickly by first assessing the 'grouping characters' of the virus so as to answer the question 'To what group does this virus belong?' and then determining the 'differentiating characters' that are best suited to distinguishing between members of that group, and that can be checked against descriptions of particular viruses, such as are given in the C.M.I./A.A.B. Descriptions of Plant Viruses or Smith (1972).

The types of test used for virus identification (Ross, 1964; Noordam, 1973) depend on the facilities available to the pathologist, and as these can vary widely no identification keys have been published, and a more flexible multi-entry point system such as punched cards or polyclaves would be required, though none has yet been produced.

Once the 'grouping characteristics' of a new

isolate have been determined, these may be compared with descriptions of groups (Brandes and Wetter, 1959; Gibbs, 1969; Harrison et al, 1971), and an intelligent guess made at the identity of the new isolate. Then host range, serological and cross-protection tests are made to check the correctness of the guess.

Serological testing is the most reliable of these specific methods, providing that suitable preparations of the unknown virus can be made. Moreover, it is suited to revealing very different degrees of affinity; often it can differentiate closely related strains, and yet it can also detect more distant relationships than the cross-protection test or host-range comparisons. Unidentified viruses with isometric particles are most conveniently tested by the double-diffusion method in agar or agarose gel, using a range of antisera to known viruses, whereas the tube precipitation test is the best for viruses with elongated particles. Both kinds of test can show either close or distant antigenic relatedness (Chapter 8), but to give the most complete picture reciprocal tests should be made, using antisera to the unidentified and the known viruses, in reactions with each of the antigens.

Cross-protection tests are somewhat less reliable, but give very satisfactory results with some viruses, notably those that cause ringspot symptoms followed by acquired tolerance. Tests made by inoculation of sap are preferable to those by graft inoculation. Isolates of known identity are of course needed for challenge inocula; these isolates should produce obvious local lesions, and with viruses that characteristically cause indistinct local lesions, such as cucumber mosaic virus in tobacco, special necrotic lesion-causing variants have been selected. In cross-protection tests, qualitative answers are more useful than quantitative; the number of lesions produced is less important than whether lesions are produced and what they look like when compared with lesions produced by the challenge virus in virus-free plants of the same age. The tests should if possible be reciprocal. Also, they are better done in leaves systemically infected with the protecting virus than in inoculated leaves as these are more likely to contain pockets of tissue not completely invaded

by the protecting virus.

There are many viruses whose particle shape and size is not known, to which antisera have not been made, and which have not been transmitted by inoculation of sap. In such instances, isolates are tentatively grouped together as strains, usually only on the basis of their effects on natural hosts and graft- or vector-inoculated indicator plants, their vectors and their mode of transmission by vectors. This has led to much confusion. Closely related strains have been considered distinct viruses and distinct viruses have erroneously been considered to be the same. Until more critical tests can be applied to such viruses, this situation is of course inevitable, but the risks in identifying viruses solely from the symptoms they induce should never be forgotten.

Chapter 13

Transmission by vectors and in other natural ways

As previously mentioned, viruses apparently cannot enter intact plants but gain access through wounds, which are sometimes produced naturally, such as when the foliage of one plant brushes against another, but are usually made by other organisms that may carry the viruses. When a virus is assisted in this way the virus-carrying organism is called a vector. Although all plant viruses may have had a vector at some time in their history, not all seem to have vectors today. Some are propagated by man in vegetatively reproduced plants; in tubers, bulbs, runners, graft-wood, etc. We shall not describe these man-dependent systems here, but there are two other kinds of natural spread without vectors to discuss before dealing with transmission by vectors.

13.1 Transmission without vectors

13.1.1 *Transmission by contact*

When the leaves of a virus-infected plant brush against those of a virus-free plant, the edges of the leaves and the leaf hairs are wounded, and the virus in the sap exuded by the infected plant may enter a wound in the virus-free one and infect it. This chance is greatest with viruses occurring in high concentrations in sap, and with plants that are readily wounded and are highly susceptible to infection. Potato virus X spreads in potato crops in this way, and the virus strains that reach the highest concentrations in infected plants are those that spread the most rapidly. Roberts (1946) found that potato virus X also spread between plants whose only contact was below ground, most probably when the healthy roots rubbed against infected ones. In woody species, the roots of adjacent plants occasionally become grafted together, and apple mosaic virus is reported to spread in apple through such unions (Hunter, Chamberlain and Atkinson, 1958). However, perhaps the most remarkable feature of virus

transmission by contact is that it is apparently so rare.

13.1.2 *Transmission through seed and pollen*

Most viruses seem not to be transmitted through true seed or pollen but at least thirty, including the potato spindle tuber viroid (Hunter, Darling and Beale, 1969), are transmitted in this way and there may be several others. With most virus/host combinations (e.g. lettuce mosaic virus in lettuce) the proportion of seedlings infected through the seed is usually small, but in some, for instance tobacco ringspot virus in soybean, most of the seed may be infected (Desjardins, Latterall and Mitchell, 1954).

Many factors affect seed transmission; for example at high or very low temperatures a smaller proportion of seed is infected than at intermediate temperatures. Among host plants, species in the Leguminosae seem prone to infection through the seed and, among viruses, the nepoviruses typically are seed-borne. The amount of transmission through seed also depends on the host genotype (Smith and Hewitt, 1938). For seed transmission to occur, plants must in general be infected before the ovules are fertilized, and thus the amount of transmission further depends on the time of infection in relation to the time of flowering and on the position of the flowers on the plant; for instance, early infection gives a high percentage of transmission of tobacco ringspot virus in soybean (Crowley, 1959), whereas barley plants inoculated 10 days before heading give a larger proportion of seed infected with barley stripe mosaic virus than plants inoculated earlier or later.

Although virus transmission from pollen to seed, and hence to seedlings, has been less studied than transmission from the mother plant to the seed, the two seem to occur with the same combinations of virus and host. Usually a similar or somewhat larger proportion of seed transmission occurs when flowers on infected plants are polli-nated with pollen from healthy plants than when healthy flowers are pollinated with pollen from infected plants. However when mixtures of pollen from infected and healthy raspberry plants are used, the infected pollen seems inefficient and most of the progeny are healthy (Lister and Murant, 1967). Most pollen-transmitted viruses do not invade the plant bearing the flowers being pollinated, but George and Davidson (1963) reported that prunus necrotic ringspot virus spreads in cherry orchards in this way, as also does raspberry bushy dwarf virus in raspberry crops (Cadman, 1965).

The viruses discussed above typically invade the seed embryo. By contrast, TMV is mostly found in and on the testa of tomato seed, where it may be inactivated by treating the seed with solu-tions of trisodium phosphate or hydrochloric acid. A small amount of seed-borne TMV is not inactivated by such treatments, and is apparently in the endosperm of the seeds, where it inactivates only slowly and may persist in stored seeds for several years. Broadbent (1965) found that con-taminated tomato seed rarely gave infected seedlings except when these were transplanted, and concluded that during transplanting the virus entered wounds made when the roots and hypo-cotyls were abraded by virus-containing testas or endosperms. Most other seed-borne viruses are carried in the embryo of the seed and may survive for several years in it. Indeed, the stability of viruses in seed is illustrated by the fact that treating bean seed at temperatures high enough to inactivate bean common mosaic virus in sap did not inactivate it in seed.

Not all viruses cause symptoms in the plants they infect through the seed; those, such as the nepoviruses, that cause severe shock symptoms in newly infected plants, which then recover and show few or no symptoms, also cause few or no symptoms in seedlings infected through the seed. In this respect the seedlings behave like cuttings. Thus it is important, when checking for seed transmission, to test seedlings by inoculating their sap to suitable indicator plants.

There is no generally agreed reason for the failure of most plant viruses to be transmitted through seed. Viruses restricted to the phloem presumably cannot enter seed because the phloem of the mother plant and embryo are not con-nected directly. With other viruses, the most plausible explanation is that they cannot replicate and/or are degraded in meristematic tissue such

as is found in developing seeds; indeed seedlings grown from both true seed and nucellar seed of infected citrus are virus-free (Weathers and Calavan, 1959). Some viruses that are not seed-transmitted have never been detected in embryos but others are not difficult to detect. For example, southern bean mosaic virus is found in the embryos of immature seeds of *Phaseolus vulgaris*, and can come from either the pollen- or ovule-parent, but is inactivated when the seed matures and is not transmitted (Cheo, 1955).

Some seed-borne viruses are readily detected by inoculating extracts of embryos or pollen to indicator plants. The characteristic tubular par-ticles of barley stripe mosaic virus can also be found in extracts of individual seeds of barley by electron microscopy or by micro-serological tests in agar gel (Hamilton, 1965), though such high virus concentrations in seed are probably excep-tional.

13.2 Transmission by vectors

13.2.1 *Range of organisms involved*

Although man may spread TMV by handling tomato plants, and rabbits or dogs may spread potato virus X in potato crops when their coats, contaminated with virus-containing sap, brush against the foliage of healthy plants, these examples are really analogous to transmission by contact as described above. The same can be said of the spread of viruses or viroids, such as that causing citrus exocortis (Garnsey and Jones, 1967), on cutting, grafting or pruning knives, and in other similar ways.

The first vectors in the strict sense that were recognized as being associated with a plant virus disease were insects, the leafhoppers *Inazuma dorsalis* and *Nephotettix cincticeps*, which were reported in Japan by Takata in 1895 and Takami in 1901 to cause rice dwarf disease and later shown to carry the disease-causing virus. No other type of insect was implicated in virus spread until Doolittle (1916) showed that the aphid *Aphis gossypii* could transmit cucumber mosaic virus and Oortwijn Botjes (1920) found the aphid *Myzus persicae* could transmit potato leaf roll virus. About 400 species of insects are now known to transmit over 200 different viruses, and of all these vectors *M. persicae*, which carries over 60 viruses, is probably the most important.

However, not all vectors are insects and some viruses are transmitted by leaf- and bud-feeding

Table 13.1 The range of known vectors of plant viruses and virus-like agents

Taxonomic group of vector	Approximate no vector species	Approximate no. viruses transmitted[1]	Types of vector relation[2]
Insecta			
Hemiptera			
Aphididae (aphids)	200	160	N, S, P
Psyllidae (psyllids)	1	1	—
Coccidae (mealy bugs)	15	2	S
Aleyrodidae (white flies)	3	25	(N), S, P
Auchenorrhyncha (leaf- and planthoppers)	60	35	S, P
Gymnocerata (tingid bugs)	2	2	P
Coleoptera (beetles)	30	20	S, P
Thysanoptera (thrips)	6	3	P
Orthoptera[3]	10	6	—
Dermaptera[3]	1	1	—
Lepidoptera (moths)[3]	4	5	—
Diptera (flies)[3]	2	3	—
Acarina			
Eriophyidae (leaf and bud mites)	8	10	(S), P
Tetranychoidea (spider mites)	2	3	—
Nematoda (nematodes)			
Dorylaimida	20	14	P
Gastropoda[3]	6	1	—
Fungi			
Chytridiales	3	6	N, P
Plasmodiophorales	3	3	P

[1] No vectors are known for more than 250 viruses and virus-like agents.
[2] N = viruses persist a few hours; S = persist days; P = persist weeks in vector.
[3] Vectors in these taxa are thought to have little biological importance.

eriophyid mites; for example, *Phytoptus ribis* transmits the currant reversion agent (Amos *et al*, 1927). More recently, vectors of viruses have been found among soil-inhabiting organisms; a dorylaimid nematode by Hewitt *et al* (1958), and a chytrid fungus by Teakle (1960). It would be surprising if there are not also other types of vector to be found.

What do all these types of vectors (Table 13.1) have in common? They all penetrate unwounded plant cells, usually when feeding on the plants, and thus have the opportunity to acquire virus from an infected plant and inoculate any healthy plants they may feed on subsequently. They also retain the virus in an infective form between acquiring and inoculating it. The very different ways in which these processes occur with different vectors and viruses make a fascinating series of biological adaptations.

In general a virus that has vectors in one major taxonomic group is not also transmitted by organisms in another group, although there are reports of tobacco ringspot virus being transmitted by nematodes (Fulton, 1962), thrips (Messieha, 1969) and spider mites (Thomas, 1969). We shall return later to this question of vector specificity, but first we shall describe the different virus-vector systems and their different transmission characteristics (Table 13.2).

13.2.2 *Terminology used in virus-vector studies*

Many terms are used by experimenters to define various treatments and phenomena encountered when describing and evaluating the relations between viruses and their vectors, particularly insect vectors. Some of the more useful terms are the following:

Acquisition access period The time for which an initially virus-free vector is allowed access to a virus source and could, if it desired, feed on that source.

Acquisition feeding period The time for which an initially virus-free vector actually feeds on a virus source. This term and the *inoculation feeding period* are sometimes used to refer to periods when a vector with stylet mouthparts probes a plant, although it is not possible to be sure that food is being imbibed during all this time.

Table 13.2 Characteristics of transmission of some viruses by their vectors

Vector	Virus	Time needed to acquire virus (minimum)	Latent period (minimum)	Time needed to inoculate virus (minimum)	Persistence in feeding vector (maximum)	Multiplication in vector
Myzus persicae (Aphididae)	Potato Y	10 sec	None	15 sec	2 h	No
M. persicae	Beet yellows	5 min	None	5 min	3 days	No
Hyperomyzus lactucae (Aphididae)	Sowthistle yellow vein	2 h	8 days	Probably <1 h	Weeks	Yes
Planococcoides njalensis (Coccidae)	Cacao swollen shoot	1 h	—[1]	15 min	4 days	Unlikely
Bemisia tabaci (Aleyrodidae)	Tomato yellow leaf-curl	30 min	21 h	30 min	20 days	—
B. tabaci	Cucumber vein yellowing	20 min	None	10 min	6 h	No
Nephotettix impicticeps (Auchenorrhyncha)	Rice tungro	30 min	None	15 min	6 days	Unlikely
Circulifer tenellus (Auchenorrhyncha)	Beet curly top	1 min	4 h	1 min	Weeks	Unlikely
Agallia constricta (Auchenorrhyncha)	Wound tumor	Probably <1 h	12 days	Probably <1 h	Weeks	Yes
Acalymma trivittata (Coleoptera)	Squash mosaic	5 min	<10 h	Probably a few min	Weeks	—
Thrips tabaci (Thysanoptera)	Tomato spotted wilt	30 min	(5 days)[2]	5 min	Weeks	Probably
Aceria tulipae (Eriophyidae)	Wheat streak mosaic	15 min	—	15 min	9 days	—
Xiphinema index (Dorylaimidae)	Grapevine fanleaf	15 min	—	15 min	Weeks	No
Olpidium brassicae (Chytridiales)	Tobacco necrosis	2 min	None	2 h	—	Unlikely
Polymyxa graminis (Plasmodiophorales)	Wheat mosaic	Probably a few days	—	4 h	Probably several days	Possibly

[1] — = not known.

[2] Figure for Frankliniella schultzei.

Inoculation (or test) access period The time for which a virus-carrying vector is allowed access to a virus-free plant, and could feed on it.

Inoculation (or test) feeding period The time for which a virus-carrying vector appears to be feeding on a virus-free plant.

Latent (or preinfective) period The time from the start of an acquisition feeding period until the vector can infect healthy plants with the virus. The minimum latent period is more often quoted than the latent period for half the vector individuals (LP_{50}).

Transmission threshold period The minimum total time that a vector needs to acquire a virus and inoculate it to a virus-free plant.

Persistence The time for which a vector remains infective after leaving a virus source. Persistence is divided into three main, but intergrading, categories: (1) *Non-persistent*, persisting for a few (usually less than four) hours at about $20°C$; (2) *Semi-persistent*, persisting for 10 to 100 hours; (3) *Persistent*, persisting for more than 100 hours, in some instances for the life of the vector.

Recently there has been an attempt to replace these somewhat arbitrary categories of persistence by categories based on the behaviour of the virus in the vector during transmission. These are:

(a) *Stylet-borne viruses;* viruses carried on the stylets and thought to include many or all non-persistent viruses.

(b) *Circulative viruses;* viruses which pass through the gut wall into the haemolymph of the vector, eventually reaching the mouthparts in the saliva, but not multiplying in the vector.

(c) *Propagative viruses;* viruses that multiply in the vector.

The main drawback to using these terms is that the mechanisms of transmission they suggest are difficult to prove and have in fact been established for very few of the viruses whose relative persistence in vectors is known.

13.2.3   *Transmission by aphids*

Aphids (Fig. 13.1) have mouthparts apparently well suited to inoculating viruses, because the stylets are so slender that they often cause little damage to the plant tissue punctured by them. The four long narrow stylets are held together by longitudinal ridges and grooves on their adjacent surfaces and consist of two outer mandibles and two inner maxillae (Fig. 13.2). The maxillae have additional grooves on their appressed surfaces which form two canals (Fig. 13.2), the anterior one for food and the other for saliva. When not in use the stylets are withdrawn into the protective central lower lip, or labium. In the head of the aphid are food and salivary pumps, which help to regulate the flow of food material and saliva. The saliva is ejected ahead of the stylets, especially during the longer probes, and coagulates to form a sheath around them; however there is also some evidence of a second more watery type of saliva which does not gel, and can be ejected and then imbibed (Miles, 1959). Work with isotope-labelled plants indicates that when *Myzus persicae* starts to feed on a suitable host plant about 10 μl of plant sap is taken up in the first hour, after which the rate of uptake increases to about 40 μl/h (Watson and Nixon, 1953), the change presumably occurring when the stylets of the aphids penetrate the phloem (Fig. 13.3), the tissue they prefer. During short probes, less sap is apparently taken up in 7 to 9 minute- than in 5 minute-access periods. This implies that some material is regurgitated in the longer period and, when the appropriate tests were made, nearly half the aphids were found to regurgitate in a 3 minute-access period more than half the sap they had acquired in the previous 5 minutes (Garrett, 1973).

Viruses transmitted by aphids can usefully be separated into groups differing in the length of time the viruses persist in the aphids (Fig. 13.4). *Non-persistent* viruses, which include potyviruses, carlaviruses, cucumoviruses and alfalfa mosaic virus, can usually be transmitted by manual inoculation of sap and many of them cause mosaic symptoms. Specificity between the virus and species of vector is not well developed, and usually one virus is transmitted by several or many aphid species. The transmission threshold period can be as short as 2 minutes, and the viruses can be efficiently acquired and inoculated during probes of only 10 to 30 seconds (Table 13.2), indicating that the virus is acquired from and inoculated to cells of the epidermis. Some workers maintain that the aphids acquire and inoculate the viruses when probing into cells (Bradley and Cousin, 1969) but others state that virus is transmitted only during probes between cells (van Hoof, 1958). However, it is agreed that aphids probe more often in the grooves of the epidermal surface that follow the transverse walls of the underlying epidermal cells than in other parts of the epidermis, but it is difficult to tell whether the stylets remain in intercellular spaces or penetrate the cells close to the transverse walls.

Transmission of non-persistent viruses is favoured by keeping the aphids off plants for an hour (or even less) before placing them on the virus source plant (Watson, 1938). This treatment

changes the behaviour of the aphids, for whereas a fasted aphid will make a series of brief probes when put on a plant, a recently fed one may take longer to make its first probe and this, on average, will last longer; and it is known that non-persistent viruses are more readily acquired during brief probes than longer ones.

Many experimental results suggest that non-persistent viruses are carried on, in or near the mouthparts. They include the brief transmission threshold periods of non-persistent viruses, and the failure of aphids to retain these viruses through the moult, when the cuticle covering the body and lining the mouthparts, foregut and hindgut of the insect is cast. However, attempts to localize more precisely where the virus is held have not yet given unequivocal results. Bradley and Ganong (1955) ingeniously manipulated the stylets of virus-carrying aphids so that different lengths protruded from the labium, and then exposed them to ultra-violet radiation. When 5 μm of the stylets was exserted, few of the irradiated aphids later transmitted the virus, although many of the control aphids that had been similarly treated but not irradiated, and others that had been irradiated without their stylets exserted, did transmit. It therefore seemed that the virus was held on the terminal 5 μm of the stylets, but it was later found that irradiation alters the aphids' probing and feeding behaviour (Bradley, 1964), and alternative explanations are thus possible. Some workers have suggested that the virus is held on the outside of the mandibular stylets, which have a series of ridges (Fig. 13.2), others that the virus is held in the food canal, and still others that it is carried in a plug of gelled saliva, which is produced where the stylets puncture the leaf (Sylvester, 1962). More recently Garrett (1973) found that the number of aphids that transmitted virus was correlated with the number that regurgitated at least 600 μm^3 of sap

(Fig. 13.5) (i.e. at least ten times the volume of the food canal in the stylets), and has suggested that virus is transmitted when regurgitated from the foregut. More evidence is needed before any of these ideas can be accepted or rejected, and in the meantime the term stylet-borne should perhaps be avoided.

Kassanis and Govier (1971b) suggest that for transmission of some non-persistent viruses the vector aphid must acquire not only virus particles but also a transmission factor, which occurs in infected tissues and can be separated from virus particles (Govier and Kassanis, 1974). Thus *M. persicae* that had fed on purified preparations of potato virus Y through a membrane failed to transmit the virus unless they had previously fed on an infected plant, which was effective even when previously irradiated with ultra-violet to inactivate virus particles in the superficial layers. The same factor also enables both potato aucuba mosaic virus and potato virus C, neither of which is normally aphid transmitted, to be transmitted by *M. persicae*.

Semi-persistent viruses are a somewhat ill-defined category, and are sometimes regarded as essentially non-persistent viruses that persist for anomalously long periods in the aphid, and are acquired from and inoculated to deeper tissues in the leaf. These viruses include beet yellows virus and its allies, which have long (1 to 2 μm) very flexuous particles, affect the phloem and cause leaf yellowing. They also include parsnip yellow fleck virus, which has isometric particles about 30 nm in diameter. Vector efficiency is not increased by keeping the aphids off plants for an hour before placing them on a virus source, and although its transmission threshold period is about 30 minutes, beet yellows virus is transmitted much more efficiently when access periods of several hours are used (Watson, 1940), possibly because it is best acquired from and inoculated to the phloem. There is no good evidence of how or where these viruses are held within the aphids, but they seem not to be retained through the moult. Semi-persistent viruses seem to have more vector specificity than non-persistent viruses, perhaps because of the longer feeding periods they require, and few species of aphid transmit them efficiently.

Persistent aphid-transmitted viruses behave quite differently in their vectors. They have minimum acquisition access and inoculation access periods of 10 to 60 minutes, usually have a latent period of 12 hours or more, and can be

Fig. 13.1 Arthropod vectors of plant viruses. **a,** Aphid, *Aphis fabae,* vector of several potyviruses. Length 2 mm. **b,** Leafhoppers, *Nephotettix cincticeps,* vector of rice dwarf virus. Left, female of long-day form; right, female of short-day form. Lengths about 5.5 and 4.5 mm, respectively. (Courtesy R. Kisimoto; Kisimoto, 1973.) **c,** Colony of mites, *Aceria tulipae,* vector of wheat streak mosaic virus, on a leaf; note eggs. **d,** Ventral aspect of *A. tulipae* adult female. Length about 0.2 mm. (**c,** and **d,** courtesy J. T. Slykhuis.) **e,** Adult female mealybug, *Planococcoides njalensis,* vector of cacao swollen shoot virus. Length about 3 mm. (Courtesy P. F. Entwistle.) **f,** Adult beetle, *Oulema melanopa,* vector of cocksfoot mottle and phleum mottle viruses. Length about 4 mm. (Courtesy J. A'Brook.)

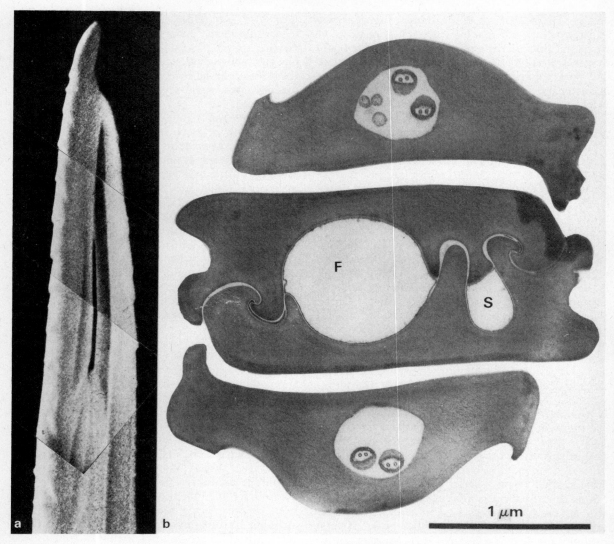

Fig. 13.2 a, Tips of stylets of *Myzus persicae* as seen in the scanning electron microscope. Note external ridges on the mandibular stylets. The enclosed maxillary stylets are just visible in the centre of the stylet bundle towards the bottom of the picture. **b,** Electron micrograph of transverse section of the stylet bundle of the aphid *Hyperomyzus lactucae*. Note the food (F) and salivary (S) canals enclosed between the interlocked maxillary stylets. The cavity in each mandibular stylet contains two nerves. (Courtesy G. T. O'Loughlin.)

retained by the aphid for at least a week (Table 13.2; Fig. 13.4). Infected plants show mainly leaf-rolling and yellowing symptoms, and many of the viruses are probably most concentrated in, if not restricted to, the phloem of the plant. Thus although some of the viruses, such as pea enation mosaic and lettuce necrotic yellows, are sap-transmissible, many others are not. Most seem to fall into one or other of two main groups. Viruses in the first group, typified by barley yellow dwarf virus (30 nm isometric particles), seem not to multiply in their vectors (Paliwal and Sinha, 1970), whereas those in the second group, which includes lettuce necrotic yellows virus (70 × 230 nm bacilliform particles), do. Both lettuce necrotic yellows virus (O'Loughlin and Chambers, 1967) and sowthistle yellow vein virus (Sylvester, 1969b; Sylvester and Richardson, 1969; 1970) infect their vector *Hyperomyzus lactucae* systemically; sowthistle yellow vein virus is also transmitted through the eggs of infective *H. lactucae* to some of the progeny; this is called 'transovarial transmission' (Sylvester,

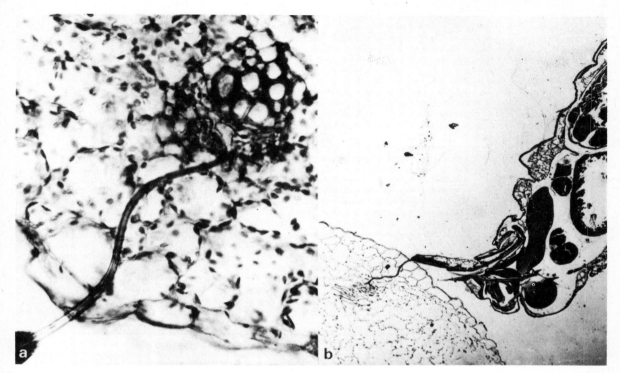

Fig. 13.3 Sections of plant tissue showing aphid stylets penetrating to the phloem, **b,** low magnification, showing aphid, labium and stylets; **a,** higher magnification view of stylets.

1969a). In the aphid, many particles of each virus, both with and without outer envelopes can be found in many cell types, with those of lettuce necrotic yellows virus occurring in the cytoplasm (Fig. 13.6), and particles of sowthistle yellow vein virus being in the nucleus, perinuclear space and cytoplasm. Sowthistle yellow vein virus also multiplies in primary cultures of cells of *H. lactucae* (Peters and Black, 1970).

Persistent viruses are mostly transmitted efficiently by only one or a few aphid species, they are retained through the moult (i.e. ecdysis does not make an infective aphid non-infective) and their transmission is not increased by fasting the aphids before placing them on the virus source plant. The latent period of barley yellow dwarf virus and its allies is less than a day and the viruses persist for several days but not necessarily for the life of the aphids. By contrast the latent period of sowthistle yellow vein virus is at least a week and persistence is more often for life (Fig. 13.4). Several persistent viruses can be detected in the haemolymph of their vectors and, when such haemolymph, or virus preparations from plants,

are injected into the haemocoele of virus-free aphids, the viruses are later transmitted. Hence it seems that circulation of the virus within the vector from gut to haemocoele to salivary gland is normally a necessary preliminary to its inoculation in saliva to virus-free plants.

13.2.4 *Transmission by leaf-, plant- and tree-hoppers*

These insects (Fig. 13.1) make up the second most important group of virus vectors (Table 13.1). Most of the vector species are leafhoppers (Cicadelloidea), but nearly twenty are plant-hoppers (Fulgoroidea), and the tomato pseudo curly top agent is transmitted by the treehopper, *Micrutalis malleifera* (Membracidae; Simons, 1962). Typically the viruses they transmit cause yellowing and/or leaf-rolling, and only a few are sap-transmissible. Some of the viruses have been purified and of these most have large isometric (70 nm diameter) or bacilliform (60 to 100 by 200 to 400 nm) particles resembling those of reoviruses and rhabdoviruses respectively (Chapter 16). The vector species mainly feed from the phloem of plants and the viruses (such as clover wound tumor virus) seem most concentrated in and

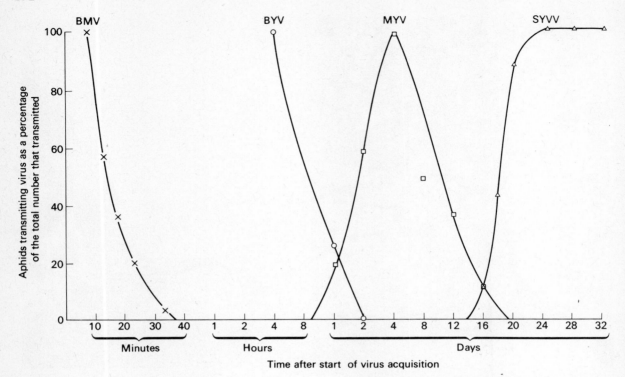

Fig. 13.4 Transmission records of aphid-borne viruses. BMV, beet mosaic, a non-persistent virus transmitted by *Myzus persicae*; 3 minute acquisition access period (Watson, 1946). BYV, beet yellows, a semi-persistent virus transmitted by *M. persicae*; 2 hour acquisition access period (Watson, 1946). MYV, malva yellows, a persistent virus transmitted by *M. persicae*; 12 hour acquisition access period (modified from Costa, Duffus and Bardin, 1959). SYVV, sowthistle yellow vein, a persistent virus multiplying in *Hyperomyzus lactucae*; 2 day acquisition access period. (Duffus, 1963.)

possibly restricted to the phloem and pseudophloem. The minimum acquisition and inoculation access times, which are a few minutes, thus probably reflect the times needed to reach and feed on these tissues.

Leafhopper vectors seem to transmit viruses in three of the ways described for aphids, but no non-persistent leafhopper-transmitted viruses are known. Most commonly the viruses multiply in their vectors, but beet curly top virus persists for long periods in *Circulifer tenellus* without seeming to multiply (Bennett and Wallace, 1938), and rice tungro virus resembles semi-persistent aphid-borne viruses in persisting for only a few days in *Nephotettix impicticeps* and not being retained through the moult (Ling, 1966). Rice tungro virus has no detectable latent period in its vector, whereas the latent period for beet curly top virus

is usually at least 4 to 6 hours (Severin, 1921) and for viruses that multiply in their vectors is more than a week.

Work with leafhoppers first showed that some

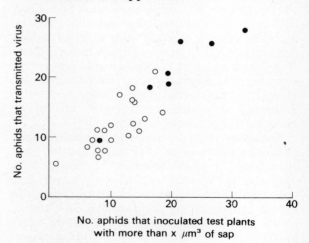

Fig. 13.5 Comparison of numbers per batch of *Myzus persicae* that transmitted cucumber mosaic virus to cucumber seedlings with numbers of similarly treated aphids that inoculated $x \mu m^3$ of sap. Acquisition access period up to 9 min, inoculation access period 3 min. ●, Expt 1 ($x = 570$), seven batches, 40 aphids per batch; ○, Expt 2 ($x = 640$), twenty batches, 30 aphids per batch. (Redrawn from Garrett, 1973.)

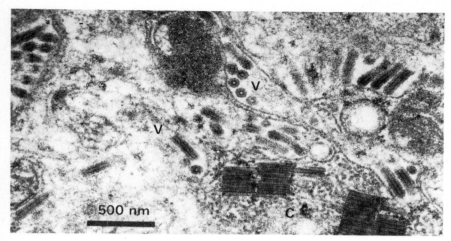

Fig. 13.6 Electron micrograph of section of mid-gut cell of *Hyperomyzus lactucae* carrying lettuce necrotic yellows virus. In the cytoplasm the bacilliform virus particles (V) are seen in longitudinal and transverse section. Bundles of cores of virus particles (C) are also sectioned longitudinally. (Courtesy G. T. O'Loughlin; O'Loughlin, 1969.)

viruses can multiply both in plants and in invertebrates. There were two main lines of evidence. One line developed from the finding that rice dwarf virus is transmitted through the eggs to about 60% of the progeny of infective female *N. cincticeps* and virus-free males, though not to the offspring of virus-free females and infective males (Fukushi, 1933). Although the mechanism of entry of the virus into the insect's eggs is not known for sure, there is evidence that virus particles are held on the surface of the symbiont cells that occur in mycetocytes, which carry the symbionts to the oocytes (Nasu, 1965). Strong evidence for multiplication in the vector was obtained by Fukushi (1940) when he found that rice dwarf virus was transmitted transovarially for six generations, despite strict precautions being taken to prevent progeny insects acquiring the virus from plants. The second and clinching line of evidence came from experiments with wound tumor virus and *Agallia constricta* (Black and Brakke, 1952). Extracts of insects that had fed on infected plants were diluted and injected through fine glass needles into virus-free insects kept on plants immune to the virus. After a suitable interval, usually 2 to 4 weeks, extracts were made from the injected insects and were then injected into a further set of virus-free insects; these serial passages were repeated many times. The infectivity of some of the insects in each injected batch was tested by putting them to feed on suitable test plants, and was found to be fully maintained throughout the experiments. By the last passage it was calculated that the original inoculum had been diluted at least 1 in 10^{18}, yet infectivity was maintained, and therefore the virus must have multiplied in the insects. In later experiments particles of the virus obtained from plants were found to be indistinguishable in the electron microscope from those obtained from vector insects (Brakke, Vatter and Black, 1954).

Viruses that multiply in leafhoppers or planthoppers persist in their vectors for long periods, often for life, and several of them are transmitted through the eggs to progeny. The frequency with which single insects transmit them is independent of the length of the acquisition access period, they have a latent period in their vectors of a week or more and they are retained through the moult.

The sequence of events occurring between virus acquisition and inoculation is most completely known for wound tumor virus in *Agallia constricta* and we shall discuss only this example. Serological assays, using the precipitin ring test, showed that soluble virus antigen first appeared in extracts of whole insects 5 days after feeding on infected plants, and was at a maximum after a further 5 days; the accumulation of virus particles followed a similar pattern (Reddy and Black, 1966; Gamez and Black, 1968; Fig. 13.7). The insects did not however start to infect plants until 12 to 35 days after acquiring virus particles, and their efficiency as vectors decreased thereafter (Fig. 13.8; Whitcomb and Black, 1961). The explanation of this delay became clear when the fluorescent antibody technique was used to locate virus antigen within the insects. Sinha (1965) detected virus antigen first in a corner of the filter chamber (between the oesophagus and midgut) 4 days after virus acquisition, but nowhere else until the twelfth day, when both the filter chamber and adjacent parts of the ventriculus contained much of it. Virus antigen was first found in the fat

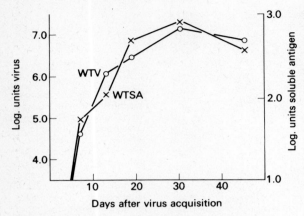

Fig. 13.7 Production of soluble antigen (WTSA) and particles (WTV) of wound tumor virus in leafhoppers, *Agallia constricta*, after feeding for 1 day on infected plants. (Data from Reddy and Black, 1966.)

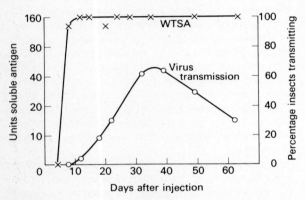

Fig. 13.8 Production of soluble antigen (WTSA) in *Agallia constricta*, and ability of the leafhoppers to transmit wound tumor virus, after they were injected with virus particles. (Whitcomb and Black, 1961.)

body, brain and Malpighian tubules on the 14th day and in the salivary glands on the 17th day. The virus therefore seems to spread, presumably from the filter chamber by way of the haemolymph to most other organs of the insect, except apparently the male reproductive organs, but including the salivary glands, from which it presumably passes with the saliva to healthy plants. Also, Shikata and Maramorosch (1965) have shown by electron microscopy of thin sections of the insect's tissues that the virus particles occur, often embedded in amorphous 'viroplasm', and sometimes aggregated in microcrystals, in the gut wall, fat body, Malpighian tubules, salivary glands, epidermis and muscle, but not in the testicles or vesiculae seminalis. More recently, methods have been developed of culturing leafhopper cells (Grace,

1973), and infecting cell monolayers with wound tumor virus (Fig. 13.9). This provides a sensitive and accurate method of virus assay (Chapters 4 and 7; Chiu, Reddy and Black, 1966; Chiu and Black, 1969).

By contrast, beet curly top virus seems either not to multiply in its vector *Circulifer tenellus* or multiplies to only a very limited extent. Unlike most other leafhopper-transmitted viruses, the ability of individual *C. tenellus* to transmit virus is related to the length of their acquisition access period (Freitag, 1936). Bennett and Wallace (1938) ingeniously assayed beet curly top virus by feeding *C. tenellus* through a membrane on extracts of other *C. tenellus*, and found that the virus content of leafhoppers first fed on infected beet and then transferred to maize, which is immune to the virus, decreased with increasing time on the maize. Leafhoppers in which the virus was no longer detectable could however be recharged by feeding them on infected beet. The virus was found in the haemolymph, alimentary tract and salivary glands of the insects. The simplest hypothesis to explain all these results is that beet curly top virus circulates but does not multiply in *C. tenellus*, but such a negative can of course never be proved.

13.2.5 *Transmission by whiteflies*

Whiteflies (Aleyrodidae; Table 13.1), particularly *Bemisia tabaci*, are the vectors of several important virus-like agents prevalent in hot countries. They mainly feed on phloem tissue, and because their larval stages are sedentary and there is no evidence of transovarial transmission, only adults spread these virus-like agents in the field. The diseases produced are of two main kinds, mosaics such as abutilon mosaic and distortions such as tobacco leaf curl. Most of these pathogens are not sap-transmissible and little is known about the properties of their particles. Some of the better studied ones are acquired by whiteflies in access periods of 30 minutes to 4 hours, have a latent period of a few hours, are inoculated in 15 minutes to 2 hours, are retained through the moult and continue to be transmitted for 5 to 25 days (Costa, 1969). By contrast, the cucumber vein-yellowing agent, which is readily sap-transmissible, can be acquired by *B. tabaci* in 20 minutes, inoculated in 10 minutes, there is no latent period and transmission by feeding whiteflies continues for only 6 hours; behaviour more

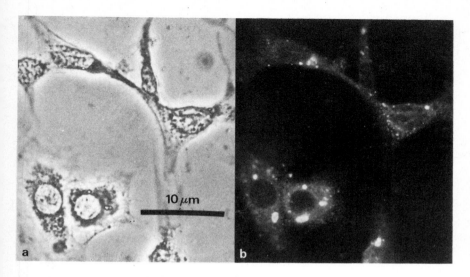

Fig. 13.9 Cultured cells of *Agallia constricta* infected with wound tumor virus. **a,** Phase contrast microscopy; **b,** stained with fluorescent antibody to the virus, showing aggregates of virus antigen in the cytoplasm. (Courtesy R. J. Chiu and L. M. Black.)

like that of semi-persistent aphid-transmitted viruses (Harpaz and Cohen, 1965).

There is no good evidence for multiplication of any of these agents in whiteflies, and the vector relations of many of them resemble those of the aphid-borne barley yellow dwarf virus. The transmission of the tomato yellow leaf curl agent by *B. tabaci* is unusual in that the ability of the insects to transmit decreases after they leave the source plants and few individuals retain the agent for 15 days, but during this period they seem less able to acquire and transmit a fresh charge of it than before or afterwards (Cohen and Harpaz, 1964).

13.2.6 *Transmission by beetles*

The particles of beetle-transmitted viruses (Table 13.1) reach a large concentration in plant sap, are relatively stable *in vitro*, and most of those described so far are isometric, about 30 nm in diameter. Many of the viruses can be placed in three groups, the tymoviruses, the comoviruses and the group typified by turnip crinkle virus. They usually cause mosaic and mottling diseases and do not have particularly wide plant-host ranges. Nearly all the vector species (Fig. 13.1) are leaf-eating members of Chrysomelidae (mostly Halticinae and Galerucinae), but echtes Ackerbohnen-mosaik-Virus is transmitted by the weevils *Apion vorax* (Apionidae) and *Sitona lineatus* (Curculionidae) (Cockbain, 1971).

The viruses can be acquired in 5 minutes, some can be inoculated immediately after acquisition and others a day later (Catherall, 1970), and

they persist in the insects for days or weeks. Markham and Smith (1949) considered that the vector species lack functional salivary glands and that virus is inoculated to plants by regurgitation of stomach contents during feeding. Selman (1973) questions this view; he describes maxillary glands and indicates that regurgitation during feeding is not normal. Also, some viruses can be found in the haemolymph of infective beetles; for example, squash mosaic virus in *Acalymma trivittata* and *Diabrotica undecimpunctata* (Freitag, 1956), and southern bean mosaic virus in *Ceratoma trifurcata* (Slack and Scott, 1971). Moreover, *C. trifurcata* injected in the haemocoele with a purified preparation of southern bean mosaic virus transmitted the virus for about the same period as virus acquired by feeding (Fig. 13.10). How the virus is conveyed from haemolymph to mouthparts is not clear but the haemolymph seems to be a reservoir of virus that may play an important role in transmission. There is no evidence that beetle-transmitted viruses multiply in their vectors.

Some beetle-borne viruses have also been transmitted experimentally by a few other insects with biting mouthparts, such as grasshoppers, but there is no evidence that these vectors are of any great importance in natural conditions.

13.2.7 *Transmission by other insects*

Several other groups of insects have been implicated in virus transmission but there are only a few examples from any one group. At least four species of thrips (Thysanoptera) transmit tomato

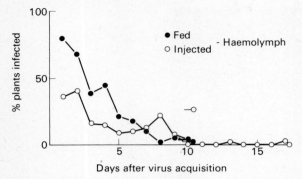

Fig. 13.10 Transmission of the cowpea strain of southern bean mosaic virus by beetles, *Ceratoma trifurcata*, each transferred to a series of plants. The beetles acquired the virus either by feeding on infected plants or by injection of virus particles into the haemocoele. The extra points at 10 days indicate the amounts of virus detected in haemolymph by injection into further beetles. (Data from Slack and Scott, 1971.)

spotted wilt virus (Sakimura, 1962), which is widely distributed, has a large plant-host range and possesses particles that are extremely labile *in vitro*. Nevertheless it persists for days or weeks in the thrips and perhaps multiples in them. Interestingly, the virus can only be acquired by the larval stage of *Frankliniella schultzei*, in which it usually has a latent period of 5 to 12 days. It is retained through the moult and inoculated by adults, in which it may persist for a few weeks (Bald and Samuel, 1931). *Thrips tabaci* can aquire the virus in 30 minutes and inoculate it in 5 minutes, and other features of transmission are the same as for *F. schultzei*. *T. tabaci* apparently also transmits tobacco ringspot virus, which has nematode vectors as well (Messieha, 1969), and *Seriothrips occipitalis* and *Taeniothrips sjostedti* have been reported to transmit cowpea mosaic virus (Whitney and Gilmer, 1974), which is more usually beetle-borne. It used to be thought that thrips fed by gashing open the epidermis and sucking up the contents of the epidermal cells, but detailed studies show that their feeding behaviour is perhaps more complex and less brutal (Mound, 1973). However, it is not clear what kind of behaviour most favours virus transmission.

Several species of mealy bugs are recorded (Dale, 1962; Fig. 13.1) as vectors of isolates of cacao swollen shoot virus, which has a moderate host range among woody plants, has particles with straight sides and rounded ends, and is sap-transmissible, though not readily. The virus can be acquired in 1 hour and inoculated in 15 minutes, although more plants are infected when

longer periods are used. Cacao swollen shoot virus persists for 3 to 4 days in *Planococcoides njalensis* and is retained through the moult (Roivainen, 1971). Other viruses have been described with particles morphologically similar to those of cacao swollen shoot virus and it will be interesting to see whether they too have vectors in the Coccoidea.

The two tingid bugs (Gymnocerata) described as virus vectors, one in Europe and the other in North America, both transmit pathogens of sugar beet that cause the leaves to curl and roll. The agent causing the European beet disease, *Krauselkrankheit*, is the better studied. It has a prolonged latent period in the bugs, persists through the winter in hibernating *Piesma quadrata*, and can be acquired by larvae but is not inoculated until the insects become adult (Volk and Krczal, 1957). It can be detected in haemolymph and when injected into the haemocoele can later be inoculated to plants; experiments using serial injections from bug to bug indicate that it multiplies in the vector (Proeseler, 1966; Schmutterer and Ehrhardt, 1964).

Still more groups of insects have been found occasionally to transmit viruses that attain high concentrations in plants, but the relevance of these reports to natural conditions is uncertain. For instance, grasshoppers and some lepidopteran larvae can act as vectors of potato virus X and TMV in experiments, although they may have to be restrained from devouring the part of the leaf they have just contaminated with virus particles. Then again, an unusual type of relation is described between sowbane mosaic, a virus occurring in large concentration in *Chenopodium murale*, and the leaf miner fly *Liriomyza langei*; the females are thought to carry the virus particles on their ovipositors, which are used to lacerate leaf tissues during feeding. *L. langei* also transmitted TMV from *Petunia hybrida* to *P. hybrida* (Costa, da Silva and Duffus, 1958), and *Cyrtopeltis modestus* (Miridae) transmitted it from tobacco to tobacco (Costa and Carvalho, 1960).

13.2.8 *Transmission by mites*

Some eriophyid mites (Eriophyidae; Fig. 13.1) are firmly established as virus vectors. The adults have elongate bodies about 0.2 mm long, possess only 4 legs, have narrow plant-host ranges, and infest the leaves, buds and other tender parts of plants. They are carried from plant to plant more readily by wind than by their own efforts and they desiccate readily. Because they are small and delicate they are difficult to manipulate in experiments, but can be moved with the aid of a fine hair mounted on a handle. They have slender stylets which puncture plant cells, and the two pads at the apex of the rostrum seem to help in bringing saliva to the stylets and in sucking up juice. Some species (e.g. *Eriophyes gracilis* on raspberry) cause symptoms not unlike those caused by viruses (Fig. 3.23), but others cause no macroscopic damage.

Four of the viruses with eriophyid vectors have filamentous particles about 700 nm long, are sap transmissible and, as their names indicate, cause mosaic and mottling diseases. Virus-vector relations have been studied in the most detail with wheat streak mosaic virus and *Aceria tulipae*. The virus can be acquired in 15 minutes, persists for up to 9 days, can be inoculated in 15 minutes and passes through the moult. Adults cannot acquire the virus for transmission but all stages can inoculate it, and the virus is not transmitted through the egg (Slykhuis, 1955; Staples and Allington, 1956; Orlob, 1966). The mid and hind gut of the mites become packed with virus particles (Paliwal and Slykhuis, 1967) but it is not known whether these can be inoculated by the mites to plants. Ryegrass mosaic virus, by contrast, is apparently retained for only 1 day by *Abacarus hystrix* (Mulligan, 1960).

Graft-transmissible virus-like agents that are transmitted by eriophyids cause currant reversion, peach mosaic, fig mosaic, pigeon pea sterility and rose rosette diseases. There is little information on the vector relations of these agents, but the currant reversion agent suppresses the production of hairs on blackcurrant bushes, making them better hosts for the mite vector, *Phyoptus ribis* (Thresh, 1964).

There are reports that two sap-transmissible viruses are transmitted by spider mites (Tetranychidae); potato Y by *Tetranychus telarius* (Schulz, 1963) and tobacco ringspot by *Tetranychus* sp. (Thomas, 1969), but Orlob (1968) could not confirm the findings with potato virus Y. Spider mites are less than 1 mm long, have 8 legs, feed by piercing cells and sucking their contents, and infest a wide range of plant species. Another member of the Tetranychoidea, *Brevipalpus obovatus*, is associated with citrus leprosis disease, and electron microscopy has revealed that leaf cells at the periphery of the characteristic lesions contain bullet-shaped virus-like particles about 100 to 130 × 40 nm in size, particularly in the nuclei (Kitajima *et al*, 1972). This agent may be the first member of a new group of mite-transmitted viruses.

13.2.9 *Transmission by nematodes*

Two groups of viruses are transmitted by soil-inhabiting plant parasitic nematodes; the nepoviruses, which have isometric particles about 30 nm in diameter, and the tobraviruses, which have straight tubular ones. The nematode vectors are in two corresponding groups of the order Dorylaimida (Harrison, Mowat and Taylor, 1961), which includes relatively few species known to be plant parasites; nepoviruses are transmitted by members of the family Dorylaimidae (*Xiphinema* and *Longidorus* spp.; adults about 3–10 mm long) whereas tobraviruses are transmitted by members of the family Trichodoridae (e.g. *Trichodorus* spp.; adults about 1 mm long). There is considerable specificity between virus and vector species, and no vectors are reported among the many plant-parasitic species in the order Tylenchida. Both groups of viruses occur naturally in many species of plants and have even wider experimental host ranges.

The vector nematodes have probing mouthparts analogous to those of aphids but consisting of a single central stylet (Fig. 13.11). In *Trichodorus* the stylet is a slightly curved tooth about 50 μm long, and is used with a rapid probing action (Wyss, 1971). In *Xiphinema* and *Longidorus* it is tubular, the distal part, or stylet (odontostyle), and the proximal part, or stylet extension (odontophore), together often measuring 200 μm long or more, with a lumen about 0.4 μm in diameter. Both groups of vectors seem often to feed on the young tissue near the root tips (Fig. 13.11), and may cause galls to develop, or growth to stop, irrespective of whether they are carrying virus. *Trichodorus* spp. usually feed on epidermal cells, moving from one to another after a few minutes, whereas *Xiphinema* spp., which have longer stylets, can penetrate more deeply, probably reaching the vascular cylinder, and may feed at one point for an hour or more. Several of the vector species have wide host ranges but *X. index* seems to feed mainly on a few species such as grapevine and fig.

Only minor differences have been found in the vector relations of the different viruses. Both nepoviruses and tobraviruses can be acquired in 15 minutes to 1 hour and inoculated in a similar period (Das and Raski, 1968; Ayala and Allen, 1968), both groups are retained by the vectors for

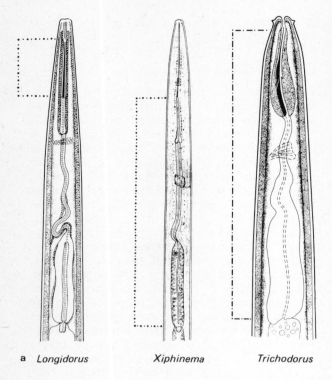

a *Longidorus*   *Xiphinema*   *Trichodorus*

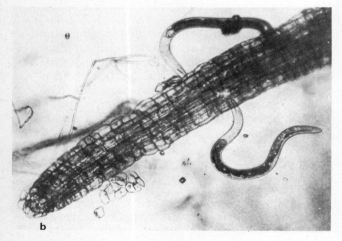

b

Fig. 13.11 a, Diagrams of the head and alimentary canal, down to the oesophago-intestinal valve, of vector nematodes in three genera. The site of retention of virus particles is indicated by the broken line to the left of each diagram. In *Longidorus*, the virus particles are found on the guide sheath surrounding the odontostyle. In *Xiphinema*, they are found lining the lumen of the odontophore and of the oesophagus, including the lumen of the muscular oesophageal bulb. In *Trichodorus*, the particles line the lumen of the entire pharynx and oesophagus. (Courtesy C. E. Taylor; Taylor, 1972.) **b,** *Paratrichodorus christiei* feeding near the tip of a cranberry root. (Courtesy B. M. Zuckerman.)

weeks and are transmitted by larvae and by adults. Although individual nematodes can transmit virus to a series of plants, there is no evidence for transovarial transmission, and no evidence that the viruses multiply in their vectors (Harrison, 1973). Transmission seems to involve the reversible association of virus particles with specific surfaces in the food canal (Taylor and Robertson, 1969). Particles of viruses transmitted by *Longidorus elongatus* become associated with the stylet guiding sheath, whereas those transmitted by *Xiphinema* spp. become associated with the oesophageal wall, and particles of tobacco rattle virus attach to the surface of pharynx and oesophagus (Figs. 13.11, 13.12). It is presumed that virus particles attach to these surfaces when sap of infected plants passes down the food canal, and dissociate when saliva passes in the reverse direction into punctured cells. A correlation between serological specificity and vector specificity among strains of two nepoviruses suggested the hypothesis that the protein surface of the virus particles plays a critical role in transmission (Harrison, 1964a), and work with virus hybrids shows that the vector specificity of raspberry ringspot virus is determined by the piece of the genome that carries the gene for virus coat protein (Harrison *et al*, 1974). Thus although the mechanism of association and dissociation is not clear, it seems to involve the surface properties of the virus particles.

The viruses are apparently not retained through the moult, and this correlates well with the effect of moulting on the sites where the viruses are retained in the nematodes. At moulting the odontostyle and the lining of the stoma and stylet guide sheath are shed with the cuticle covering the nematode's body, and hence virus particles retained in the buccal capsule of *Longidorus elongatus* are presumably cast with the cuticle. Similarly, those retained by *Xiphinema* are also removed because the oesophageal lining sloughs off during moulting and passes posteriorly through the oesophageal/intestinal non-return valve (Taylor and Robertson, 1970a).

13.2.10 *Transmission by fungi*

Species in two major groups of root-infecting fungi act as vectors of some soil-borne viruses. In the Chytridiales, *Olpidium* spp. transmit tobacco necrosis, satellite and cucumber necrosis viruses, and also the agents causing tobacco stunt and lettuce big vein diseases; in the Plasmodio-

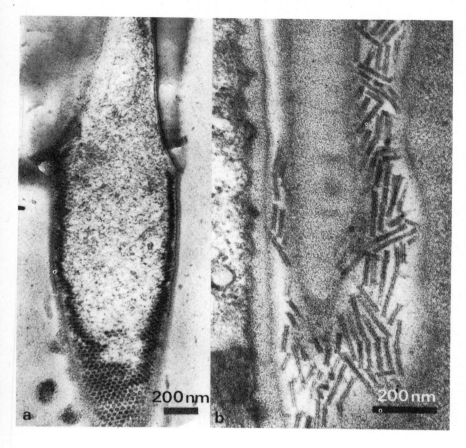

Fig. 13.12 Electron micrographs showing virus particles in sections of vector nematodes. **a,** oblique longitudinal section through lumen at the junction of the odontostyle (above) and odontophore (below) of *Xiphinema diversicaudatum*. Particles of arabis mosaic virus line the odontophore lumen but not the odontostyle lumen. (After Taylor and Robertson, 1970a.) **b,** Nearly longitudinal section of the oesophagus of *Paratrichodorus pachydermus* carrying tobacco rattle virus. Virus particles of the two characteristic lengths can be readily seen lining the oesophageal lumen. (After Taylor and Robertson, 1970b.) (Courtesy C. E. Taylor and W. M. Robertson.)

phorales, *Polymyxa graminis* transmits soil-borne wheat mosaic virus, and *Spongospora subterranea* transmits potato mop-top virus. In addition there are reports of transmission of a few other viruses (Teakle, 1972), and there are several further soil-borne viruses which seem likely to have fungus vectors.

Olpidium spp. have a simple life history. The zoosporangia, which are mostly in cells close to the surface of roots, produce exit tubes and liberate the uniciliate zoospores into the liquid surrounding the root (Fig. 13.13a). The zoospores swim and drift in films of soil water to another root, attach to its surface, withdraw their cilia and produce a thin outer cyst wall. After about 2 hours each zoospore produces an infection canal, which can be seen penetrating the wall of the nearest root cell, and the protoplasm of the zoospore moves through the canal into the cell, leaving the cyst wall outside. In 2 to 3 days a thallus grows in the cell and develops into a new zoosporangium with one or more exit tubes. It is thought that some zoospores fuse in pairs before infecting and

that these zygotes produce the thick-walled resting sporangia (Fig. 13.13c) that can withstand drying, and later germinate to give zoosporangia, which in turn produce zoospores. The plasmodiophoromycete vectors have a rather similar life history (Karling, 1968) to that of *Olpidium*.

Olpidium transmits different viruses or virus-like agents in one or other of two ways. In the first, the particles of tobacco necrosis (Teakle, 1962), satellite (Kassanis and Macfarlane, 1968) or cucumber necrosis (Dias, 1970) viruses are rapidly acquired by zoospores from solution and carried into root cells: in control experiments little virus infection occurs when the roots are immersed in virus preparation alone. The particles are found attached to the whole zoospore surface (Temmink, Campbell and Smith, 1970; Fig. 13.14) but it is thought that those on the surface of the cilium are withdrawn with the cilium into the zoospore protoplasm and then transferred with the protoplasm to a root cell. Growth of the *Olpidium* thallus in the root cell is not necessary for virus multiplication; heating roots for 10 seconds at 50°C, 2 to 24 hours after inoculation with virus-carrying zoospores, kills *O. brassicae* without preventing infection and replication of tobacco necrosis virus (Kassanis and Macfarlane, 1964). Thus these viruses seem to have only a very transient association with *Olpidium*, and they are not retained in resting spores.

By contrast the agents of lettuce big vein

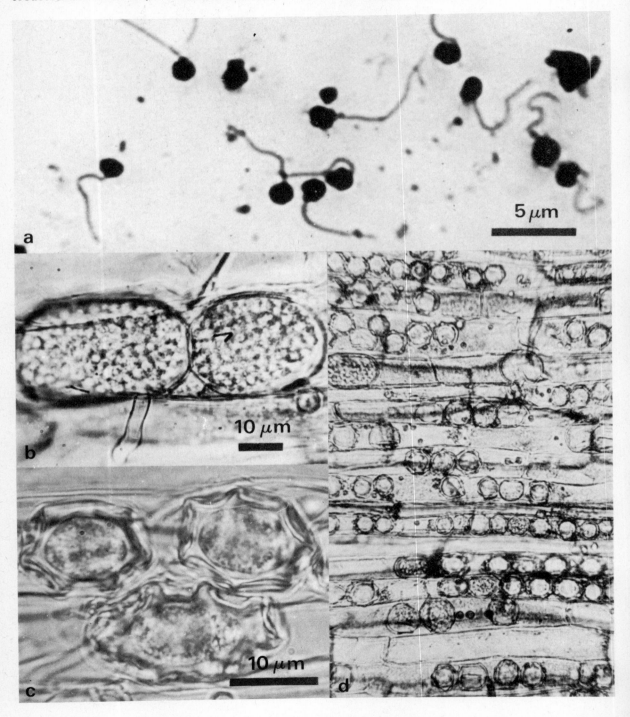

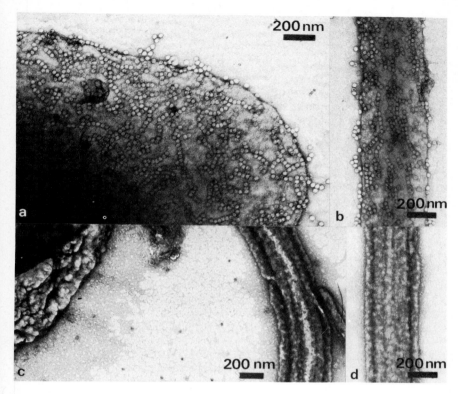

Fig. 13.14 Electron micrographs of *Olpidium brassicae* zoospores exposed to tobacco necrosis virus. **a,** Virus particles attached to surface of zoospore and **b,** its flagellum, using a vector isolate of the fungus. **c,** Lack of virus particles on surface of zoospore and **d,** flagella of a non-vector isolate of the fungus (Temmink, Campbell and Smith, 1970; courtesy R. N. Campbell.)

(Campbell, 1962; Campbell and Grogan, 1964) and tobacco stunt (Hidaka and Tagawa, 1962) are apparently held internally, and are retained by resting spores. They are acquired by the fungus while it is growing in plant cells and are not acquired by zoospores from solutions. The same sort of behaviour is shown by soil-borne wheat mosaic and potato mop-top viruses in their plasmodiophoromycete vectors (Rao and Brakke, 1969; Jones and Harrison, 1969). Resting spores remain viable and can probably retain the viruses for several years, and the viruses are transmitted to roots or tubers by the biciliate zoospores. Neither virus is proved to multiply in its vector, but may well do so.

With the discovery of so many viruses infecting fungi (Chapter 16), a few of them having particles seemingly similar to those of viruses infecting angiosperms, there seems a real possibility that fungi may constitute reservoirs of viruses that infect higher plants. Indeed it has been claimed that TMV has been obtained from conidia of the powdery mildews *Sphaerotheca lanestris* and *Erisyphe graminis* (Yarwood, 1971; Nienhaus, 1971), and TMV seems to infect the phycomycete *Pythium sylvaticum* (Brants, 1971). Also it should be remembered that although a remarkable range of organisms has been implicated in virus spread, the list is unlikely to be complete because the vectors of many viruses and virus-like agents are still not known, and there are therefore excellent prospects of additions.

13.2.11 *Specificity of transmission by vectors*

In the preceding part of this chapter we have described how viruses are transmitted by vectors, but failure to transmit a given virus also poses many problems. We would like to know, for example, why the vectors of most viruses are confined to one taxonomic group, why TMV is not efficiently transmitted by Hemiptera, and why different strains of a virus may differ greatly in vector transmissibility. The answers to these and related questions are mostly not yet known, and the best we can do at present is to indicate

Fig. 13.13 The vector fungus, *Olpidium brassicae*. **a,** Zoospores. **b,** Zoosporangia (one with an exit tube) in root cells. **c,** Resting spores in root cells. **d,** General view of resting spores and zoosporangia in root cells at low magnification. (Courtesy J. A. Tomlinson.)

some of the factors which seem to be involved and the ways in which they may act.

Transmission is the outcome of a sequence of processes; acquisition of virus particles by the vector, survival of infectivity during the association of virus and vector, and inoculation of virus particles to plants followed by the initiation of infection. Failure of any one of these processes will result in failure to transmit, and we shall consider them in order. In general, the study of virus strains that differ in transmissibility has provided more information on the causes of vector specificity than comparisons of unrelated viruses because the strains are the more likely to differ in behaviour in only one critical process.

A non-vector may fail to acquire virus because it does not feed on the appropriate tissue. For example, superficial feeders such as thrips will not reach the phloem and pseudophloem where particles of wound tumor virus occur. Similarly, plant species that are hosts of the virus may not be hosts of the non-vector. Squash mosaic virus is transmitted by at least five species of chrysomelid beetle and turnip yellow mosaic virus by at least twelve others, but there seems no plant species that is a host for both viruses and any of the beetles. By contrast turnip yellow mosaic virus shares several plant hosts and several beetle vectors with turnip crinkle virus. A different kind of phenomenon is the basis of specificity of transmission by the chytrid fungi, *Olpidium brassicae* and *O. cucurbitacearum*, of tobacco necrosis and cucumber necrosis viruses, respectively. Electron microscope observations show that particles of tobacco necrosis virus but not of cucumber necrosis virus become attached to zoospores of *O. brassicae* suspended in virus preparations, whereas the opposite is observed with *O. cucurbitacearum* (Temmink, Campbell and Smith, 1970); only the virus whose particles attach is transmitted.

In other instances, virus particles are taken up by non-vectors, but not transmitted. Beet curly top virus and TMV can be recovered from the bodies of non-vector aphids (*Myzus persicae*) after feeding on infected plants, and arabis mosaic and strawberry latent ringspot viruses can be extracted, probably from the intestine, of the non-vector nematode *Longidorus elongatus* (Taylor, 1968). Viruses that are transmitted by *L. elongatus* or other nematodes have particles that become attached to particular surfaces in the buccal capsule or oesophagus, and viruses whose particles do not attach are not transmitted (Table 13.3). Here specificity of transmission is correlated with specificity of attachment, and attachment seems to depend on some property of the protein coat of the virus particles (Harrison *et al*, 1974). A similar phenomenon is perhaps the basis of the inability of aphids to transmit potato virus C, even though they are able to transmit the closely related potato virus Y; plants infected with potato virus C seem to lack a transmission factor which is needed in addition to particles of the virus (§13.2.3; Kassanis and Govier, 1971b) and may be involved in attachment.

There are other barriers to be passed by viruses whose particles circulate in insects. In classic work, Storey (1938) found that individuals of a race of the leafhopper *Cicadulina mbila* that could not transmit maize streak virus, were able to transmit after their intestines had been punctured with a needle; in this species the ability to transmit, and hence presumably the movement of virus through the gut wall, is inherited as a single dominant, sex-linked gene (Storey, 1932). Some other leafhopper- or planthopper-transmitted viruses behave similarly and are efficiently acquired and transmitted by nymphs but not by adults, though adults become efficient vectors when their guts are punctured. Examples are the European wheat striate mosaic agent in *Javasella*

Table 13.3 Specificity of virus transmission by nematodes, and site of retention of virus particles within the nematode. (From Harrison, Robertson and Taylor, 1974.)

Nematode species	Virus Raspberry ringspot (strain S)	Arabis mosaic	Grapevine fanleaf	Tobacco rattle
Longidorus elongatus	++(G)	0(0)	—	0(0)
Xiphinema index	0(0)	0	++(E)	—
X. diversicaudatum	0(0)	++(E)	0(0)	—
Trichodorus pachydermus	—	—	—	++(P, E)

Symbols for transmission: ++ = much, 0 = none.
Symbols for retention site: (E) = oesophagus, (G) = guide sheath, (P) = pharynx, (0) = none, — = not tested.

pellucida (Sinha, 1960) and wound tumor virus in *Agallia constricta* (Sinha, 1963). Perhaps adults transmit inefficiently because virus does not multiply adequately either in the wall of the filter chamber or in some other part of the insect's gut. Indeed, some isolates of wound tumor virus that have been maintained in plants but not been transmitted by vectors for several years cannot multiply even in nymphal *A. constricta*, and are not transmitted (Black, Wolcyrz and Whitcomb, 1958); the particles of one such strain contain the usual seven polypeptide species, but lack one of the twelve pieces of double-stranded RNA found in normal *A. constricta*-transmitted isolates (Reddy and Black, 1973b). Incidentally, the existence of such strains indicates that different factors are involved in the ability of the virus to infect insect cells and plant cells.

The gut wall is not the only barrier to the circulation of virus particles in insects. Rochow (1969) found that two vector-specific isolates of barley yellow dwarf virus (one efficiently transmitted by *Rhopalosiphon padi* but not by *Macrosiphum avenae*, and the other efficiently transmitted by *M. avenae* but not by *R. padi*), could both be detected in the haemolymph of either species after feeding on infected plants. Also, after injecting preparations of each virus isolate into the haemocoele of each species, only the normally transmissible isolate was transmitted. When, however, *R. padi* were fed on plants infected with both isolates, or injected with virus preparations from such plants, both isolates were transmitted. Treating the preparations with antiserum to the *M. avenae*-transmitted virus isolate had no effect on transmission, nor did *R. padi* transmit the *M. avenae* isolate when injected with a mixture of purified preparations of the two isolates (Rochow, 1970). These results are interpreted to indicate that heterologous coating (§12.2.5) occurs in the mixedly infected plants, and that the coat protein determines vector specificity, possibly by affecting the ability of the virus particles to move into the salivary glands.

One instance is known of a non-vector preventing infection of plants by means of a vector. Kassanis and Macfarlane (1965), working with tobacco necrosis virus, and vector and non-vector strains of *Olpidium brassicae*, found that when roots were put into a suspension of virus particles together with zoospores of a vector strain of *O. brassicae*, and then immediately into a suspension of zoospores of a non-vector strain of *O. brassicae*, fewer virus infections resulted than when the roots were not exposed to the non-vector.

These examples illustrate the range of stages in transmission at which vector specificity can be determined. In an increasing number of examples the virus coat protein seems to play a critical role in vector specificity, especially with viruses that do not multiply in their vectors. Virus coat protein therefore seems often to have a function in vector transmission in addition to its well-known protective role.

13.2.12 *Transmission dependent on a second virus*

In work on tobacco rosette disease, Smith (1946) discovered a remarkable phenomenon. He showed that the disease was caused by the combination of two apparently unrelated viruses, tobacco mottle and tobacco vein-distorting viruses. Tobacco vein-distorting virus is not sap-transmissible but is transmitted in the persistent manner by the aphid *Myzus persicae*, whereas tobacco mottle virus is sap-transmissible and is only transmitted by *M. persicae*, and then in the persistent manner, if the source plants also contain tobacco vein-distorting virus. Transmission of tobacco mottle virus therefore requires the aid of a helper virus, in this instance tobacco vein-distorting virus.

Some other viruses behave similarly. For example carrot red leaf virus serves as the helper for transmission of carrot mottle virus by the aphid *Cavariella aegopodiae* (Watson, Serjeant and Lennon, 1964), anthriscus yellows virus as helper for transmission of parsnip yellow fleck virus by *C. aegopodiae* (Murant and Goold, 1968), and either potato virus Y or other potyviruses as helper for transmission of potato aucuba mosaic virus by *M. persicae* (Kassanis, 1961; Kassanis

Table 13.4 Effect of potato virus Y (PVY) on the transmission of potato aucuba mosaic virus (PAMV) by aphids, *Myzus persicae* (Kassanis and Govier, 1971a)

Virus in source plants		Transmission of
First plant[1]	Second plant[1]	PAMV[2]
Healthy	PAMV	0/14
PAMV	PVY	0/13
PVY	PAMV	13/25
PVY+PAMV	—	15/25

[1] Source plants were tobacco cv. Xanthi-nc.
[2] Numerator is the number of *Capsicum annuum* test plants that became infected, denominator is the number exposed to infection; 15–20 aphids per test plant.

and Govier, 1971b). In each instance the persistence of the assisted virus in its vector is the same as that of the helper virus, but whereas carrot red leaf is a persistent virus, anthriscus yellows virus is semi-persistent and potato virus Y non-persistent. Also, in each instance the assisted virus is sap-transmissible whereas, of the helper viruses, only the potyviruses can be transmitted by inoculation of sap.

Potato virus Y and potato aucuba mosaic virus do not need to be acquired from the same plant, providing that aphids feed on the source of the helper first (Table 13.4), and then potato aucuba mosaic virus can even be acquired by aphids feeding on purified virus through a membrane (Kassanis and Govier, 1971b). The anthriscus yellows virus-parsnip yellow fleck virus system behaves similarly (El Nagar and Murant, 1973). In these instances it seems possible that plants infected with the helper virus contain a virus-induced transmission factor distinct from the particles of the helper virus, and with properties that determine the vector relations of both helper and assisted viruses. However such a factor is not yet proved to exist.

Chapter 14

Virus ecology

14.1 Introduction

Individual plants have limited lives, so viruses must spread from one plant to another in order to survive. In the last chapter we described the various ways in which vectors transmit viruses, and in this chapter we go on to discuss spread in nature.

In well-established plant communities, such as old pastures or woodland, virus diseases are rarely obvious, presumably because the most virulent strains, and also the plants most susceptible to damage by viruses, have been eliminated. When the community is changed, for example by cultivation or by introducing new plant species, the ecological relationships are changed and virus diseases often become evident. It is for example probable that the cacao swollen shoot viruses and their mealy-bug vectors were indigenous in trees and shrubs in West Africa long before cacao was introduced from tropical America in the nineteenth century, but when the new crop was grown widely the viruses spread to it, with disastrous consequences. Crops are quite unnatural communities in that they usually consist of very large numbers of only one type of plant all at a similar stage of development – a situation which often favours the rapid spread of the fastest multiplying strains of both virus and vector. Because of the economic incentive, there has been much more emphasis on the study of the ecology of viruses in crops than in natural plant communities, and the discussion in this chapter will be similarly biassed. Because so much work has been done, it is impossible here to give more than a broad outline with some illustrative examples; further details may be found in the articles listed in section 2 of the appendix.

14.2 Sources of viruses and virus vectors for crops

14.2.1 *Planting material*

The vegetatively derived progeny of virus-infected plants is usually also virus-infected, whereas most, but not all, sexually-produced progeny is virus-free. Such infected plants in the planting material used to establish a new crop are a potent source of virus for the crop. For example Tomlinson (1962) found that five crops of lettuce grown from ordinary commercial seed, 2.2% to 5.3% of which was infected with lettuce mosaic virus, gave crops in which 25% to 96% of the plants were infected, whereas six crops grown from specially selected seed (less than 0.1% infected) gave crops with 0.5% or less infected plants at the end of the growing season. Similarly, seed-borne viruses such as tomato black ring virus can reach crop sites in naturally dispersed weed seeds (Murant and Lister, 1967).

Vectors too can be carried on planting material. For example, the nematode *Xiphinema index* probably largely owes its present worldwide distribution to being transported in soil with the roots of grapevine rootstocks from Europe.

14.2.2 *Virus-carrying vectors or virus-infected plants surviving from a previous crop on the same site*

Many of the soil-borne viruses persist in their vectors in the soil, and are transmitted to a new crop when it is planted. For example the nematode *Trichodorus pachydermus* still transmits tobacco rattle virus after being kept in soil free from plants for more than one year.

Comparable behaviour is found with soil-borne wheat mosaic and potato mop-top viruses, which are transmitted respectively by the plasmodiophoromycete fungi *Polymyxa graminis* and *Spongospora subterranea*. These viruses have narrow host ranges and survive in resting spores of the fungi, probably for many years (McKinney, 1923; Rao and Brakke, 1969; Jones and Harrison, 1969; 1972). Some nematode-borne viruses, such as tomato black ring virus, persist only a few weeks in their vector, *Longidorus elongatus*, but persist longer in nature in infected seeds. Thus when dormant weed seeds germinate in spring, virus-free *L. elongatus* reacquire tomato black ring virus by feeding on the roots of the young infected weed seedlings (Table 14.1).

Infected crop plants, weeds, or their seeds surviving from a previous crop are also sources of many other viruses, such as cucumber mosaic virus (Tomlinson and Carter, 1970). Doncaster and Gregory (1948) found that potato ground-keepers (plants growing from tubers left in the ground after harvest) may number several thousands per acre in wheat grown the year after a potato crop, and that some potato plants could still be found after five years of successive cereal crops. Groundkeepers that were infected with aphid-borne viruses were a potent virus source; for example when one part of a potato crop was grown on land carrying groundkeepers which came from a potato crop grown 2 years previously it contained 96% virus-infected plants at the end of the growing season, whereas in another part of the same crop without groundkeepers only 9% of the plants became infected.

14.2.3 *Viruses brought by vectors from sources outside the crop site*

Many crops are infected by virus-carrying vectors from several different outside sources. For example, in Europe, plants in sugar-beet crops are frequently infected with one or more of three aphid-borne viruses; beet yellows, beet mild yellowing and beet mosaic viruses. These viruses are transmitted by aphids, mainly *Myzus*

persicae, which has a wide host range. Both yellowing viruses infect and overwinter in several ways; in sugar-beet crops being kept for a second year to produce seed; in other widely grown forms of *Beta vulgaris* such as fodder beet, mangold and beetroot; in spinach; and in wild beet growing on the foreshore around the coasts. Also, beet mild yellowing virus occurs in many common weeds such as *Capsella bursa-pastoris*, *Senecio vulgaris*, *Stellaria media* and *Chenopodium album* (Russell, 1968). The relative importance of these sources is not known, and doubtless changes from place to place and from one year to the next. The weed hosts are ubiquitous; near towns beetroot and spinach are common; in the country beet-seed crops, fodder beet and mangolds are more important; and near the coast wild beet is sometimes important. Fodder beet and mangolds are grown for cattle food, and are stored during the winter in clamps often at the sides of fields. The clamps provide suitable conditions for aphids to overwinter, and when the alate (winged) aphids leave these clamps in the spring they take viruses with them. Figure 14.1 is an aerial photograph showing where beet mild yellowing virus has spread from a mangold clamp into an adjacent sugar-beet crop.

Mature crops are frequently a source of virus-carrying vectors for nearby young crops, especially when an overlapping sequence of crops is grown, as in tropical countries. This may happen even in temperate climates, for example in Canada when winter wheat is sown too early in the autumn and germinates before the maturing crops

Table 14.1 The ability of *Longidorus elongatus* to infect bait plants with tomato black ring virus (TBRV) before and after feeding on seedlings of different indigenous weed species that had been infected through the seed. (After Murant and Lister, 1967.)

Source of field soil	Initial numbers of *L. elongatus* in 500 g soil	Initial infectivity of the soil[1]	Weed species allowed to grow[2]	Percentage of the weed seedlings infected with TBRV	Final numbers of *L. elongatus* in 500 g soil	Final infectivity of the soil[1]
Invergowrie, Perthshire	25	5/30	All weeds	Not tested	22	8/30
			Stellaria media only	19	18	3/30
			Spergula arvensis only	8	18	0/30
			No weeds	0	20	0/30
Glendevon, Kinross-shire	70	0/30	All weeds	Not tested	134	5/30
			Stellaria media only	22	97	20/30
			Chenopodium album only	4	170	5/30
			No weeds	0	62	0/30

[1] Numerator is the number of pots of soil in which turnip bait plants became infected. Denominator is the total number of pots tested.
[2] Weed seeds occurring in the soils were allowed to germinate, and all except the species indicated were removed daily for 3 months, and then the final tests were made.

100 m

Fig. 14.1 Aerial photograph of a crop of sugar beet growing near Bury St Edmunds, England in 1965. The field contained a mangold clamp (lower left corner), the contents of which were not completely used until after emergence of the sugar-beet seedlings. Viruliferous aphids spread from the mangolds to the beet, and they and their progeny infected patches of plants with beet mild yellowing virus. (Courtesy G. H. Brenchley.)

of spring wheat have been harvested. The winter crops then often become severely infected with wheat streak mosaic virus, which is transmitted by mites that are blown from the maturing crops or are carried by volunteer wheat plants growing on summer fallowed land (Slykhuis, 1953, 1955).

Virus and vector may come from the same source. For example those viruses which are only acquired by their vectors during long periods of feeding will only be transmitted by vectors that have spent such a period on a virus-infected source plant. Most of these viruses are also best transmitted to healthy plants during long feeds, so that they are mainly spread by species that colonize both the source plant and crop plant. For example, in Europe bean leafroll virus is trans-

mitted between and within lucerne, bean and pea crops by the pea aphid (*Acyrthosiphon pisum*), which colonizes all these. However, an exception to this rule is lettuce necrotic yellows virus, which multiplies in its aphid vector *Hyperomyzus lactucae*. In Australia this virus is common in the weed *Sonchus oleraceus*, which is also a host of the vector. The aphids transmit the virus from *S. oleraceus* to lettuce but do not colonize the lettuce, and secondary spread within the lettuce crops is probably negligible (Stubbs and Grogan, 1963).

Another interesting example of a virus that spreads into but probably not within a crop is provided by parsnip yellow fleck virus, the commonest virus infecting parsnips in Britain. In laboratory tests, this virus is transmitted by the aphid *Cavariella aegopodii* only when source plants also contain anthriscus yellows virus (§13.2.12). The two viruses occur together in wild umbelliferous species but only parsnip yellow fleck virus can infect parsnip. Hence transmission takes place mainly between wild umbellifers, and from these to parsnip (Murant and Goold, 1968).

Virus and vector do not always come from the same source. For example most viruses that are transmitted by aphids in the non-persistent manner have many aphid species as vectors (Kennedy, Day and Eastop, 1962), and are transmitted best by aphids that probe the plants briefly. When aphids alight on, or crawl onto, a plant they usually probe the plant briefly to test whether it is suitable as a host, and in doing this they may acquire or inoculate non-persistent viruses. Thus non-persistent viruses may be transmitted by many of the aphid species entering the crop, even those that do not settle and colonize the crop; indeed non-colonizers may be more important than colonizers because they fly from plant to plant in search of a host species, and do not settle. For example, in a field experiment in England, red clover vein mosaic virus spread among white clover plants in a pasture when *Cavariella* spp. were caught on sticky traps, but not when aphids known to colonize white clover (e.g. *Acyrthosiphon pisum*) were caught; in subsequent glasshouse experiments *Cavariella aegopodii* and *C. theobaldi* transmitted red clover vein mosaic virus from clover to clover as efficiently as *A. pisum* (Varma, 1967). Thus the importance of an aphid species as a vector of a non-persistent virus is not necessarily related to its ability to colonize hosts of the virus, but may be more dependent on the abundance and activity of the vector.

The abundance of a particular vector species entering a crop varies from year to year and throughout the year. The number of insects surviving the winter and entering crops in the spring depends on weather conditions during the winter. In south-western U.S.A. the beet leafhopper overwinters in desert areas, where rain falls mostly in winter. When the desert plants are affected by drought in late spring the leafhoppers move to irrigated sugar-beet crops, infecting them with beet curly top virus and starting an epidemic which spreads over considerable distances (Fig. 14.4).

In severe winters in Britain most adult aphids are killed and only eggs survive, whereas in mild winters large numbers of adults survive on weeds, in crops and root clamps, and in sheltered gardens. After mild winters alate adult aphids are produced earlier and in larger numbers than after severe winters, and this difference is reflected in the incidence of viruses. Thus the incidence of beet yellowing diseases in British sugar-beet crops is closely correlated with the number of *Myzus persicae* entering the crops in the spring,

which in turn depends on the weather of the preceding months, so that yellows incidence can be predicted with some accuracy just from weather records (Fig. 14.2).

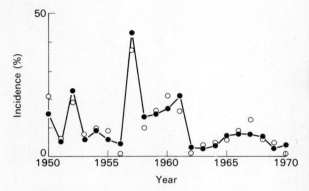

Fig. 14.2 The mean incidence of beet yellows disease in sugar-beet crops in Britain in August, from 1950 to 1970 (unbroken line and discs), compared with the disease incidence predicted from weather records of the same years (circles). The predicted values were calculated (Watson, 1966) from the multiple regression equation relating the incidence of yellows in August with the number of days of frost in the previous January, February and March, and the deviations from average April temperatures. (Courtesy Marion A. Watson.)

Insects may migrate when for one reason or another their plant hosts become unsuitable; for example many aphid species will only colonize immature shoots or senescing leaves, so that they are found on different plants at different times of the year. Figure 14.3 shows the annual life cycle of *M. persicae* which has a wide host range, transmits many viruses efficiently and has restless habits, and is therefore one of the most important virus vectors. When insects become crowded they leave plants. Small colonies of aphids usually produce apterous (wingless) adults only, but when the plants become unsuitable or the colony crowded, alate adults are produced, and these then migrate. Similarly, when conditions become less favourable, leafhoppers may produce long-winged instead of short-winged adults.

14.3 Pattern and amount of spread of virus in crops

The speed with which viruses spread varies greatly, and depends on two main groups of factors, those affecting spread between plants, and those affecting spread within each plant (Chapter 11).

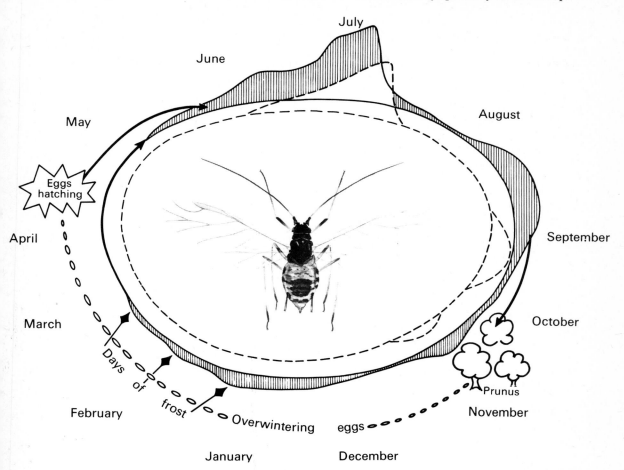

July

June

May

August

Eggs
hatching

April

September

March

Days

of

frost

Overwintering

eggs

October

Prunus

November

February

January

December

Fig. 14.3 Diagram of the annual life cycle in Britain of the aphid *Myzus persicae*, an important virus vector. The hatched areas represent the numbers of the aphid on herbaceous hosts. These individuals are all alate or apterous females, reproducing by parthenogenesis. They reach a maximum population during the summer, feeding on many different plant species including potatoes and sugar beet. The population decreases greatly during the autumn and winter, and may be completely killed by prolonged frost. During the autumn, alate aphids landing on *Prunus* spp. start colonies, which produce sexual forms; these mate and eggs are laid. The eggs are resistant to frost, and hatch in the spring to give colonies which produce alate females that migrate to the summer hosts. The aphid population is partially controlled by various predators and parasites (ichneumons, ladybirds, hover flies and fungi), whose population changes (broken line) follow that of the aphid.

14.3.1 *Vector movement*

One important factor determining how far viruses spread from a source is the distance that can be covered by the vector. Soil-inhabiting vectors such as nematodes (Harrison and Winslow, 1961) and fungi move only slowly, and the viruses they transmit therefore usually spread only small distances, unless soil containing these vectors is moved, for example when it is cultivated, or dust is carried by the wind. Most nematode-transmitted viruses are also transmitted through pollen and seed (Lister and Murant, 1967), and may be spread much more widely and quickly in these ways than by their vectors, which spread only 0.3 to 0.6 m per year. The viruses which spread furthest and fastest are those transmitted by wind-borne or flying insects; flying aphids may be carried by a wind for tens or hundreds of kilometres in one day.

In practice, the mobility of vectors depends on many factors, and especially on the environmental conditions. The factors that affect the movement of nematodes in soil are discussed by Wallace (1963); the most important is water. Thus the soil must contain more than a critical amount of water before *Trichodorus* will transmit tobacco rattle virus (Cooper and Harrison, 1973b), and this amount depends on the pore structure of the soil. Fungus vectors also depend on the soil water for mobility and, in the U.S.A., infection with soil-borne wheat mosaic virus occurs particularly in the poorly drained parts of fields. Soil pH too can be important, and potato tubers rarely become infected with potato mop-top virus or its fungus vector, *Spongospora subterranea*, at or below pH 5 (Fig. 15.9; Jones and Harrison, 1968).

The main environmental factors affecting the activity of aerial vectors are air movement and temperature. An example of the interplay of the meteorological factors affecting the behaviour of aphids is discussed in §14.3.2. With mites, the situation is simpler because they are carried passively in the wind (Slykhuis, 1955), and the direction, amount and distance of spread of wheat streak mosaic virus from a source by *Aceria tulipae* seem related to the speed and prevailing direction of the wind.

14.3.2 *Patterns and gradients of disease incidence*

Much can be deduced of the ecology of a virus from the distribution of infected plants in a crop and the time at which the plants become infected. When plants are virus-infected at emergence from the soil, it is likely that the planting material was infected or possibly that the plants were infected by a soil inhabiting vector. When these infected plants are distributed at random in the crop or are only in the rows containing one lot of planting material, it is even more likely that the planting material was infected. However when plants in patches develop symptoms at about the same time after planting, and when such patches are related to soil differences, this suggests the virus is soil borne. Such patches usually occur in much the same place in successive crops, although the affected areas may slowly enlarge.

Most of the important viruses of annual crops are transmitted by aerial vectors that move from plant to plant. Often a few plants in the crop are infected first and vectors then spread virus from these to other plants in the crop; this often results in the diseased plants being in expanding patches or clumps (Fig. 14.1), in which individual plants may be in different stages of symptom development. These patches are not in the same place in successive crops.

When virus is spreading from a source outside a crop, it is often possible to see a gradient in the incidence of infection when the crop is near the source (Fig. 14.1). The detailed interpretation of such gradients has been discussed by Gregory (1968). Far from the source the gradient may be so flat that the diseased plants seem to be at random. Where there is a gradient, the incidence is usually inversely and exponentially related to the distance from the virus source and will often closely fit the formula

$$\log I = a + bx \quad \text{or} \quad \log I = a + b \log x,$$

where I is the disease incidence at distance x from the source and a and b are constants; b usually has a negative value. When I is small, as for example when the virus is first spreading, the gradient (b) is often large and negative, but after further spread, especially secondary spread, then b decreases. At the centre of a patch of infected plants few healthy ones remain, so that more vector activity is needed to infect these few than the same number elsewhere in the crop, where most of the plants are healthy; thus when I is large the gradient becomes artificially flattened because of 'multiple infection'. Therefore before comparing disease gradients the effect of multiple infection should be removed by adjusting the data using a transformation, which has been tabulated by Gregory (1948).

Disease gradients depend on many factors, such as the mobility of the vector (Fig. 14.4). Viruses spread by leafhoppers, which are strong fliers, usually have shallower gradients than those spread by aphids, mites, mealy bugs or nematodes. The disease gradients of a particular virus spreading in one type of crop will differ in different places and from one year to the next. Also for various reasons one must be cautious when extrapolating disease gradients such as those in Fig. 14.4 to estimate the incidence of virus at greater distances because, for instance, aphids settle more readily on some hosts than others and thus gradients will be steeper on acceptable hosts and will be flattened over non-hosts. Moreover gradients will be influenced by obstructions; for example those of viruses spread by aerial vectors will be affected by hedges and woodlands (Lewis, 1966), and by the prevailing wind direction.

One vector species may give different gradients with different viruses. For example, non-persistent aphid-transmitted viruses usually give steeper gradients than persistent ones, not just because they persist in the aphid for different times but also because each is spread best in different

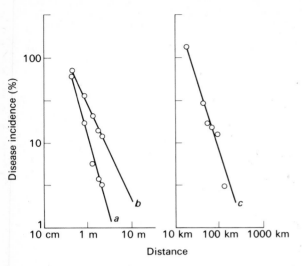

Fig. 14.4 Incidence of virus infection at different distances from a virus source. The incidence of potato leafroll virus (*b*) and potato virus Y (*a*) at distances from infector plants in a plot experiment at Rothamsted, England in 1943; and the incidence of beet curly top virus (*c*) in bean crops at different distances from the winter breeding grounds of the leafhopper *Circulifer tenellus* in Idaho. (Data from Gregory and Read (1949) and Annand *et al* (1932), and transformed to compensate for multiple infections.)

meteorological conditions (Cockbain, Gibbs and Heathcote, 1963). Winged aphids are restless when they first become mature, and in warm conditions with low wind speeds fly from their host plant. At the start of a flight they fly upwards strongly, and if weather conditions are suitable (e.g. turbulent with convective mixing) they may fly for several hours and be carried long distances. When, however, there is either little wind and stable air conditions (Fig. 14.5), or high wind, they will not rise very far and may be deposited on nearby plants, and though they may probe the plants briefly they will not settle and usually soon fly again. After a long flight, or a series of shorter flights, the urge to fly is satisfied, and when the aphid has landed on a suitable host, it will settle and feed. Non-persistent viruses are transmitted most frequently by aphids that are probing or feeding for only short periods on infected plants, and are therefore spread most when, because of weather conditions, the aphids fly only short distances (large *b*). In contrast persistent viruses are best spread by aphids that feed for longer times, which they will only do after longer flights in weather conditions favouring dispersal (small *b*). This is illustrated in Fig. 14.4, which shows the gradients of potato leafroll virus (persistent – *b* negative and small) and potato virus Y (non-

persistent – *b* negative and larger) spreading in the same experimental plots.

The total incidence in the crop of those viruses which produce steep disease gradients and which spread into the edges of fields from outside sources will be proportional to the total length of the edges of the crop. Thus under similar conditions large fields will contain a smaller proportion of diseased plants than small ones, because large crops, unless of unusual shape, have a smaller ratio of edge to total area (van der Plank, 1949a, b).

Usually the distance between rows in a crop is greater than the distance between plants in the row; thus it is not surprising that sometimes viruses spread from plant to plant along the row more often than from row to row. Where a virus has spread from plant to plant within the crop there will be more diseased plants next to one another than expected by chance. One way of detecting this was devised by van der Plank (1946), who showed that if, in a sequence of n plants, d are diseased, then the number of pairs of adjacent diseased plants (p) expected by chance will be $d(d-1)/n$, with a standard error (when n is large) of $\sqrt{p}$. In Fig. 14.6 (data for 1948), for example, 401 plants out of 1728 are diseased; among these 275 pairs can be found instead of the 93 expected by chance, suggesting that the virus has spread from plant to plant. Van der Plank's method is

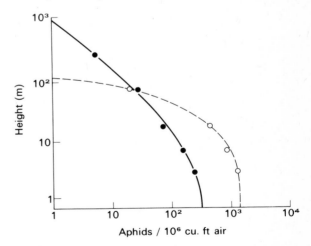

Fig. 14.5 Numbers of aphids caught in suction traps at different heights. In the morning (unbroken line and discs) a 16 km/h wind was blowing, distributing the aphids throughout the profile by turbulent mixing, thus ensuring long flights by the aphids. In the afternoon of the same day (broken line and circles) the wind speed and mixing were less, and aphids accumulated near the ground, presumably making shorter flights than in the morning. (Courtesy L. R. Taylor.)

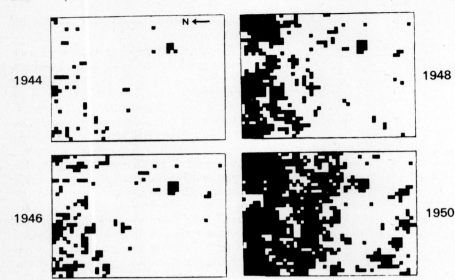

Fig. 14.6 The spread of cacao swollen shoot virus in cacao (*Theobroma cacao*) in a field experiment in Trinidad (Dale, 1953). Each black square represents one infected tree. Trees were spaced on a 12 ft (3.7 m) square lattice, and were in the shade of *Erythrina peoppigiana* trees. (Courtesy W. T. Dale; see also Fig. 14.7.)

most accurate when the proportion of diseased plants is small.

14.3.3 *The progress of epidemics*

The rapidity with which virus incidence increases in crops varies greatly. When the source of infective vectors is outside the crop, and there is no spread within the crop, then the increase in incidence will be proportional to the number of infective vectors entering the crop, and a graph showing the incidence at different times would, in theory, resemble the 'simple interest' graphs of economists, but will be flattened at large incidences of infection because of the effect of multiple infections. More usually the epidemic in the crop is mainly dependent on secondary spread from plant to plant within the crop, so that the rate of increase is related to the number of plants already infected, and is therefore exponential and of the 'compound interest' type. At large incidences of infection the progress of the epidemic often slows (Fig. 14.7), either because the plants age and become resistant to infection, or because the vector population decreases. To compare epidemics statistically or even graphically it is best to transform the data to obtain a linear relationship between time and infection incidence. This is done by converting the proportion of diseased plants d to log d, or log $[d/(1-d)]$ or to probits (Fig. 14.7).

14.4 Factors affecting spread of virus within the crop

14.4.1 *Agronomic and host factors*

Typically, plants are most susceptible to virus

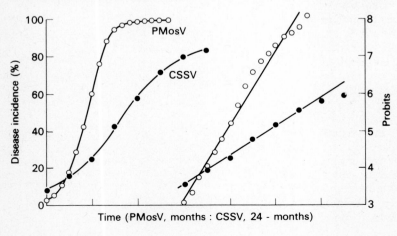

Fig. 14.7 The progress of epidemics of viruses in crops; pea mosaic virus (circles) in sweet peas (*Lathyrus odorata*) and cacao swollen shoot virus (discs) in cacao (*Theobroma cacao*). The rate of progress of the epidemics can be compared better when the percentages (left) of infected plants are transformed into probits (right). Note that the time scale for pea mosaic virus has units of months, and that for cacao swollen shoot virus units of two years. (Data from Hull and Selman 1965 and Dale, 1953.)

infection when they are young, and are often most suitable as hosts for vectors at the same time; as they grow and age the plants become resistant to infection. Therefore crops planted so that they are young at the time when virus-carrying vectors are moving among the plants will have a larger incidence of virus than those planted either earlier or later. This effect is common and is illustrated in Table 14.2, which shows the incidence of pea

Table 14.2 Incidence of pea mosaic virus in plots of field bean (*Vicia faba*) sown on different dates at Rothamsted, England

Date examined	Plots sown on		
	March 9	April 27	May 3
July 1	0.8[1]	0.8	2
July 14	0.8	7	8
August 7	2	35	41

[1] Figures are the percentage of plants showing symptoms.

mosaic virus in plots of field beans sown on different dates.

Viruses also spread more slowly in older crops not only because plants resist infection as they age but also because the multiplication and systemic spread of the virus are slower in older plants, which are therefore poorer sources of virus for vectors than young plants. Similarly, different cultivars of a crop may differ in susceptibility to infection and as virus sources (Chapter 15).

Adequately manured plants growing well are more susceptible to infection, and a better source of virus for vectors, than nutrient deficient ones but, fortunately for the farmer, well-manured virus-infected crops usually outyield poorly fed virus-free ones. Large plants also are more likely to be visited by vectors than small ones purely because of their size; for example in one bed of cauliflower seedlings 30% of the large seedlings were infected with cauliflower mosaic virus but only 15% of the middle-sized and 5% of the small ones (Broadbent, 1957).

There are many recorded instances of the incidence of viruses in crops being affected by plant spacing. Some viruses, such as the mealy-bug-transmitted cacao swollen shoot virus, are spread principally by vectors that crawl from plant to plant, so that the virus spreads mainly where the plants are touching (Thresh, 1958). The incidence of viruses brought into the crop by flying insects is also influenced by planting density, because the number of virus-carrying insects flying into a given area of crop is independent of the number of plants in that area. When the area contains few plants a larger proportion will be infected than when it contains many. In African crops of groundnut there is a further effect of close planting that is incompletely understood. The incidence of groundnut rosette virus in plots of close-spaced plants is much less than would be expected from the incidence of the virus in plots of wide-spaced plants (A'Brook, 1964, 1968). In the plots of close-spaced plants the ground is obscured by the crop much sooner than in the plots of wide-spaced plants, and it has been suggested that the continuous canopy may inhibit aphid vectors from alighting (A'Brook, 1973a). Aphids that have flown for some time are attracted to colours at the yellow and red end of the spectrum and are repelled by blue, so that flying aphids may be more strongly attracted by groundnut plants surrounded by reddish soil than by a continuous canopy of mature shining blue-green leaves. This is presumably also the reason why fewer gladiolus and squash plants became infected with viruses when the ground around the plants was covered with light-reflecting aluminium foil than when it was bare (Smith *et al*, 1964; Chapter 15).

When a crop contains more than one species of plant this will influence virus spread; often the different plant species grown together in a mixed crop are not hosts for the same viruses and vectors, and viruses spread less in such crops than in crops of a single species. For example in Britain lucerne is often grown in alternate rows with a grass, usually cocksfoot, and in such mixed crops the spread of alfalfa mosaic virus is decreased. In one experiment, the incidence of alfalfa mosaic virus in plots of pure lucerne was three times that in the lucerne plants in plots of lucerne plus cocksfoot. Virus may not spread at all among plants that are a minor component of the vegetation and are covered and surrounded by other plants. For example, aphid-borne viruses do not spread among potato groundkeepers growing in cereal crops because the groundkeepers rarely become infested with aphids; thus the groundkeepers provide a record of the virus incidence in the crop from which they were derived several years before (Doncaster and Gregory, 1948). Similarly, a crop of sugar beet can be kept virus-free by growing it under a cover crop of a different species (Fig. 15.5), even in localities where beet viruses are widespread, and this is a convenient way of producing virus-free sugar-beet crops to be grown for a second year to provide seed (Hull, 1954; Chapter 15).

14.4.2 *Virus and vector factors*

There may be important differences in the ecology of different virus strains. For example, outbreaks of the Scottish strain of raspberry ringspot virus are only found where the soil contains the nematode *Longidorus elongatus*, whereas those of the English strain are associated with *L. macrosoma*. Each virus strain is more efficiently transmitted by the species with which it is naturally associated, and indeed *L. macrosoma* rarely or never transmits the Scottish strain (Harrison, 1964a; Taylor and Murant, 1969). Similarly, different isolates of barley yellow dwarf virus differ greatly in vector specificity. Some are transmitted by one species, which may, for example, be *Macrosiphum avenae*, or *Rhopalosiphon padi*, whereas others are transmitted by two or more species (Rochow, 1969). The differences in vector specificity of isolates of barley yellow dwarf virus may be associated with differences in their virulence for plants. Plumb (1971) found that in the U.K., barley yellow dwarf virus appeared first each year in cereal crops in the south and west. The first isolates to appear in these crops were mostly virulent and were carried by *Rhopalosiphon padi*, whereas most of the isolates arriving late in the year were avirulent and were carried by *Metapolophium dirhodum* or *Sitobion avenae*.

In general, the strains of a virus that survive in nature are those that multiply sufficiently to be readily acquired by vectors yet do not damage their hosts so severely that infected plants fail to survive and constitute a reservoir of virus. For example, naturally occurring strains of the mite-transmitted ryegrass mosaic virus seem mostly to be of intermediate virulence, although both avirulent and lethal strains can easily be selected (Wilkins and Catherall, 1974).

Where a virus has more than one vector species, the different species may differ in efficiency. For example beet yellows virus has been experimentally transmitted by more than two dozen aphid species, yet in British sugar-beet crops only two aphids are at all common. *Aphis fabae* is the commoner and when abundant causes damage as a pest, whereas *Myzus persicae* occurs in much smaller numbers. However surveys over several years of the incidence of sugar-beet yellows disease and of the two aphid species showed that the virus incidence is closely correlated with the numbers and earliness of arrival in the crop of *M. persicae*, and is quite independent of the numbers of *A. fabae* which is more sedentary and in experiments is much the less

efficient vector (Watson and Heathcote, 1966). Another complication is that different species may be responsible for virus spread in different places. For example, whereas *Toxoptera citricidus* seems to be the most important vector of citrus tristeza virus in S. Africa and S. America, *Aphis gossypii* is the most important in some parts of the U.S.A., where *T. citricidus* does not occur. Similarly, different clones of one aphid species may also differ in their efficiency as vectors of a virus. For example, Cockbain and Costa (1973) found that clones of *Acyrthosiphon pisum* differed in their ability to transmit pea enation mosaic virus; this difference was inherited in clones for at least two years, and their efficiency as vectors of pea enation mosaic virus was not correlated with their efficiency as vectors of bean leafroll virus. When differences in vector efficiency are found it is advisable to check whether the different clones are indeed all of the same species, for observations of this sort have on several occasions proved to result from mis-identification.

Many other factors also affect the proportion of individuals in a natural vector population that transmit a virus. These include the stage (or instar) of the vector (Chapter 13), the availability of appropriate sources of virus and the environmental conditions, especially temperature (Chapter 15).

In most conditions, though not all, viruses decrease the competitive ability of their host plants, and this increases the need of the viruses to spread to virus-free plants to survive. Indeed some viruses affect their host plants and/or vectors in ways which increase the chances of the virus being spread.

Many aphid-transmitted viruses cause the host plants to become chlorotic, which makes them more attractive to flying aphids seeking a host. Baker (1960) found that aphid species commonly found on sugar beet in England prefer red-beet plants infected with beet yellows virus to comparable healthy ones; aphids preferred the leaves showing the severest symptoms, and on these *Myzus persicae*, which is the most important vector of the virus, lived half as long again and produced three times as many progeny as on healthy leaves.

The morphology of plants, and hence the microclimate at their surface, is altered by viruses, sometimes to the advantage of the vectors. For instance the thrips vector of tomato spotted wilt virus prefers the curled leaves of infected plants of *Emilia sonchifolia* to the flatter leaves of healthy plants. Likewise the beet leafhopper, *Circulifer tenellus*, which prefers dry conditions, thrives better on stunted beet with curly top disease than on healthy beet, which maintain more humid conditions (Carter, 1973), and the currant gall mite, *Phytoptus ribis*, colonizes blackcurrant bushes with reversion disease far more readily than the more hairy healthy bushes (Thresh, 1967).

14.5 Finding which are the important vectors in nature

If one wishes to control the spread of a virus, then it is important to know how it is being spread. For most plant viruses this means trying to find the vector, which can be difficult because plants in natural conditions usually provide food and lodging for many different organisms, and many of these may be suspect. The type of vector can often be guessed either from the pattern of spread of the virus or, when the affinities of the virus are known, the vector type can often be predicted by analogy with the known vectors of allied viruses (see Chapters 2, 12 and 13). Usually the vector of a virus is identified by trying to correlate virus spread with the presence or activity of various potential vectors and then using the most likely vector on its own in transmission tests.

Two main types of experiment have been used to obtain clues to the identity of the vectors of soil-borne viruses. In the first, soil samples are collected from around infected plants and around healthy plants in the same field, and their fauna and flora compared to see whether the 'infectivity' of the soil is specifically correlated with the presence and number of any one species of organism. Figure 14.8 shows an outbreak of strawberry latent ringspot virus in a raspberry crop, where the distribution of the nematode vector *Xiphinema diversicaudatum* coincided with that of virus-infected plants and infective soil (Taylor and Thomas, 1968). In the second type of experiment, infective soil is treated or fractionated in various ways and attempts made to correlate the infectivity of each soil fraction with the presence or activity of one organism (Harrison and Cadman, 1959). For example, air drying abolishes the infectivity of soils containing all known nematode-borne viruses because the vector nematodes are killed by this treatment, whereas soils containing some fungus-transmitted viruses are still infective after air drying because these viruses are within

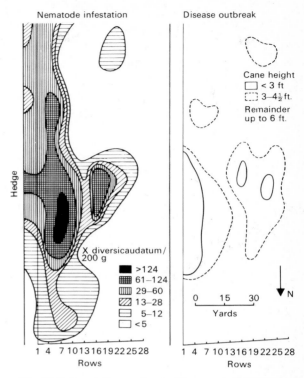

Fig. 14.8 Population contours of the nematode *Xiphinema diversicaudatum* in soil in relation to an outbreak of strawberry latent ringspot virus in a plantation of raspberry cv. Malling Jewel. The plants were stunted by the virus infection. (Courtesy C. E. Taylor; Taylor and Thomas, 1968.)

the fungus resting spores, which resist desiccation.

Aerial vectors, such as insects, are usually easier to manipulate and test as vectors than soil-inhabiting vectors. Insect populations on plants are readily estimated, and the number of flying insects can be assessed by trapping them in various ways (Fig. 14.9; Heathcote, 1957, 1958; A'Brook, 1973b). However, despite these advantages it is often difficult to find the identity of an aerial vector or vectors, because there are often many related species that could be implicated. Also, aerial vectors are very mobile, may only visit plants briefly, and many leave long before the plants show symptoms and plant pathologists arrive. Several other difficulties can readily be seen from the examples discussed in §14.4.2.

For various reasons, correlations between virus spread and any one type of estimate of the population of a potential vector can mislead. Firstly, the populations of several species are often similarly affected by weather. Second, the numbers of insects caught in traps depend on the number flying, not on the number visiting plants.

Fig. 14.9 Three types of aphid trap: **a,** the sticky trap; a yellow cylinder, to which is clipped a plastic sheet thinly coated with tree-banding grease. (Doncaster and Gregory, 1948; Heathcote, 1957.) **b,** The water trap; a dish, usually yellow, containing water to which has been added a drop of detergent and insecticide (Moericke, 1949). **c,** The suction trap; air is blown by an electric fan at the top into a conical sieve, insects collect at the bottom and are divided into batches by discs dropped at regular intervals into the cylinder at the end of the cone (Johnson, 1950). (Courtesy G. D. Heathcote and L. R. Taylor.)

Third, the number of winged insects on crop plants only indicates how many have stayed on the plants, not how many have visited, which for non-persistent aphid-borne viruses is probably more important. Finally there is the difficulty of assessing the relative importance as vectors of flying and crawling forms of one species.

Obviously the best way of finding which vector is most important will vary greatly with different viruses, and it is usually best to use a combination of approaches. For example, Schwarz (1965) showed that in South Africa the incidence of citrus tristeza virus in lime bait plants exposed for monthly periods over three years in citrus orchards was correlated with the number of *Toxoptera citricidus* found on the bait plants (Fig. 14.10) and not with the number caught in yellow

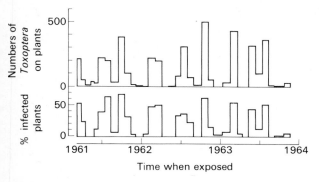

Fig. 14.10 The relation between the incidence of citrus tristeza virus in West Indian lime plants exposed for one month to infection in citrus orchards in Eastern Transvaal, South Africa, and the numbers of the aphid *Toxoptera citricidus* found on them. (Data from Schwarz, 1965.)

water traps, nor the numbers of *Aphis gossypii* either on plants or in traps. Presumably the water traps gave an estimate of the number of aphids flying, not the number that settled on and infected the bait plants.

14.6 Viruses which spread without vectors

In this chapter we have mainly discussed the behaviour of vector-borne viruses, but there is also some information about viruses that spread without a vector in the usual sense.

Labour-intensive crops, particularly those grown in glasshouses, are sometimes infected with viruses transmitted by contact or on tools, hands, etc. This is because intensive cultivation systems often involve frequent manipulation of the plants and this favours viruses that spread by contact. For instance tomato mosaic virus, a strain of TMV, is spread in tomato crops in glasshouses mainly on the tools and hands of workers tending the plants. This virus can survive for years in infective tissue or in sap dried on clothing, tools or woodwork in the glasshouse, as well as in plant debris in the soil. It also infects the testa and endosperm, but not the embryo, of tomato seed and, when seedlings are transplanted, it is often trans-

mitted from the remnants of seeds to the young plants (Broadbent, 1965). Infection of a few seedlings in a crop by soil- or seed-borne virus is thus followed by rapid spread of virus from these seedlings to the remainder when the plants are handled.

Potato virus X spreads in potato crops by contact between infected and healthy shoots or roots (Roberts, 1948), and some other viruses whose particles occur in tissues in large concentration behave analogously. Todd (1958) found that the incidence of potato virus X in a potato stock approximately doubled each year, and showed that the virus could also be transmitted by contact with contaminated clothing or contaminated fur of rabbits or dogs. Mowing machines seem largely responsible for the spread of white clover mosaic and cocksfoot mottle viruses in swards. Again both are viruses whose particles occur in large concentrations, but cocksfoot mottle virus also has an insect vector, the beetle *Oulema melanopa* (Serjeant, 1967).

A few other viruses spread in a totally different way. Prunus necrotic ringspot virus is transmitted by cherry pollen to pollinated cherry trees (George and Davidson, 1963), and raspberry bushy dwarf virus behaves similarly in raspberry. Plants that are prevented from flowering do not become infected with either virus.

14.7 Ecological systems

In nature all viruses need a succession of hosts to survive but, as we have already seen in this chapter, the way in which the succession is achieved differs greatly from one virus to another, and depends on the type and behaviour of the vector. Viruses that spread without the aid of a vector in the usual sense may be pollen-borne to pollinated plants, or may attain large particle concentrations in plant tissue and spread by contact. However there are also viruses such as apple mosaic, which seem rarely to spread but survive largely because they are perpetuated in nursery stocks that are vegetatively propagated by man. Whether all these viruses had vectors at an earlier stage in their evolution, or have them now in other situations, is debatable but it is clear that these viruses mainly owe their present distribution, directly or indirectly, to man's activities.

Among vector-borne viruses, those that are confined to short-lived plants have efficient transmission systems, whereas those that infect but do not kill long-lived perennials can survive without spreading rapidly. Many of the more 'successful' viruses, however, survive in perennial reservoir hosts, and spread from these to annuals when the conditions are favourable.

In general, vector activity is more important than vector numbers in determining the amount of virus spread although, as already mentioned, types of vector behaviour that favour the transmission of one virus may be unfavourable for spread of another. Vectors inhabiting the soil usually move over only short distances in search of hosts, and the survival of the viruses they transmit seems to involve either the ability to infect roots of a wide range of plants (e.g. tomato black ring virus) or a method of surviving in the vector for long periods (e.g. potato mop-top virus). Among viruses with aerial vectors these kinds of property are less important, and some with narrow host ranges in nature and limited persistence in their vectors (e.g. potato virus Y) are nevertheless relatively successful.

The discussion in this chapter has been largely confined to the spread of viruses into and within crops. Some of these viruses (potato leafroll and potato Y viruses in Britain) depend wholly on crops for their survival. Others, such as barley yellow dwarf virus, depend only partly on crops and some, for example tobacco rattle virus in Britain, infect crops only incidentally. Finally, there are viruses that are not known to occur in crop plants. These have been little studied and, though their ecology is subject to the same general principles as that of other viruses, it would be surprising if their survival and spread did not involve some novel ecological systems.

14.8 Effects of virus infection on other plant pathogens

The ecological niche provided by a plant to its parasites and pathogens is altered by virus infection. We have mentioned in §14.4.2 some of the ways in which vectors are affected, but non-

vectors, such as leaf-infecting fungi, are affected too. For example, in British potato crops, plants infected with potato leafroll virus are often more heavily infected with late blight (*Phytophthora infestans*) than virus-free plants. However when the fungus is inoculated under laboratory conditions, the virus-infected leaves are found to be less susceptible to infection than the healthy ones. The effect observed in the field occurs because the rolled leaves of virus-infected plants retain rain and provide a more favourable microclimate for infection by the fungus than the flat leaves of healthy plants (Richardson and Doling, 1957). Similarly, broad bean plants infected with bean leafroll virus are more heavily infected in the field with the chocolate spot fungus (*Botrytis fabae*) than virus-free plants, but in this instance not only does the microclimate of the rolled leaves favour fungus infection but also the virus-infected plants are more susceptible to experimental infection than the healthy ones. Yet another example is provided by sugar-beet plants infected with yellowing viruses. In Central and Eastern Europe these show an enhanced susceptibility to the fungus *Cercospora beticola* and in Britain to *Alternaria* spp. In Britain *Alternaria* infection is especially common on beet infected with beet mild yellowing virus and the fungus is responsible for much of the loss in yield associated with virus infection (Russell, 1966b); thus the yield of virus-infected plants is increased when fungicidal sprays are used or genes for *Alternaria* resistance are incorporated into cultivars. Viruses causing mosaic symptoms, as well as those causing yellowing and leaf rolling symptoms, can affect the incidence of leaf-infecting fungi; thus in North America, potato plants infected with potato virus Y are more heavily infected with early blight (*Alternaria solani*) than comparable virus-free plants (Hooker and Fronek, 1960).

By contrast with these examples, there seems little evidence from experiments that infection in nature with other kinds of pathogens affects the susceptibility of plants to virus infection.

Chapter 15

Ways of preventing crop losses

15.1 Effects of virus infection on yield

The yield lost by crop plants when they become infected varies greatly. Infected plants may yield nothing, either because they are killed or because the part of the plant that is harvested is lost. Usually, however, infected plants survive but yield less than healthy ones, the loss depending on various factors such as the strain of virus, the species (or variety) of plant, and when the plant became infected (Fig. 15.1). Of these the last is

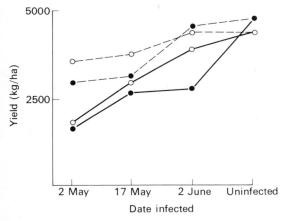

Fig. 15.1 The yield of plots of barley (cv. 'Proctor'; circles) and wheat (cv. 'Koga II'; discs) infected on different dates with either a virulent (unbroken line) or an avirulent (broken line) strain of barley yellow dwarf virus. (Data from Watson, 1959.)

often the most important, and the yield loss is often proportional to the area under the line in the 'time/incidence' graph of an epidemic (Fig. 14.7). For example in England, plants infected with yellowing viruses in crops of sugar beet yield about 4.5% less than virus-free ones for each week that they are infected, so that the yield loss can be predicted from estimates of the disease incidence at different times (Watson, Watson and Hull, 1946).

Viruses not only decrease the weight of crop but also its quality; for example tomato aspermy virus in chrysanthemum, plum pox virus in plum, and spraing caused by tobacco rattle virus in

potato decrease the sale value of the flowers, fruit and tubers, respectively. The effect on quality may be subtle; for example, King Edward potatoes infected with potato paracrinkle virus not only yield about 10% less weight of tubers than healthy plants, but also produce tubers of less uniform size (Bawden and Kassanis, 1965).

The yield of a crop partly infected with virus is usually larger than would be expected from the size and number of infected plants in the crop, because the infected plants use less space, nutrients and water than virus-free plants, and hence adjacent virus-free plants grow larger and partly compensate for the poor growth of the infected plants (Reestman, 1972). For example, in severe winters in Britain, cocksfoot plants (*Dactylis glomerata*) infected with cocksfoot streak virus are killed, but even when 30% of the plants die, the remaining plants grow larger and fully compensate for the dead plants by the end of the following summer (Catherall and Griffiths, 1966).

Viruses also change the growth habit of their host plants. Using an elegant simulated sward technique, Catherall and Griffiths (1966) have shown that cocksfoot plants infected with cocksfoot streak virus produce a larger amount of vertical growth and fewer tillers than virus-free plants, and start growing earlier in the spring. Infected plants therefore compete more successfully with virus-free plants, and cause greater yield loss, in hay crops than in frequently-cropped pastures. By contrast perennial ryegrass plants (*Lolium perenne*) infected with barley yellow dwarf virus produce a smaller amount of vertical growth and more tillers than virus-free plants, and therefore compete more successfully with virus-free plants when cut or grazed frequently than when grown for hay (Catherall, 1966).

15.2 Ways of preventing losses: the rationale

So far no chemical has been found which will eliminate virus from infected plants. Some chemicals (e.g. purine and pyrimidine analogues) decrease virus multiplication, but these do not eradicate the virus and often damage the plant more than the virus would have done. Therefore virus diseases are mainly avoided by prevention rather than cure, that is by growing crops from virus-free seeds or stocks, and trying to stop viruses entering and spreading through the crops. The aim is to stop or delay the start of a virus epidemic in the crop and, if an epidemic starts, to decrease its rate of progress so that the area under the line of a time/incidence graph such as Fig. 14.7 is minimized.

The rate of spread of virus in a crop can be decreased both by slowing its rate of spread within individual plants, using for example plant varieties in which the virus multiplies poorly, and also by slowing spread between plants through the use of pesticides, cover crops, plant varieties resistant to infection, etc. The method or methods that will be most effective in a particular situation depend on many factors, both ecological and economic. Thus the cost of using a plant variety that escapes or resists infection will usually be much less than the cost of applying a chemical to kill virus vectors. Plant breeding or changes in agronomic practices can be useful for even the lowest value crops, such as pastures, whereas pesticides can only be used for high-value crops.

However it is obviously likely that a combination of various control measures will more effectively prevent crop losses than one or a few measures. This approach is fashionably called *integrated control*.

15.3 Obtaining virus-free planting material

15.3.1 *Simple methods*

Virus-free planting material often is obtained from virus-free crops, which have been inspected carefully at regular intervals and from which all infected plants have been removed (*rogued*). Rogueing is most effective for eliminating viruses that cause noticeable symptoms. Viruses or virus strains that cause few or no noticeable symptoms will not be eliminated and may even be selected by rogueing; also, some seed-borne viruses, such as the nepoviruses, usually cause no obvious symptoms in plants infected through the seed. For these reasons it is desirable to supplement inspection and rogueing by laboratory tests. In the Netherlands, Germany and parts of the U.K., for example, leaf samples from potato 'seed' stocks are tested serologically on a large scale for potato X, potato S and potato paracrinkle viruses. Stocks are then classified for sale according to the incidence of infection.

Where a vegetatively propagated cultivar is not totally virus-infected, tests can be made to find the virus-free plants to be used as mother plants, but where the cultivar is totally infected special techniques, described later in this chapter, are needed to obtain virus-free plants. However, in a few instances some of the vegetative progeny of systemically infected plants escape infection; some viruses spread only slowly or sporadically through their hosts, and thus some virus-free progeny can be obtained from, for example, potato plants infected with potato mop-top virus, or gladiolus plants infected with cucumber mosaic virus.

An interesting technique is used to obtain virus-free plants of various *Citrus* species. *Citrus* seedlings are of two types; one type is sexually produced, and the other is produced from the nucellus and thus derived from the female parent. Nucellar seedlings are rarely infected with the viruses that occur in the parent plant, which they often closely resemble agronomically, and so after selection for agronomic features they can be used as virus-free mother plants (Weathers and Calavan, 1959).

Some crops, for example, brassicas, tobacco, rice and most woody perennials, are planted using young plants that have been grown in special nursery beds. These beds of young susceptible plants, often in warm sheltered situations, are ideal for the spread of diseases, and great care must be taken to stop the spread of viruses at this stage.

15.3.2 *Heat therapy*

In 1889 Kobus reported that the growth of sugar cane ratoons obtained from canes suffering from sereh disease was much improved when the ratoons were put in water at 50°C for 30 minutes before planting. Kunkel (1936) showed that peach plants could be freed from yellows disease agent by growing them at 34 to 36°C for several weeks, and about half the known viruses of vegetatively propagated horticultural species can be eliminated from plants by this treatment (Fig. 15.2). In this way valuable cultivars of raspberry, strawberry, grapevine and many other species have been revitalized (Hollings, 1965; Nyland and Goheen, 1969).

The plant may be treated when dormant by

Fig. 15.2 Leaf of ornamental *Abutilon* infected with the virus-like agent of Abutilon variegation disease **(a)**, and **(b)** a leaf from a cutting cured of infection by keeping the mother plant for 4 weeks at 36°C. (Courtesy B. Kassanis.)

Fig. 15.3 Virus-free plantlet of potato (cv. Golden Wonder) grown from an excised meristem of a plant infected with potato virus A. (Courtesy B. Kassanis and A. Varma.)

dipping it briefly in water at about 55°C, but this treatment kills many plants. A safer method, commonly used, is to keep the actively growing plant at 35 to 40°C in air for several weeks. Few viruses with rod-shaped or filamentous particles, but many with isometric particles, have been eliminated from plants in this way. The ability of a virus in the plant to withstand high temperatures is not related to its *in vitro* stability; for example, tomato spotted wilt virus, which is very labile *in vitro* and is inactivated in 10 minutes at about 45°C, survives in plants at 36°C, whereas plants infected with tomato bushy stunt virus, which can withstand 80°C for 10 minutes *in vitro*, lose much virus when kept at 36°C, and most cuttings taken from them after three weeks of heat treatment are virus-free (Kassanis, 1954).

In countries with very hot climates, plants may perhaps be freed of certain viruses naturally, every year. Thus in the Patna plain of India, potato tubers are stored during the summer at temperatures up to 36°C for as long as 6 months. Thirumalachar (1954) found that this treatment freed tubers of potato leafroll virus, whereas the virus survived in all tubers kept in the modern way, in refrigerated stores.

Even those viruses which are not eliminated may multiply or spread more slowly in plants at high temperatures, so that shoots produced by the plant while at the high temperature may be virus-free, and tip cuttings may produce virus-free plants.

15.3.3 *Apical meristem and tip culture*

The apical meristems of systemically infected plants may contain little or no virus, and Morel and Martin (1952, 1955) reported that virus-free dahlia and potato plants could be grown from apical meristems excised from infected plants. The meristem tips are dissected from plants in sterile conditions, and grown into small plants on nutrient agar (Fig. 15.3) or on a filter paper wick dipping into a nutrient solution. The nutrients supplied include sugars, salts and plant growth hormones (Stone, 1963; Kassanis and Varma, 1967). When sturdy enough, the plantlets are transferred to soil and grown normally. Some plants such as carnations, chrysanthemums and potatoes can be propagated in this way without too much difficulty, but others, especially monocotyledons, are much less easy to culture from meristem tips because these either die, produce disorganized callus tissue, or grow very slowly. When the plant is heat-treated before use, a larger portion of the shoot tip may be taken. This has the advantage that the shoot tips are more robust than apical meristems, and grow into plants sooner.

Virus-containing apical meristems or tips

sometimes grow into virus-free plants; for example, carnation meristems containing carnation mottle virus will produce virus-free plants, presumably because the virus is inactivated after the meristems are excised (Hollings and Stone, 1964).

Apical meristem or tip culture is now widely used to produce virus-free stocks of vegetatively propagated plants (Fig. 15.4), and is often combined with heat treatment or chemotherapy. One example of its practical use in Britain was when Kassanis (1957b) freed the King Edward variety of potato from potato paracrinkle virus, with which all stocks of this cultivar were infected. In the field, the virus-free clone yields 10% more than infected clones, and has rapidly replaced infected stocks (Bawden and Kassanis, 1965).

A similar but little used method of obtaining virus-free plants of vegetatively propagated species involves the culture of callus derived from infected plants. The callus cells are cultured on sterile nutrient agar, and kinetin used to induce them to grow and produce shoots, which are then cut off and rooted. Virus-free potato plants have been obtained in this way (Svobodova, 1966) but a disadvantage is that the plants derived from callus cells may have chromosome abnormalities.

15.3.4 *Chemotherapy*

No chemical has been found that will consistently and specifically eradicate virus from a systemically infected plant, but the multiplication of viruses is usually decreased when infected plants are sprayed, watered or infiltrated with chemical analogues of the purine and pyrimidine bases of nucleic acids. These unnatural bases, for example 2-thiouracil or 8-azaguanine, upset some part of the nucleic acid metabolism of the infected plant. Different analogues may not however work in the same way; nor does one analogue have the same effect on all viruses. For example, 2-thiouracil not only decreases the amount of TMV nucleoprotein produced by an infected plant, but also is incorporated into the RNA in the virus particles, and decreases their specific infectivity (Matthews, 1956). By contrast 2-thiouracil is apparently not incorporated into turnip yellow mosaic virus RNA, but induces infected plants to produce unusually large numbers of RNA-free protein shells of the virus (Francki and Matthews, 1962b).

These base analogues, and other chemicals such as malachite green, may be used, like heat therapy, to increase the success of the apical meristem and tip culture technique, but their value for this purpose is not well established.

Fig. 15.4 Carnation plants cv. Joker. Plant on right infected with carnation ring-spot and carnation mottle viruses, plant on left from the same stock but freed from these viruses by heat treatment and meristem-tip culture. (Courtesy M. Hollings and O. M. Stone.)

15.4 Controlling sources of viruses and vectors outside the crop

In Chapter 14 we described some of the many sources of viruses and vectors, and it is obvious that infection of crops can be decreased by controlling the vectors at their sources. However viruses can spread over great distances and often overwinter in wild plants or pastures far from the crops in which they cause most damage, and it is difficult to get farmers and landowners to take concerted and costly measures to control virus sources on their land for the benefit of crops elsewhere. However, large-scale control programmes have been tried with some success; thus the desert foothills in California have been sprayed from the air to kill the overwintering leafhopper vectors of beet curly top virus. Such programmes are expensive; for example in 1951, about 70 000 hectares of desert were sprayed at a cost of about $·4 per hectare.

Young crops may be infected by vectors leaving the maturing crops of the previous season, and a crop-free period will then control the virus, provided the period is longer than the survival time of the vector or the virus it is carrying. For example a crop-free period will control mite-borne wheat streak mosaic virus in Canadian wheat crops (Slykhuis, 1955). Similarly, in Britain there has been a campaign to ensure that the mangolds in clamps, which are a potent source of over-

wintering aphids and sugar-beet yellowing viruses (Fig. 14.1), are sprayed with either insecticide or maleic hydrazide (to inhibit shoot growth) before clamping, or are used in the spring before sugar-beet crops are sown. However some vectors, such as leafhoppers, can survive and retain the viruses they are carrying for long periods, and the mere removal of virus-infected source plants may not control spread.

Seed crops of biennial species such as sugar beet can, in their second year, be a source of pathogens for newly planted crops. Thus it is essential that the plants for seed (steckling beet) are kept virus-free in their first year of growth. This was originally achieved in Britain by growing the stecklings for one year in the far north and west, away from the aphids and viruses in the main beet growing area of the south-east, and then transplanting stecklings to the south-east for their second year. However, the stecklings are now more simply and effectively protected by growing them under cereal cover crops during the first year in the south-east; they can then be grown on *in situ* (Fig. 15.5).

In places with cold winters, many aphids overwinter as eggs on woody plants; thus it was found that in some places the ornamental *Prunus* trees grown for aesthetic reasons alongside roads in the Netherlands were excellent overwintering hosts for *Myzus persicae* (Fig. 14.3) (Hille Ris

Fig. 15.5 b, Sugar-beet seed crop in its first year of growth with a barley cover crop to protect it from virus yellows. (Courtesy G. D. Heathcote.) **a,** Tobacco planted between rows of barley on the slopes of Mount Unzen, Japan. The barley cover crop protects the tobacco seedlings from the aphid-borne cucumber mosaic virus.

Lambers, 1955), and the trees had to be removed.

In many countries, quarantine measures and import restrictions have been introduced to control the movement of plants and seeds, and hence of viruses and virus vectors, from one country or province to another. They are most likely to be effective in geographically isolated countries, such as Australia and New Zealand, which are free from many of the pests and diseases that are common in the Old World. The advent of air travel has however made it much more difficult to make government regulations of this type effective, and national frontiers are not of course respected by airborne vectors.

15.5 Stopping spread into and within the crop

15.5.1 *Agronomic practices*

Any change in the management of crops may influence the spread of viruses and, by intelligent changes, virus spread can be greatly decreased.

Plants are usually most susceptible to virus infection when young, and their natural resistance to infection increases with age. Thus the spread of a virus can be greatly decreased by planting the crop so that the seedlings emerge either after or well before the period when the vectors are most actively transmitting the virus. For example, in Britain it is best to sow broad beans early in the spring, instead of later, to avoid aphids and viruses (Table 14.2). In some instances, the ability of virus-carrying vectors to infect plants decreases during high summer temperatures; thus in Israel, the spread of the leafhopper-borne maize rough dwarf virus is greatly decreased by delaying the time of sowing (Harpaz, 1961).

The density of plants in a crop and the area of a crop both affect the incidence and spread of viruses (Chapter 14). Loss caused by virus can often be decreased by planting at high density, though of course at very high plant densities the inter-plant competition may decrease yield. With viruses that spread from hedgerows into crops there will be a smaller total virus incidence in a few large fields than in a comparable area made up of small fields. Partly for these reasons, field crops usually have a smaller incidence of virus than plants in market gardens.

High planting density can be combined with rogueing to control viruses. Thus van der Plank and Anderssen (1945) in South Africa found that when a tobacco crop was planted with regularly spaced seedlings 50% became infected with tomato spotted wilt virus, whereas when, instead of single seedlings, groups of two or three were planted, the incidence of virus was 9% and 1% respectively, and the virus-infected plants could be removed when the plants were thinned.

Rogueing is also frequently used successfully for controlling spread in perennial and tree crops. It is used for example on a large scale in West Africa to control the spread of cacao swollen shoot virus in cacao (Thresh, 1958). Rogueing is also used to control viruses in Scottish seed potato crops because, in the districts where the best of these are grown, virus spread is mostly from sources within the crop.

A common agronomic practice in many parts of the world is to grow seed-potato crops in cool windy regions where aphid vectors arrive late in the season and do not fly often. In places where the conditions are somewhat more favourable for aphids, such as the Netherlands, the potato foliage is destroyed before the crop is fully mature, so that viruses transmitted to the shoots may be prevented from reaching the tubers.

Plants in mixed plant communities usually have a lower virus incidence than those in comparable pure stands. Thus alfalfa mosaic virus is less prevalent in cocksfoot/lucerne mixtures than in lucerne alone. This phenomenon is the basis of the practice of growing sugar-beet stecklings, during their first year, under a cereal cover crop to protect the stecklings from infection by beet yellowing viruses (Fig. 15.5b). Similarly, tobacco plants in southern Japan are commonly grown between rows of barley (Fig. 15.5a), and are at first enclosed under polyvinyl sheeting, to protect them from aphids carrying cucumber mosaic virus. The barley is harvested when the tobacco is barely half grown.

Special measures are often taken to protect virus-free mother plants used for vegetative propagation. For example, virus-free raspberry and strawberry mother plants are grown in sterilized soil in gauze-houses designed to exclude aerial vectors.

Yet other methods are used to control viruses, such as the tomato mosaic strain of TMV, that are mainly spread by people when they tend the plants. The main initial sources of this virus for glasshouse tomato crops, however, are contaminated debris in the soil, and seed (Broadbent and Fletcher, 1966). Integrated control measures for controlling the spread of tomato mosaic virus therefore involve many simple forms of hygiene, and the use of genetically resistant tomato cultivars. The hygiene precautions include treating seed with detergent or acid to inactivate virus particles at the surface; growing the plants *in situ* from seed without transplanting; sterilizing the glasshouse soil and woodwork before each crop; not smoking, because many smoking tobaccos contain TMV; and washing frequently in virus-inactivating fluids the hands, tools and clothing of all who work in the glasshouses. Similar considerations apply to keeping tobacco crops free of TMV, but the control measures that are used differ in detail and emphasis. In particular, it is important to use steam-sterilized or chemically-sterilized soil for the seedbeds.

15.5.2 *Certification schemes*

Certification schemes are administrative procedures designed to ensure, as far as possible, that viruses are prevented from infecting those crops that are grown to provide seed or planting material for other crops. They are mainly used for vegetatively propagated species. Crops or planting material meeting stringent standards are given a certificate indicating their health status; lower grades of certificate may be awarded when slightly less stringent criteria are met. The Certification Scheme used in Scotland for 'seed' potatoes (Todd, 1961) is a good example. Although this scheme was started to designate pure stocks of cultivars immune to wart disease, and virus diseases were not seriously considered until 1932, its main function in recent years has been to decrease the incidence of virus infection by encouraging farmers to propagate the healthiest stocks.

Clones are propagated for four years from the progeny of a single plant derived from a stem cutting. Each year they are inspected in the field, and serological tests are made to check freedom from potato X, S and M viruses. Subsequently, clones may be combined and are eligible for the highest grade of certificate (Virus-tested stock). In 1972, the standard required for the more widely grown Foundation Seed grade was as follows: 99.95% plants true to varietal type, no potato virus Y, and not more than 0.01% plants infected with potato leafroll virus, 0.05% with potato virus X, 0.25% with black leg disease or 0.02% with witches' broom disease. Also, the land used for the crop must be free from potato root eelworm, and must not have carried potatoes for at least five years. The continuing success of the scheme

is shown by the figures; the proportion of the 'seed' acreage reaching Foundation Seed standard increased from 21% in 1967 to 43% in 1972.

Other comparable schemes are operated in other crops, particularly for soft fruit, tree fruit and ornamental species. The principles used in these are the same; selection of healthy mother stock; propagation in specified ways; and testing and/or inspection to check continued freedom from virus infection. Certification schemes are most successful in controlling viruses that do not spread rapidly from plant to plant and are essentially confined to the species in question.

15.5.3 *Varieties of plants*

Most commercial varieties of plants have been deliberately or unwittingly selected to minimize the incidence and effect of diseases, including those caused by viruses, and the cheapest way of further avoiding losses is to breed and select new varieties that suffer even less. Indeed, in conditions where viruses are spreading very rapidly, such as in many tropical countries, breeding disease-escaping cultivars may be the only effective way of avoiding crippling epidemics. Most work has been done to find or produce cultivars that resist infection by viruses, but where this has been impossible cultivars that tolerate virus infection are sought.

There are several instances where resistant or tolerant cultivars have been used very successfully. For example, in the western U.S.A., where beet curly top virus is widespread, the sugar-beet industry has been saved by introducing tolerant cultivars (Fig. 15.6). Another success has been the use of specially selected wheat cultivars in areas of the U.S.A. where wheat was previously devastated by soil-borne wheat mosaic virus; some of these cultivars are not infected in the field although they are susceptible to infection by inoculation of sap, whereas others are infected but tolerate the infection and are not severely affected (McKinney, 1948).

There have been other successes but also many disappointments for plant breeders. Many thousands of strains, varieties and crosses of sugar beet have been tested for resistance to beet yellowing viruses, but without success, though tolerant cultivars that also escape infection in the field seem promising; not only are they tolerant of infection but, even when infected, are poor hosts for the aphid vectors and are a poor source of virus for the aphids (Russell, 1966a; Lowe and

Fig. 15.6 Successively bred sugar-beet cultivars tolerant of infection with beet curly top virus. Plots a, the susceptible European cultivar 'Rand G Old Type'; b, the first curly top resistant cultivar, 'U.S.1'; c, 'U.S.33'; d, 'U.S.12'; e, 'U.S.22'; f, the most tolerant cultivar, 'Improved U.S.22'. (Courtesy U.S. Department of Agriculture.)

Russell, 1969). Similarly in groundnuts, crops of the cultivar Mwitunde contain a smaller incidence of groundnut rosette virus than those of other cultivars, even though Mwitunde is equally susceptible in laboratory tests; this is because it is a poor host for the aphid vector and a poor source of virus for the vector (Evans, 1954).

Resistance can also take the form of hypersensitivity or immunity to infection, and the genetical control of these reactions in some crop plants, such as potato (Cockerham, 1955, 1970), has been studied in some detail. Indeed, major genes for hypersensitivity or immunity are found in some old cultivars; for example, King Edward potato is hypersensitive to common strains of potato virus X and Lloyd George raspberry cannot be infected with common strains of raspberry ringspot virus, even by grafting. There is a danger that mutable pathogens like viruses will produce new variants able to overcome the resistance of such cultivars, and resistance-breaking strains of both potato virus X (Cockerham, 1943) and raspberry ringspot virus (Murant, Taylor and Chambers, 1968) have been known to occur for many years; they have not however become common. By contrast the use of tomato plants containing a single gene for tolerance to TMV resulted in the selection and rapid spread of virus strains virulent for the plants (Pelham, Fletcher and Hawkins, 1970). The behaviour of most tolerant cultivars however seems different, perhaps because tolerance can be determined by many minor genes, and, with these cultivars, plant breeding for virus control is more apt to be a continuing process. For example Fig. 15.6 shows the reaction, to modern strains of beet curly top virus, of successively produced tolerant sugarbeet cultivars. A danger in using tolerant cultivars is that infected tolerant plants may be a more important source of virus for further spread than severely affected plants. This may have been the reason why beet curly top virus seems to have become more important in crops other than sugar beet after tolerant beet cultivars were introduced (Bawden, 1950).

The fact that a plant cultivar escapes virus infection in one place does not ensure its escape in another, particularly when distance between the two places is large compared with the range of the vector. Thus although the relative incidence of the aphid-borne bean leafroll virus in different broad bean cultivars was similar in different parts of Britain (Aldrich, Gibbs and Taylor, 1965), all the 26 varieties of peas that escaped infection with the nematode-borne pea early-browning virus when grown in soil from one site in England became infected in soil from another site only fifty miles from the first (Harrison, 1966).

The incidence of viruses in crops is also affected by varietal characters unrelated to their intrinsic susceptibility or their reaction to viruses or vectors. For example plant colour can have an effect because, after they have been flying for some time, aphids are attracted to yellow. Thus more aphids, and plants with lettuce mosaic virus, are found in plots of yellow-leaved varieties of lettuce than in comparable plots of green or browner-leaved varieties.

15.5.4 *Protective inoculation with avirulent strains of viruses*

The finding that plants infected with one strain of a virus usually fail to develop additional symptoms when inoculated with other strains of the same virus (§11.6.1) suggests that it might be possible to protect a crop against naturally occurring virulent strains of a virus by deliberately infecting the crop plants with avirulent strains. Posnette and Todd (1955) showed the potential of this approach for protecting cacao plants against the virulent strains of cacao swollen shoot virus that are widespread in Ghana; in a field experiment only 14% of trees protected with a mild strain of the virus developed severe symptoms when graft inoculated with a severe strain compared with 76% of the unprotected trees.

Protective inoculation has been used to prevent the spread of virulent strains of passionfruit woodiness virus in passionfruit (Simmonds, 1959) and of tomato mosaic virus in tomato crops (Rast, 1972), but it is not popular because of the intrinsic danger of broadcasting any mutable pathogen however innocuous it may seem. Another hazard of this method is that a virus that is avirulent for one species may have normal virulence for other host species, and when broadcast has more opportunity to spread to other species. Furthermore the protection given by this technique operates only against other strains of the same virus, and other viruses infecting the protected plants are likely to produce more than the usual amount of damage, because unrelated viruses can have additive or even synergistic effects.

15.5.5 *Crop protection chemicals*

Various pesticides may be used to kill or inhibit the activity of plant-virus vectors, but for various reasons pesticides have only been successfully used to control a few viruses. Firstly, most pesticides are expensive, and can therefore only be used with valuable crops. Secondly, many vector species are so efficient that almost all vector individuals must be killed to give a worthwhile control of a virus. Thirdly, pesticides rarely kill rapidly enough to prevent incoming virus-carrying vectors from transmitting virus to treated crop plants. For these reasons pesticides seem to be most useful for controlling viruses that are acquired and transmitted by their vectors only during long feeds, and for viruses that are transmitted mainly within the crop instead of coming from outside.

Insecticides such as nicotine and pyrethrum have been used for several hundred years, but since 1945 interest in pesticides has greatly increased and a variety of complex synthetic chemicals has been developed. Pesticides are formulated in various media to dilute them to useful concentrations, and to ensure that they contact and are retained by the vector or by the plant host of the vector. They may be sprayed onto plants, or made into a fog or smoke that is blown around the plants, or mixed with talc or vermiculite and applied as a dust. Seeds may be steeped in pesticide before sowing, or may be sown together with pesticide in dust or granule form. Soils are treated with volatile pesticides by injection; less volatile chemicals can be incorporated into the soil in dust or granules during cultivations.

(a) *Natural products* The best-known natural products used as insecticides are nicotine, pyrethrum and rotenone. Nicotine is obtained from the leaves of various species of tobacco, principally *Nicotiana rustica*. It is a nerve poison and is very toxic for animals, but has a low toxicity for most plants. Pyrethrum is produced by various species of *Chrysanthemum*, especially *C. cinerariaefolium*. It quickly immobilizes insects at low concentrations, but only kills at higher concentrations. Unfortunately it is very unstable in sunlight. It has a very low toxicity for mammals and plants. Rotenone is obtained from various legumes, especially *Derris* and *Lonchocarpus* spp. It is more selective than nicotine both in the insects and other animals it kills.

These materials in general have little effect on virus spread in field crops and are therefore little used on a field scale; however they are used to a limited extent to control aphid vectors in crops grown under glass.

(b) *Synthetic insecticides* There are two main groups of synthetic insecticides; the chlorinated hydrocarbons and the organophosphorus compounds.

The chlorinated hydrocarbons, of which DDT is the most famous, are contact poisons with high toxicity for insects, moderate to high toxicity for mammals and low toxicity for plants. They are insoluble in water and are therefore usually applied to plants in oil-in-water emulsions, or as dusts. They are very stable and persistent, and will survive in soil or in animals for several years. This is a major drawback because they accumulate in tissues, are passed along food chains, and can have drastic effects on some species, particularly predatory birds. Other insecticides in this group include benzene hexachloride, of which the purified active γ-isomer is used and called 'lindane', and also various cyclodienes called 'aldrin', 'chlordane', 'dieldrin', 'endrin', and 'heptachlor'. Chlorinated hydrocarbons have been used to control virus spread, but with limited success. They do not move systemically through the plant, and repeated application is needed to protect young emerging foliage; this is costly and may damage the crop. In Britain, the dissemination of potato leafroll and Y viruses, which spread mainly within the crop, was shown to be greatly decreased by repeated DDT sprays (Broadbent, Burt and Heathcote, 1958); the spread of potato leafroll virus, which is best transmitted after long acquisition and inoculation feeding periods, is more effectively controlled than that of potato virus Y, which can be transmitted in brief feeding periods by migrating aphids. By contrast, spraying sugar beet with DDT gave variable results and in some experiments even increased virus spread (Dunning and Winder, 1965).

The organophosphorous compounds were developed from nerve poisons studied for military purposes in the 1940s. They seem to inhibit acetyl cholinesterases in the nervous system. Most are of low toxicity for plants, and some are absorbed through, and transported systemically to, all parts of the plant. They can be applied as sprays, dusts or granules to the plants or seeds. Most are very toxic for mammals but are unstable, especially in alkaline solutions; they may persist in the plant for as little as one or two days ('Phosdrin') or several weeks ('Disyston' or 'Menazon'). One of the most widely used is demeton-methyl ('Metasystox'), which has been successfully used on a large scale in Britain to control beet yellowing viruses in sugar-beet crops (Hull and Heathcote, 1967). In most years only relatively few viruliferous aphids enter these crops, and most of the virus spread is from plant to plant within each crop. Thus the spread can be greatly decreased by killing the aphids in the crop. In Britain the field staff of the British Sugar Corporation inspect crops at frequent intervals in the spring, and advise all farmers in a district to spray their sugar beet soon after most of the migrant viruliferous aphids have entered the crops, and before their progeny have moved to other plants within the crop. Some of the many experiments on which this spray-warning scheme is based are summarized in Fig. 15.7, which shows that demeton-methyl persists in plants for about 2 weeks, and is best applied about 2 weeks before the peak *Myzus persicae* population on the plants. After very mild winters and warm springs, *M. persicae* migrates to the crop early in the spring, and a second spray is occasionally useful. The insecticide does not stop aphids entering the crop and infecting a few plants, but it arrests the epidemic within the crop for 2 to 4 weeks, and though virus may spread subsequently, it will usually then spread more slowly because the plants are more resistant to infection and the aphids relatively few. The result of spraying is that fewer plants are infected, and they are infected for a shorter period of their growth, so that the yield is decreased less.

The carbamyl oximes, such as 'Temik', are another group of compounds used as pesticides, but although they have given encouraging results in experiments they may prove too expensive and toxic for general use.

(c) *Synthetic nematicides and fungicides* Various chemicals will kill soil organisms, and stop the spread of viruses transmitted by them. Almost all these materials are toxic to plants and must be applied to soil, and removed, before planting. Infection with soil-borne wheat mosaic virus (transmitted by the fungus *Polymyxa graminis*) has been stopped in small-scale experiments by treating the soil with carbon disulphide or formaldehyde. Also, the establishment in soil of *Spongospora subterranea* carrying potato mop-top virus can largely be prevented by treating virus-infected potato seed tubers, bearing powdery scabs, with formaldehyde or organic mercurial fungicides. However, the compounds used most frequently in practice are various chlorinated hydrocarbons such as methyl bromide, chloropicrin, D-D mixture (dichloropropene + dichloropropane) and Telone (D-D mixture enriched with dichloropropene); these all kill most nematodes, plus some fungi and other soil organisms; methyl bromide is a general soil sterilant and kills even weed seeds. Most of these chemicals are volatile. Therefore best results are obtained by sealing the soil surface by watering or light rolling

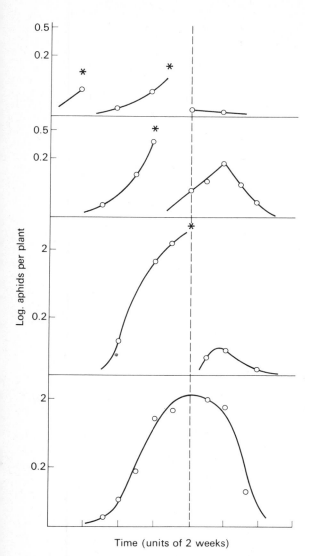

Fig. 15.7 Populations of *Myzus persicae* on sugar-beet plants in plots growing in England and sprayed with the organo-phosphorus insecticide, demeton-methyl. The graphs summarize results from 58 field experiments done in 1957–60. The aphid populations (vertical logarithmic scale) at different times (horizontal scale) are shown relative to the peak aphid population (vertical broken line) on the unsprayed plots (bottom graph). The asterisks indicate the time of application and number of sprays applied. (Data by courtesy of R. Hull.)

after the chemicals have been injected. Methyl bromide is a gas, and must be kept in the soil by covering it with a plastic sheet or other impermeable cover.

Several nematode-transmitted viruses have been effectively controlled using these chemicals, and in experiments the degree of control of arabis

mosaic and tobacco rattle viruses is proportional to the kill of the vector species (Harrison, Peachey and Winslow, 1963; Cooper and Thomas, 1971).

Other chemicals have also proved effective. Pentachloronitrobenzene, which is commonly used as a soil fungicide, also kills some nematodes and stops *Longidorus elongatus* infecting strawberry crops with tomato black ring and raspberry ringspot viruses (Murant and Taylor, 1965). Organocarbamates and organophosphates have also been used experimentally. For instance, Cooper and Thomas (1971) found that methomyl would protect potato crops from infection by tobacco rattle virus in the year of application, although the treatment did not kill as many *Trichodorus* nematodes as did D-D mixture, and its effect did not persist for a second year. Chemicals of this general type have the advantages that they can be applied in granular form, and that they are systemic in plants and act as insecticides too.

All these chemicals are costly and are therefore most used with valuable crops; for example, in Britain D-D mixture is used successfully to control nematode-borne viruses in strawberry crops (Harrison, Peachey and Winslow, 1963) (Fig. 15.8), and in Japan chloropicrin is used on tobacco seed beds to prevent infection with the tobacco stunt agent, which is transmitted by the fungus *Olpidium brassicae*. Soil treatments of this sort are effective for plants with shallow root systems but are likely to be of less use in deep porous soils with deep-rooting plants like grapevine.

15.5.6 *Miscellaneous materials*

Bradley (1963) and others (Loebenstein *et al*, 1966; Vanderveken and Vilain, 1967; Vanderveken, 1968) have found that aphids carrying non-persistent or semi-persistent viruses often fail to infect plants that have been sprayed with a water/oil emulsion. For example Bradley, Moore and Pond (1966) in Canada found that oils decreased the spread of potato virus Y in field-grown potatoes by 57 to 87%, and in an experiment in Israel the incidence of cucumber mosaic virus in plots of cucumber sprayed with 5 to 10% oil emulsions at low volume was only 10 to 20% of that in unsprayed plots (Loebenstein *et al*, 1966). The attraction of using these non-toxic materials to interfere with virus transmission is great because they are less likely to cause pollution of the environment than pesticides, but such

Fig. 15.8 Control of arabis mosaic virus in strawberries. Foreground, infected plants in control plot; background, healthy plants in plot whose soil was injected with D-D mixture (50 g/m²) before planting, to kill the vector nematodes. *Xiphinema diversicaudatum.* (Courtesy B. D. Harrison, J. E. Peachey and R. D. Winslow.)

materials need to be more effective than those now available if they are to be used widely.

Another treatment involving a relatively non-toxic material is the application of sulphur to soil containing the fungus-transmitted potato mop-top virus. When the soil acidity is increased in this way to pH 5, there is little spread of the virus (Fig. 15.9) or of its vector *Spongospora subterranea*, but the fungus is not killed, and spread again occurs when the soil pH is increased (Jones and Harrison, 1968). The spread of potato mop-top virus can also be greatly decreased by adding

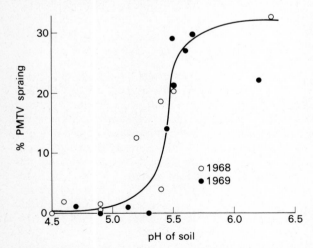

Fig. 15.9 Relation between incidence of spraing caused by potato mop-top virus in tubers of 'Arran Pilot' potato and soil pH. The plots were adjusted to different pH values by applying sulphur. Circles, 1968; discs, 1969. (Courtesy R. A. C. Jones and B. D. Harrison.)

zinc compounds to infested soil, but virus-carrying resting spores of *S. subterranea* survive and such large amounts of zinc are needed that the treatment is of questionable practical value (Cooper and Harrison, 1973a).

Controlling weeds often also helps to control viruses, unless the weeds are a source of predators of virus vectors or are serving as a barrier to virus spread. Herbicides may be used, for example, to kill potato groundkeepers in cereal crops, thereby protecting subsequent potato crops. Herbicides can also be used to delay the infection of crops with viruses that occur in weeds or their seeds (Taylor and Murant, 1968), but weed control sometimes increases the incidence of virus in the crop. Thus when weeds were controlled for two years at a site infested with nematodes carrying tobacco rattle virus, the potato tubers produced in the second year contained more spraing than those from untreated plots. Evidently the nematodes fed more often on the potatoes when the weeds were not available (Cooper and Harrison, 1973b).

15.6 Conclusions

This chapter and the preceding one illustrate some of the many different ways in which viruses survive and spread in nature, and show that viruses are difficult to control and virtually impossible to eradicate. The aim in control is to find the weakest points in the ecological system of the virus and to direct control measures at these. Thus there is little value in using crop rotation to control soil-borne viruses but much in seeking resistant varieties, and there is little value in treating crops with insecticides to prevent aphid-borne viruses being brought in from outside but much to control their spread within crops. It is obvious that an integrated control programme will be more effective than one or a few control measures. A good example of this approach is shown by the measures used successfully to prevent the spread of cucumber mosaic virus in tobacco crops in southern Japan. These are (1) spraying with insecticides the nearby overwintering vegetable crops and weed reservoirs of virus and vector, (2) early planting, (3) planting between rows of barley (Fig. 15.5), and (4) growing the tobacco seedlings under polyvinyl sheeting until they are old enough to have lost their extreme susceptibility to infection. Another example is seen in the control of virus yellows in British sugar-beet crops (Hull, 1965). Many of the overwintering sources of viruses and vectors such as weeds, mangold clamps and beet-seed crops are removed and the spread and effect of viruses in the crops minimized by using insecticides and tolerant varieties. Hull (1965) reported that 'Sugar-beet yields have increased over the last three decades by about $2\frac{1}{2}\%$ per annum and the various measures to control yellows have contributed to this. When yellows was rife, sunny, dry years gave only small yields of sugar beet; now such years give the greatest yields. The widely fluctuating yields in south Lincolnshire in the years when yellows spread unchecked from seed crops have now been replaced by consistently large yields'. Unfortunately it is probable that such encouraging results will only be obtained with valuable, well-organized, contract-grown crops such as sugar beet and tobacco.

Chapter 16

Viruses of organisms other than higher plants. Origins of viruses

In the preceding fifteen chapters of this book we have mainly discussed viruses of higher plants, and here we will outline some of the features of other viruses. We will also discuss the interesting but unresolved problem of where viruses come from.

16.1 The hosts of viruses

Several of the major phyla of organisms are not known to be associated with viruses (Table 16.1),

Table 16.1 Incidence of viruses in different phyla

| | | Phylum | Number of species[1] that: | |
			are infected	can act as vectors
Prokaryotes		Bacteria and blue-green algae	+++	−
Eukaryotes	Plants	Algae	+	−
		Fungi	++	+
		Pteridophytes	+	−
		Gymnosperms	+	−
		Angiosperms	+++	+
	Animals	Protozoa	+	−
		Nematodes	(+)	++
		Arthropods	+++	+++
		Molluscs	+	−
		Vertebrates	+++	+

Viruses have not been reported from lower plants such as diatoms, myxomycetes, bryophytes and cycads, nor from many phyla of lower animals such as the sponges, coelenterates, platyhelminths, rotifers, polyzoa, brachiopods, annelids and echinoderms

[1] Number of species with which viruses are known to be associated: − none, + 1–10 species, ++ 10–100 species, +++ more than 100 species.

and most known viruses are from only four of the phyla. Is there really this uneven distribution of viruses among different phyla, or are there reasons why we know so much more about the viruses of some phyla? One reason for the uneven distribution could be that only certain organisms lead lives that favour virus spread between individuals. For example, most plants are sedentary and their cells are surrounded by resistant cell walls; also, many primitive plants have no vascular systems. Thus viruses may not have prospered in plants until vascular plants, especially seed plants, evolved together with specialized plant-feeding arthropods, such as aphids, that could act as vectors. However, the apparent scarcity of viruses in lower animals seems unlikely to have a similar explanation, because many viruses of higher animals infect respiratory and gut surfaces via a contaminated environment, and it is not obvious why comparable viruses are not widespread in lower animals. More probably, the uneven distribution of known viruses mostly reflects the uneven concentration of activity of virologists. The viruses predominantly studied are those that cause diseases of man, and domesticated animals and plants, although more recently there has been increasing interest in using viruses to control unwanted organisms. Thus the known incidence of viruses in a phylum principally mirrors the economic importance of that phylum to man.

16.2 Viruses of prokaryotes

Twort (1915) and d'Herelle (1917) first and independently found transmissible agents that multiplied, could pass through bacteria-proof filters and caused lysis of bacteria. Ever since their discovery, these bacteriophages (or phages) have been extensively studied, initially because of their possible use for controlling pathogenic bacteria, but more recently because they are ideal subjects for those biologists and biochemists wishing to study the basic processes of life (Stent, 1960).

Many hundreds of different phages have been described, and whenever a bacterium is isolated, phages for that bacterium can usually be found too. Most phages are named by the code lettering or numbering used when they were isolated (e.g. T2 or ϕX174) though many of those that are 'male' specific in that they infect using male sex pili have code names containing f (e.g. f2, fr). Phages can be grouped conveniently by the shape of their particles (Bradley, 1967; Ackermann and Eisenstark, 1974) and the type of nucleic acid in those particles.

The largest group of phages are those with large genomes of double-stranded DNA, and complex particles with a tail. Characteristic of one group of the tailed phages are the closely related T-even phages [cryptogram D/2: 137/40: X/X: B/0] of *Escherichia coli*; incidentally T is for type, not tailed nor Twort. These have extremely complex particles (Fig. 16.1) which consist of an angular polyhedral head of about 100 nm diameter and a complex helically-constructed contractile tail of the same length. Fibres on the end of the tail stick to the wall of the bacterial cell, the distal end of the tail rests against the wall, the tail sheath contracts, the inner tube of the tail penetrates the host cell wall and the phage DNA is injected into the bacterium. After phage replication has proceeded for about 20 minutes at 37°C, the host cell lyses and several hundred progeny virus particles are released. The linear double-stranded genome of the T-even phages is large, and some of its cytosine is hydroxymethylated, presumably to thwart unfriendly enzymes in the host. The genetics of the T-even phages, particularly T4, have been studied in great detail and have provided much information on the way that genetic information is organized and its expression controlled, and there is now much interest in understanding the way that the particles of T4, which contain more than a dozen proteins, are assembled through a series of spontaneous and controlled stages along various converging assembly lines (King and Laemmli, 1973; King and Mykolajewycz, 1973).

Other tailed phages have non-contractile tails of different lengths. For example the particles of phage λ (D/2: 32/50: X/X: B/0) have a rounded polyhedral head about 60 nm in diameter with a long flexuous tail about 160 nm long. The genome of λ has short single-stranded regions of DNA at both 5′ termini, and these have an antiparallel complementary sequence so that when phage λ infects *E. coli* its genome is converted into a circular molecule.

Phage λ is one of many phages that not only can replicate in and lyse its host but also, under other conditions, can integrate into the host's DNA 'chromosome' and then be replicated with it. This phenomenon is called lysogeny, and phages which can lysogenize bacteria are called temperate phages in contrast to virulent phages, such as the T-even phages. The integrated phage DNA, or prophage as it is called, may be excised from the host DNA and become lytic again; the way in which integration and excision is controlled is a fascinating story (Hershey, 1971). Some prophage genes are expressed in lysogenized bacteria; for example, the product of one gene of phage λ suppresses expression of the other λ genes and ensures that the phage DNA remains in the lysogenic state. Other prophage genes may affect the lysogenized bacterium more obviously; for example, *Corynebacterium diphtheriae* does not cause diphtheria unless it is lysogenized with phage β (D/2: 22/*: X/X: B/0), which carries the gene for the diphtheria toxin (Euchida, Pappenheimer and Greany, 1973).

The large DNA phages are an extremely diverse and obviously successful group, and many differ from those described above. Some have bizarre particles, the heads of which are covered with spikes, others have tails covered with curly filaments (Bradley, 1967), and one, phage PBS1 of *Bacillus subtilis*, has on its tail three helically coiled tail fibres that specifically adsorb to the flagellae of its host (Raimondo, Lundh and Martinez, 1968). Not all the phages of this group have obvious tails; one such from *Spirillum itersonii* has a genome and head that are similar in size and type to those of phage λ, but no obvious tail (Clark-Walker and Primrose, 1971), though some particles show one modified vertex with short fibres (Fig. 16.1c). The large DNA phages also differ widely in the composition of their particles; for example, the tail-less one from *Pseudomonas*, unlike other phages, has particles that contain lipid (13%; Camerini-Otero and Franklin, 1972), and phage PBS1 of *B. subtilis* has a DNA genome containing uracil and not thymine (Takahashi and Marmur, 1963).

Temperate phages are excised from the bacterial DNA and become 'vegetative' (i.e. replicate as phages) at a small but constant rate, which can increased by treating lysogenized bacteria with, for example, ultra-violet or X-ray irradiation, or the antibiotic Mitomycin C. Excision from the host's DNA is sometimes inaccurate, so that a piece of bacterial DNA is attached to the phage DNA. These fragments of bacterial DNA are carried to the next bacterium lysogenized by the phage, where they may be expressed. Genes carried in this way are said to have been transduced. Indeed the distinctions between phages, prophages and plasmids, which are small extrachromosomal packets of genes transmitted

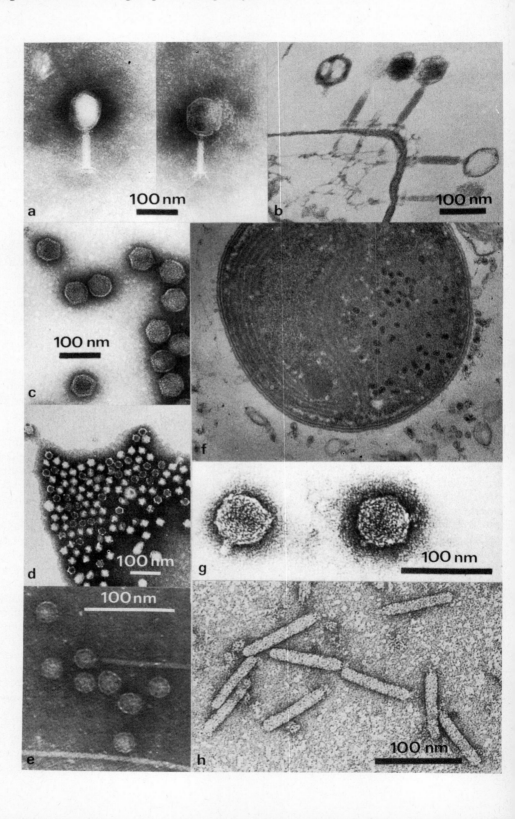

through bacterial sex pili (Meynell, 1972), are difficult to make, and perhaps all are ways that bacterial genes have evolved of grouping together for their common good. Treatments which induce temperate phages to excise, frequently also result in the production of macromolecules that are smaller than phage particles and kill sensitive bacterial cells but do not reproduce; these are the bacteriocins (Reeves, 1972). In the electron microscope some bacteriocins resemble parts of phages, such as their tails, and these bacteriocins are probably the products of incomplete pro-phages.

Not all phages have genomes of double-stranded DNA. Two extensively studied phages with circular single-stranded DNA genomes are ϕX174 (D/1: 1.7/25: S/S: B/0) which has angular icosahedral particles about 30 nm in diameter (Fig. 16.1d), and fd (D/1: 1.7/12: E/E: B/0) which has filamentous particles 870 nm long and 5.5 nm in diameter. fd infects by attaching to the end of the f sex pili of male *E. coli*, but it does not lyse infected cells like most phages; infected bacteria continue to grow and divide while releasing virus.

The ribophages [R/1: 1.1/28: S/S: B/0] are a well-known group of phages with isometric particles about 25 nm in diameter (Fig. 16.1e); like fd phage they infect through male sex pili, and the best-known of them are called Qβ, MS2 and R17. Their replication is discussed in Chapter 11; much is known of the nucleotide sequence of their genomes, the amino-acid sequence of the three proteins that their genome specifies, and the way that its synthesis and translation is controlled (Hindley, 1973).

Viruses also infect filamentous blue-green algae

(Cyanophyta) (Brown, 1972; Padan and Shilo, 1973). All that have been studied have large genomes of double-stranded DNA, and particles (Fig. 16.1) that closely resemble, both in structure and composition, the large DNA tailed phages of bacteria. They have been studied mostly for their potential as biological control agents to arrest the 'blooms' of blue-green algae that pollute some freshwater lakes (Padan and Shilo, 1973). The similarity between the phages of bacteria and blue-green algae underlines the close relationship between these two groups. By contrast some of the viruses so far isolated from mycoplasmas (Razin, 1973), are unlike any found in bacteria. For example one, called MV-L1, that infects *Achole-plasma laidlawii* (Gourlay, Bruce and Garwes, 1971; Milne, Thompson and Taylor-Robinson, 1972) has bacilliform particles, which are 16 nm wide and 90 nm long (Fig. 16.1h), and are possibly polar. Another virus from the same mycoplasma, MV-L2 (Gourlay, 1971) has irregu-lar rounded particles 50–100 nm in diameter which, unlike those of MV-L1, are disrupted by ether. There is little direct information on the genomes of these viruses, but some evidence that both are DNA.

16.3 Viruses of lower plants

16.3.1 *Viruses of algae*

An increasing number of virus-like particles are being found in algae as a result of recent renewed interest in the ultra-structure of these plants (Brown, 1972; Mattox, Stewart and Floyd, 1972). The particles most commonly found are large angular polyhedra about 150–250 nm in diameter that resemble most closely the iridoviruses of insects (§16.4.1). These particles have been found in green algae such as *Oedogonium* spp. (Fig. 16.2a; Picketts-Heaps, 1972) and *Uronema gigas* (Mattox, Stewart and Floyd, 1972), and in the brown alga *Chorda tomentosa* (Toth and Wilce, 1972). Other types of particle have also been found, such as the crystalline arrays of iso-metric particles about 40 nm in diameter in the nuclei of the green alga *Coleochaete scutata* (Mat-tox, Stewart and Floyd, 1972); these contain DNA (H. Marchant, personal communication). Similar particles occur in the freshwater red alga *Sirodotia tenuissima* (Lee, 1971).

Although these different sorts of particle are virus-like, so far none has been transmitted to particle-free isolates of the algae. By contrast,

Fig. 16.1 Particles of viruses of prokaryotes. **a,** Particles of T4 bacteriophage. Note the contractile tail sheath and baseplate with pins and long thin tail fibres. (Courtesy J. E. Gardiner.) **b,** Section through the wall of a cell of *Bacillus subtilis* recently infected with SP50 phage. Note the DNA fibrils injected by the phage particles. (Courtesy R. G. Milne.) **c,** Particles of Sil phage from *Spirillum itersonii*, a large DNA phage with tailless particles. Note the particle above the bar marker has a modified vertex that may function as a tail. (Courtesy G. D. Clark-Walker; Clark-Walker and Primrose, 1971.) **d,** Particles of ϕX174 phage. Note the prominent vertices. (Courtesy J. E. Gardiner.) **e,** Particles of Qβ, a ribophage, that infects through the sex pili of the bacterium. **f,** Section through a cell of the blue-green alga, *Plectonema*, infected with LPP-1 virus. **g,** Particles of a virus closely related to LPP-1. (Courtesy M. J. Daft and W. D. P. Stewart.) **h,** Particles of MV-L1 from the mycoplasma, *Acholeplasma laidlawii*. (Courtesy R. N. Gourlay.)

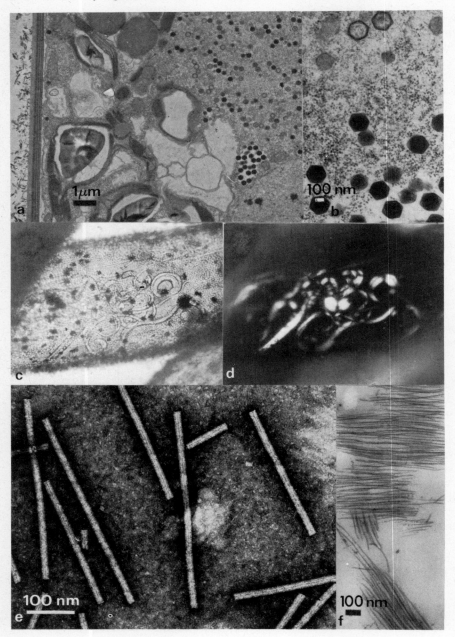

Fig. 16.2 Viruses of algae. **a,** Section through part of a germling cell of the green alga, *Oedogonium*. Note the cell wall along the left margin, starch-filled chloroplasts in centre left, nucleus lower right, and the cytoplasm filled with virus-like particles, top right. **b,** Cytoplasm as in **(a)**, at a greater magnification, showing virus-like particles. (Courtesy J. D. Pickett-Heaps; Pickett-Heaps, 1972.)

c, Cell of *Chara corallina* containing inclusions of *Chara corallina* virus. **d,** As in **(c)** but photographed in plane-polarized light to show birefringence of the inclusions. **e,** A mixture of the particles of TMV (300 nm long) and *Chara corallina* virus (530 nm long). **f,** Section through an inclusion in the cell shown in **(c)** and **(d)** (Gibbs *et al*, 1975.)

virus-like particles found in the green alga *Chara corallina* have been purified, transmitted by injection to particle-free cells, and shown to replicate there, thus confirming that they are the particles of a virus (Gibbs *et al*, 1975). The *C. corallina* virus (R/1 : 2.3/5 : E/E : A/*) has particles which closely resemble those of TMV except that they are about 530 nm long (Fig. 16.2e). These particles, which are readily purified, contain a single-stranded RNA of 2.3×10^6 molecular weight and a single species of protein that has a size and composition resembling the coat protein of TMV; they are serologically related to the particles of some but not all strains of TMV. *Chara* cells inoculated with the virus become chlorotic 4 days after infection, and the tip cells of some naturally-infected plants contain elongated birefringent inclusions (Fig. 16.2d) similar to those found in TMV-infected plants; sections of the inclusions show them to consist of parallel alignments of virus particles (Fig. 16.2f).

16.3.2 *Viruses of fungi*

Viruses and virus-like particles occur in a wide range of fungi. In §13.2.10 we discussed the relations between certain plant viruses and their fungus vectors, and here we are mainly concerned with some of the best known of those viruses that parasitize fungi and seem to have no other hosts.

(a) *Basidiomycetes* Sinden and Hauser (1950) described two diseases of the cultivated mushroom, *Agaricus bisporus*, and during the following ten years many similar mushroom diseases with unknown causes were found throughout the world. Gandy (1960, 1962) showed that one such disease could be transmitted artificially, and Gandy and Hollings (1962) and Hollings (1962) showed that the disease is closely associated with the presence of virus-like particles in the fungus (reviewed by Hollings and Stone, 1969; 1971). However the story is still not complete, because some apparently healthy fruiting bodies (sporophores) contain virus-like particles, and seemingly diseased ones may contain no detectable virus-like particles.

The diseases of cultivated mushrooms are variously known as watery stipe, La France, brown, mummy or X-diseases. Diseased sporophores are small, malformed and either shrivelled or watery, depending on the humidity. The disease may cause a total loss of crop. Healthy mycelium grows vigorously on malt agar and is white and fluffy with coarse radiating strands, whereas when diseased it grows slowly and is brown and velvety with no strands (Fig. 16.3). The viruses can be transmitted experimentally by hyphal anastomosis or by injecting purified virus preparations into very young sporophores. In commercially grown mushrooms the viruses are spread in infected mycelium (Gandy, 1960), in spores (Schisler, Sinden and Sigel, 1967; Dieleman-van Zaayan, 1968) and perhaps occasionally by insects such as the phorid fly, *Megacelia halterata* (Hollings and Stone, 1971).

Extracts of infected mushrooms contain several types of virus-like particles; in the U.K. and the Netherlands, isometric particles about 34 nm in diameter are commonly found (Hollings and Stone, 1971; Dieleman-van Zaayen and Tem-

minck, 1968), isometric particles 25 nm in diameter are frequent, and bacilliform particles of 19×50 nm and isometric particles 50 nm in diameter are uncommon. In Hollings' early work isometric particles 29 nm in diameter were common but not later. The relationships, if any, between the different particle types are not known, though the 25 nm particle, or the 29 nm particle, may occur alone in specimens whereas the bacilliform particles were only found mixed with 25 nm or 29 nm particles. In thin sections of the fruiting body or the mycelium, the 25 nm, 34 nm and bacilliform particles have been found free in the cytoplasm or in membrane-bound vacuoles; only the 34 nm particles have been found in spores, in the cytoplasm (Dieleman-van Zaayen and Igesz, 1969; Dieleman-van Zaayen, 1972).

Many methods are effective for purifying the particles of the mushroom viruses (Hollings and Stone, 1971) and also those of the viruses of other fungi, but the most universally successful seems to be the method of Kitano, Haruna and Watanabe (1961), in which a phosphate buffered extract of the mycelium is clarified using mixed ethoxy- and butoxyethanols, and the particles then concentrated and purified by centrifugation.

Gandy and Hollings (1962) reported that a combination of at least two cycles of heat treatment (33°C for 2 weeks) and mycelial tip culture could sometimes eliminate virus from infected cultures, but Dieleman-van Zaayen (1970) could not confirm this. However, not all cultures of commercially grown clones of *Agaricus bisporus* are virus-infected, and healthy spawn is available for growers. Various other hygiene measures help to prevent spread of the viruses in mushroom houses (Dieleman-van Zaayen, 1970, 1971; Hollings and Stone, 1971); these include filtering air at intakes and exits of the houses, steam sterilizing the interior of houses (methyl bromide seems less effective) and the growth medium, saturating exposed wooden surfaces with 2% sodium pentachlorophenate, and treating tools, machinery, clothing etc. with 5% formaldehyde.

There are also reports of virus-like particles in some other Basidiomycetes. Blattný and Králík (1968) found that extracts of the sporophores of *Laccaria laccata* suffering from a transmissible disease contain isometric particles 28 nm in diameter. Isometric particles about 40 nm in diameter and with a sedimentation coefficient of 150 S were found in axenic cultures of a few races of the rust fungus, *Puccinia graminis tritici*, and in spores of other rusts including *Uromyces phaseoli* and *Cronartium ribicola* (Mussell *et al*, 1973; H. W. Mussell, personal communication). Furthermore Koltin *et al* (1973) reported the presence of transmissible isometric particles 130 nm in diameter in plaques and sectors of poor growth in cultures of *Schizophyllum commune*; one wonders how many of the sectors of poor growth frequently found in fungus cultures are caused by viruses.

(b) *Ascomycetes* Virus-like particles have been obtained from the fungus *Gaeumannomyces*

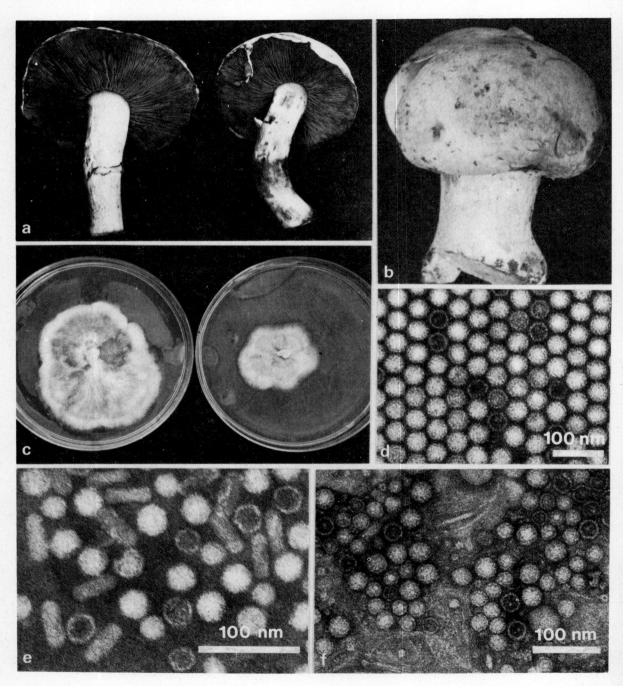

(Ophiobolus) graminis, which is the cause of take-all disease of wheat and barley. When these crops are grown consecutively, the incidence of take-all increases for three to five years and then decreases; a phenomenon known as take-all decline. Lapierre *et al* (1970) reported that in France there was a correlation between take-all decline and the presence of isometric particles of 29 nm diameter in the fungus; isolates of 'declined' fungus were weakly pathogenic, rarely produced perithecia, and in culture developed sectors and were difficult to maintain. By contrast Rawlinson *et al* (1973) found in British isolates of the fungus two types of isometric particles with diameters of 35 nm and 27 nm, which sedimented homogeneously with coefficients of 148*S* and 110*S* respectively but were of heterogeneous densities between 1.29 and 1.37. Both types of isometric particle contained a single protein species of 70 000 molecular weight, but there was no correlation between the presence of either particle and the cultural characteristics or pathogenicity of different isolates of the fungus. Lapierre *et al* (1970) and Rawlinson *et al* (1973) agree that cultures of the fungus grown from the ascospores of particle-containing parents contain no particles.

Virus-like particles also occur in *Peziza ostracoderma*, which is a common contaminant of mushroom beds, and is probably the perfect stage of *Botrytis crystallina*. Some extracts of apothecia contain isometric particles 25 nm in diameter, and others contain helically-constructed tubular particles 350 nm long and 17 nm wide. These particles resemble those of TMV, but did not infect plants usually susceptible to TMV, did not react with antiserum to type strain TMV, and the dimensions of their helix differed slightly from those of TMV (Dieleman-van Zaayen, Igesz and Finch, 1970).

Fig. 16.3 Viruses of fungi. **a,** Healthy (left) and infected (right) mushroom (*Agaricus bisporus*) sporophore, showing 'watery-stipe' symptoms typically induced by mushroom virus 1 plus mushroom virus 2 in humid conditions. (Courtesy M. Hollings.) **b,** Mushroom sporophore infected with mushroom viruses 1, 3 and 4, showing grey areas on the stipe and cap, some distortion and stunting. (Courtesy M. Hollings.) **c,** Cultures of mushroom mycelia on malt agar; left, healthy; right, infected, showing decreased growth rate but normal colour and texture. (Courtesy M. Hollings.) **d,** Preparation of virus-like particles from *Colletotrichum lindemuthianum.* (Courtesy C. Rawlinson.) **e,** Preparation of particles of mushroom viruses 1 (25 nm diameter), 2 (29 nm diameter) and 3 (bacilliform). (Courtesy M. Hollings.) **f,** Preparation of virus-like particles of two sizes from *Gaeumannomyces graminis.* (Courtesy C. Rawlinson.)

Of the few virus-like diseases of yeasts (Lindegren, Bang and Hirano, 1962; Volkoff and Walters, 1970), the most fully investigated is the killer trait of certain strains of *Saccharomyces cerevisiae*. Killer strains produce extracellular proteins that are toxic to most non-killer strains and cause alterations in their cell membrane function (Bussey and Sherman, 1973) similar to that caused by bacteriocins to sensitive bacteria (§16.2). Purified preparations of killer factor contain toxic proteins apparently of a wide range of sizes, some over 2×10^6 daltons, but most much smaller (Bussey, 1972). However, there seems to have been no detailed electron microscopic examination of the killer factor. Killer strains contain virus-like isometric particles 30 nm in diameter (Herring and Bevan, 1974). These particles contain two species of double-stranded RNA with molecular weights of 2.5×10^6 and 1.4×10^6, containing around 50% G + C (Bevan, Herring and Mitchell, 1973; Vodkin, Katterman and Fink, 1974). Work with mutants suggests that the RNA of 1.4×10^6 molecular weight controls toxin production.

Incidentally Lhoas (1972) and Border (1972) reported having infected mating *S. cerevisiae* with viruses from *Aspergillus niger* and *Penicillium stoloniferum*; these viruses have double-stranded RNA genomes and will be described below.

Finally we should mention reports of the isolation of TMV from the conidia of oak powdery mildew (*Sphaerotheca lanestris*) and barley powdery mildew (*Erysiphe graminis*) (Yarwood, 1971; Nienhaus, 1971); the significance of these observations is not known.

(c) *Phycomycetes* The lower fungi have mainly been associated with viruses as vectors of viruses of flowering plants (§13.2.10), but there have been some reports of virus-like particles in them. These include the interesting record of large enveloped isometric particles, resembling those of herpesviruses, in the saprophytic fungus *Thraustochytrium* sp. (possibly *motivum*) (Kazama and Schornstein, 1973), the smaller isometric particles in the oyster pathogen *Labyrinthomyxa marina* (Perkins, 1969), and the larger iridovirus-like particles in *Aphelidium* (Schnepf, Soeder and Hegewald, 1970). Also, Brants (1971) reported that *Pythium sylvaticum* can be infected with TMV.

(d) *Fungi Imperfecti* Virus-like particles have been found in many imperfect fungi (see reviews by Hollings and Stone, 1969; 1971), though interest has centred mostly on those in various species of *Penicillium* and *Aspergillus*. This interest followed reports (Shope, 1948, 1953; Powell *et al*, 1952) that media in which these fungi had grown had antiviral activity when injected into animals at the same time as, or soon after, they were infected with various animal viruses. The antiviral substance, which became known as statolon, was found to be active because it induced the production of interferon (Kleinschmidt, Cline and Murphy, 1964; Kleinschmidt and Ellis, 1968). The interferon-inducing compound in statolon was shown to be double-stranded RNA (Lampson *et al*, 1967; Kleinschmidt *et al*, 1968) contained in isometric virus-like particles (Kleinschmidt and Ellis, 1968; Banks *et al*, 1968).

Penicillium stoloniferum (strain ATCC 14586) yields at least three types of isometric virus-like particles; two types of about 34 nm in diameter that are serologically unrelated and can be separated electrophoretically, and occasionally a few particles about 40 to 45 nm in diameter (Banks *et al*, 1968; Buck and Kempson-Jones, 1970). Preparations of each of the electrophoretic types of 34 nm particle can be shown to be comprised of several components. For example, one electrophoretic type includes seven components with sedimentation coefficients from $66S$ to $113S$, and that contain either no nucleic acid or single or double-stranded RNAs (either alone or in combination) with molecular weights of 0.47×10^6 to 1.11×10^6 (Bozarth, Wood and Mandelbrot, 1971; Buck and Kempson-Jones, 1973). The double-stranded RNA in these particles contains 53% G + C, and the particles contain a specific RNA-dependent RNA polymerase (Lapierre, Astier-Manifacier and Cornuet, 1971). Infected cultures of the fungus show more uneven and slightly slower growth than either naturally occurring uninfected cultures, or cultures that have been freed of the particles by heat treatment and hyphal tip culture (Banks *et al*, 1968). Single spores from an infected parent yield cultures that contain both types of 34 nm particle, indicating that they do not exclude one another (Bozarth, Wood and Mandelbrot, 1971). Lhoas (1971a, b) reported that the viruses could be transmitted through hyphal anastomoses, and that protoplasts of the fungus mixed with virus suspensions became infected with one of the viruses with 34 nm particles. If the protoplast technique is reproducible it might be useful for analysing the functions of the various nucleic acid components of these viruses.

The isometric particles about 40 nm in diameter (Banks *et al*, 1969a) found in *Penicillium chrysogenum* are also well studied; they occur either free or in vacuoles in the cytoplasm but not in the nuclei of infected cells (Yamashita, Doi and Yora, 1973) and have a sedimentation coefficient of $150S$. Each particle contains one molecule of double-stranded RNA, but preparations of the particles yield three such RNAs with molecular weights of 1.9×10^6 to 2.2×10^6 and containing 51% G + C, and also a smaller double-stranded RNA with a molecular weight of 1.4×10^5 and a G + C content of 38% (Cox, Kanagalingam and Sutherland, 1970, 1971; Wood and Bozarth, 1972). *P. chrysogenum* produces penicillin, but there is no correlation between the presence or absence of the virus-like particles and the yield of penicillin (Lemke, Nash and Pieper, 1973). The particles from *P. chrysogenum* are serologically related to those from *P. brevi-compactum* (Wood, Bozarth and Mislivec, 1971).

Similar isometric particles have been found in many other species of *Penicillium* and *Aspergillus* but, with the exception mentioned above, none are serologically related (Banks *et al*, 1969b; Wood, Bozarth and Mislivec, 1971; Hollings and Stone, 1971), and they differ from each other in other properties. For example those of *P. cyaneo-fulvum* though about 33 nm in diameter sediment at $157S$ (Banks *et al*, 1969b), and those from *Aspergillus foetidus* have RNAs in the size range 1.4×10^6 to 2.8×10^6 daltons (Ratti and Buck, 1972), like those found in *P. brevi-compactum* and *P. chrysogenum*. Isometric particles have also been found in *Mycogone perniciosa*, the fungus that causes wet bubble disease of mushrooms (Lapierre *et al*, 1972), and *Piricularia oryzae*, the rice blast fungus (Spire, Ferault and Bertrandy, 1973).

It is important to note that relationships between these particles will be difficult to establish by serological tests alone, because the double-stranded RNAs elicit antibodies (Moffitt and Lister, 1973) that react with double-stranded RNAs in general.

16.3.3 *Viruses of other plants*

Very few viruses of primitive vascular plants have been recorded. Among the pteridophytes, nine species of fern, including *Polypodium aureum*, were reported as being susceptible to TMV which, however, attained only a small concentration in them (Cheo, 1972). Virus-like particles that could not be transmitted to angiosperms were found in mottled and distorted plants of hart's tongue fern (*Phyllitis scolopendrium*); the straight tubular particles measure about 135 or 320×22 nm and are, moreover, transmissible by inoculation of sap to sporelings of *P. scolopendrium*, which develop the disease symptoms (Hull, 1968).

Among the gymnosperms, an aphid-transmitted virus confined to spruce was reported by Cech, Králík and Blattný (1961), but there seems to be no more recent information on this agent. A few viruses of angiosperms are known to be able also to infect gymnosperms. For example, *Pinus sylvestris* roots became infected when mechanically inoculated with tobacco necrosis virus (Yarwood, 1959), and the roots of *Chamaecyparis lawsoniana, Cupressus arizonica* and *Picea sitchensis* became infected when exposed to nematodes carrying arabis mosaic, tobacco ringspot and tomato black ring viruses, respectively (Harrison, 1964b; Fulton, 1969). These three last are nepoviruses and, interestingly, all four viruses have soil-inhabiting vectors.

16.4 Viruses of animals

Viruses have been found in animals from many different phyla, though principally in arthropods and vertebrates (Andrewes and Pereira, 1972; Gibbs, 1973a; Fenner *et al*, 1974).

16.4.1 *Viruses of arthropods*

Viruses are associated with many different groups of arthropods. In Chapter 13 we described the various types of association between viruses of higher plants and the arthropods that act as their vectors, and in some instances are alternate hosts. There are similar associations between parasitic arthropods, such as mosquitoes, ticks and fleas, and viruses of vertebrates such as birds and mammals; these animal viruses are known collectively as the *a*rthropod-*bo*rne viruses or arboviruses (§16.4.2). Finally, there are viruses that seem confined to arthropods, and these viruses will be discussed in this section.

The best-known arthropod viruses are the polyhedrosis viruses that are found especially in phytophagous larvae of Lepidoptera, Hymenoptera, Diptera and Coleoptera. These viruses are named after the polyhedra found in infected insects; the polyhedra are crystalline inclusions of virus-specified proteins that contain occluded virus particles, which are protected from inactivation. There are three quite unrelated groups of polyhedrosis viruses; the baculoviruses, the cytoplasmic polyhedrosis viruses and the entomopoxviruses.

The baculoviruses [D/2: 100/10: U/U: I/0], once called the nuclear polyhedrosis and granulosis viruses, are perhaps the commonest polyhedroses. Baculovirus polyhedra that are eaten by insects are digested and liberate the virus particles. These infect midgut cells and then spread throughout the larva replicating in cell nuclei to such an extent that finally all that remains of the larva is a skin filled with a suspension of polyhedra, which are dispersed by water splash when the skin has ruptured. Each polyhedron contains one (granuloses) or more (nuclear polyhedroses) complex rod-shaped particles (Fig. 16.4a), which themselves contain a large double-stranded DNA genome that may be circular. Virus particles within polyhedra retain their infectivity for several years, but when the particles are liberated, either by chemical treatment or digestion in an insect, they soon lose their infectivity. The baculoviruses are being increasingly used for the 'biological control' of insect pests.

The cytoplasmic polyhedrosis viruses [R/2: $\Sigma 13–18/16–30$: S/S: I/0] resemble the baculoviruses only in their ecology. They replicate in the cytoplasm, and have, in each polyhedron, many isometric particles each of which contains the complete multipartite double-stranded RNA genome. The particles and their genome resemble closely in size, composition and morphology those of clover wound tumor virus, and the reoviruses of mammals.

The entomopoxviruses [D/*: */*: X/X: I/0] have rounded inclusions which contain large particles (Fig. 16.4c) that resemble very closely in structure and composition those of the poxviruses of vertebrates. Thus three apparently unrelated groups of insect viruses use polyhedra as an additional protective coat to help them survive the chancy journey from one rupturing host to the susceptible midgut cells of another host (Harrap, 1973).

Other insect viruses do not induce the production of protective inclusions; such viruses differ in ecology from the polyhedroses. Many of them are spread directly by mutual feeding among social insects, or by contamination of the environment of swarming or generally abundant insects. Many of the viruses, such as bee acute paralysis virus (R/1: 2/30: S/S: I/0), have isometric particles, which are about 25 nm in diameter, have no obvious subunits, and closely resemble those of the enteroviruses and other picornaviruses of vertebrates (Bellett, Fenner and Gibbs, 1973). Other viruses are spread from parent to offspring congenitally, such as for example sigma virus of *Drosophila*, a rhabdovirus.

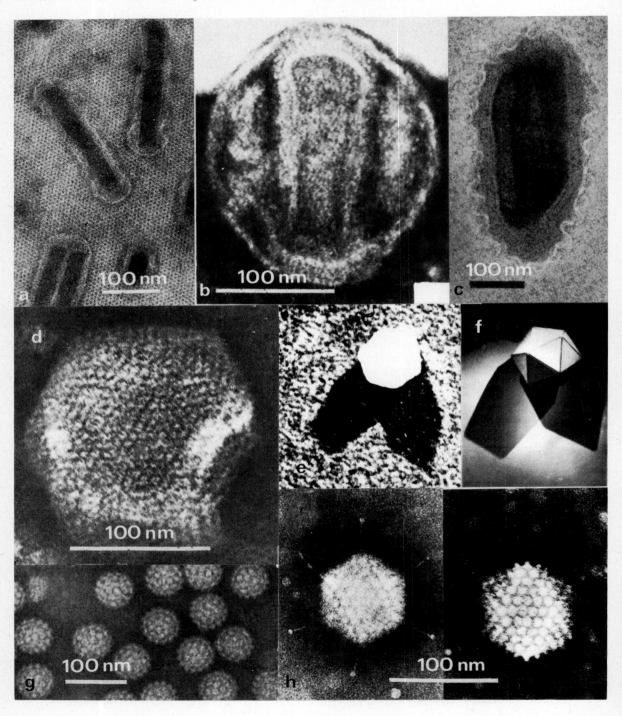

Finally there are those insect viruses whose ecology suggests that the insect may not be the principal host, but has been infected incidentally. Most notable of these are the iridoviruses [D/2: 130/15: S/S: I/*]. Many iridoviruses found in insects have complex icosahedral particles about 130 nm in diameter, and one of these was the first virus particle shown to be an icosahedron (Fig. 16.4). When these particles are sufficiently concentrated, either artificially by centrifugation or naturally in larval fat bodies, they form crystals which appear a beautiful iridescent blue because of Bragg reflection of light. Some iridoviruses, such as those found in mosquitoes, have larger particles which, presumably because of their size, form non-iridescent crystals. Iridoviruses were first found in insect larvae, but similar particles have since been found in reptiles, amphibia, molluscs, protozoa and green algae, and the particles of tick-borne African swine fever virus are somewhat similar.

Few viruses are known from other groups of arthropods, perhaps because they have been little studied, although there are reports of viruses in, for example, crabs (Bonami and Vago, 1971; Bang, 1971) and mites (Shaw and Beavers, 1970).

16.4.2 *Viruses of vertebrates*

Very many viruses have been obtained from vertebrates, in particular from mammals and birds, but fortunately for the virologist most fall

into well defined groups, several of which are described below.

The group of vertebrate viruses with the most complex genomes are the poxviruses [D/2: 160–200/6: X/X: V/0,Di,Ac,Si], of which there are at least five subgroups in addition to the entomopoxviruses (§16.4.1). The poxviruses infect a wide range of birds and mammals, and include such famous viruses as vaccinia and smallpox, which was recorded in Roman times. Most are transmitted by contact or through the gut, but some are transmitted by arthropods such as, for example, rabbit myxoma virus, which causes myxomatosis disease and is transmitted by fleas and mosquitoes. Poxviruses cause skin lesions or more general diseases. They multiply in the cytoplasm and many form cytoplasmic inclusions, like the entomopoxviruses. They have rounded brick-shaped particles (Fig. 16.4b) about 250 to 300 nm long, which have a very complex structure and are constructed of at least seventeen different proteins, some of which are enzymes.

The herpesviruses [D/2: 80/7: S/S: V/0] are another widespread group of viruses. They have rounded icosahedral particles with tubular subunits and a lipid envelope, and are thus sensitive to ether. At least two viruses of this group, namely human genital herpes virus (Type II) and Epstein-Barr virus, which is the cause of infectious mononucleosis or glandular fever, are likely causes of human cancers.

The adenoviruses [D/2:22/13:S/S:V/0] are widespread in mammals and birds, usually causing colds and sore throats. They have elegant angular icosahedral particles (Fig. 16.4h) that are about half the diameter of those of the iridoviruses, and have filaments at their vertices. Some cultures of adenoviruses contain the smaller particles of the adeno-associated (=adeno satellite) viruses (D/1: 1.5/25: S/S: V/*), which are unable to replicate properly in the absence of adenovirus, and in this respect resemble the satellite virus of tobacco necrosis virus. When the single-stranded DNA in the particles of an adeno-satellite virus is extracted, it becomes double-stranded because the DNA in half the particles is complementary in sequence to that in the others.

The papovaviruses [D/2: 3–5/7–15: S/S: V/0] have a wide range of hosts, and usually cause warts. Their particles resemble those of cauliflower mosaic virus in appearance (Fig. 16.4g) and in containing a genome of circular double-stranded DNA which, when isolated, is infective;

Fig. 16.4 Particles of animal viruses with DNA genomes. **a,** Baculovirus. Section of a polyhedron of *Bombyx mori* nuclear polyhedrosis virus showing the rod-shaped particles scattered through the crystalline matrix of the polyhedron. (Courtesy G. K. Filshie and C. Beaton.) **b,** Poxvirus. Detergent-treated particle of vaccinia virus, showing the complex structure of the particle. (Courtesy K. B. Easterbrook.) **c,** Entomopoxvirus. A section of an inclusion body of the entomopoxvirus of *Othnonius batesi*, showing the verrucose surface and coiled inner component of one particle of the virus. (Courtesy B. K. Filshie.) **d,** Iridovirus. Particle of Tipula iridescent virus treated to show the subunits. (Courtesy N. G. Wrigley; Wrigley, 1970.) **e,** A particle of Tipula iridescent virus, frozen-dried, and shadow cast in two directions; **f,** a model icosahedron illuminated by two lights. Note the similarity in the flat topped and pointed shadows cast by both particle and model; evidence that the virus particle is an icosahedron. (Courtesy K. M. Smith; Williams and Smith, 1958.) **g,** Papovavirus. Particles of rabbit papilloma virus. (Courtesy E. A. C. Follett.) **h,** Adenovirus. Particles of adenovirus type 5, showing the fibres at the vertices, and the close-packed subunits. (Courtesy R. C. Valentine.)

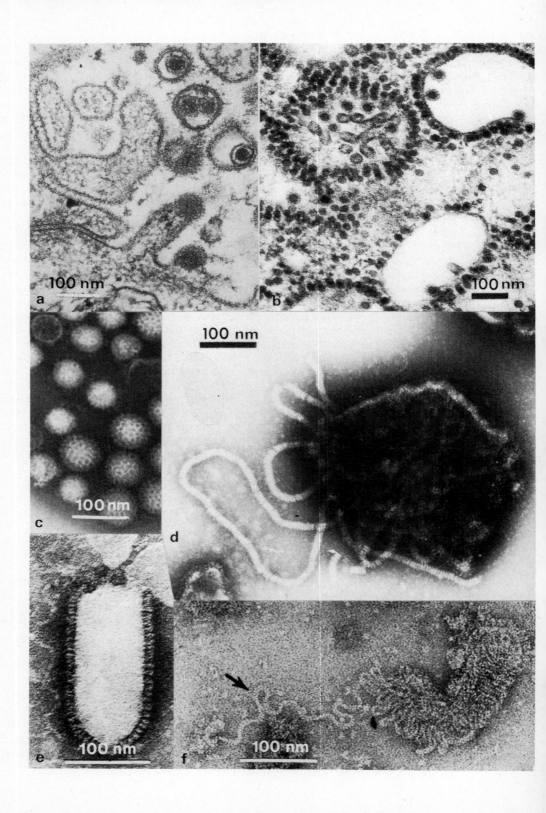

they replicate in the host's nuclei. Both the adenoviruses and papovaviruses can cause cancers when transmitted experimentally to unusual hosts; however, it is unlikely that they are a cause of cancers in their natural hosts.

The smallest viruses of vertebrates that have a DNA genome are the parvoviruses [D/1: 1.2–1.8/35: S/S: V/0]. The parvoviruses are a disparate group in that they contain the adenosatellite viruses, as well as viruses that are autonomous, such as Kilham's rat virus.

The vertebrate viruses with RNA genomes are at least as diverse as those with DNA genomes, and include not only those that use RNA as genetic material at all times, but also the leukoviruses, which have a RNA genome that is transcribed into DNA at one stage of the life cycle.

The leukoviruses [R/1: Σ10–13/1–2: S/*: V/0] are a large group of viruses with pleomorphic enveloped particles about 100 nm in diameter, that resemble in appearance the particles of tomato spotted wilt virus. Many cause leukaemia (hence the group name) in birds and rodents. The genome of the leukoviruses, like that of the orthomyxoviruses and several plant viruses, is divided into several parts. There has been much interest in these viruses since it was shown that their particles contain a RNA-dependent DNA polymerase, otherwise called a reverse transcriptase. This enzyme probably transcribes the genome into complementary DNA, which in turn is transcribed into progeny RNA molecules, but may also be incorporated into the genome of the host, and hence cause cancer.

Fig. 16.5 Particles of animal viruses with RNA genomes. **a,** Leukovirus, group B. Section of part of an adeno-carcinomatous mouse cell showing particles of mouse mammary tumour virus as they bud from the cell surface. (Courtesy D. H. Moore.) **b,** Togavirus, alphavirus group. Section of a chick cell infected with Semliki Forest virus, showing particles of the virus around cytoplasmic vacuoles, and acquiring their outer membranes as they bud into these vacuoles. (Courtesy K. B. Tan.) **c,** Reovirus. Particles of reovirus showing the double layer of subunits. (Courtesy D. W. Verwoerd.) **d,** Paramyxovirus. Particle of mumps virus treated with ether. The outer lipid-containing layer of the particles has disrupted and released the helically-constructed nucleocapsid. (Courtesy A. J. Gibbs.) **e,** Rhabdovirus. Particle of vesicular stomatitis virus, showing the subunits of the outer layer. (Courtesy R. W. Simpson.) **f,** Rhabdovirus. A disrupted particle of vesicular stomatitis virus: the arrow showing the 5 nm wide filament of subunits from which the helically-constructed nucleocapsid is made; the filament may itself be helically constructed. (Courtesy R. W. Simpson.)

Less devious is the life cycle of the reoviruses [R/2: Σ15/15: S/S: V/0,Di] which, like clover wound tumor and the cytoplasmic polyhedrosis viruses, have a multipartite double-stranded RNA genome contained in multishelled isometric particles that are about 80 nm in diameter (Fig. 16.5c). The reoviruses are widespread in birds and mammals, even bats. Each of the ten or more parts of their genome specifies a single protein.

The remaining groups of vertebrate viruses have genomes of single-stranded RNA. Some of them, namely the togaviruses and picornaviruses, have genomes which are messenger RNAs and may be translated directly, whereas the ortho- and paramyxoviruses and the rhabdoviruses have genomes that are complementary in sequence to their messenger RNAs.

The togaviruses [R/1: Σ4/6: S/S: V,I/Di,Ac] are a very large group of viruses that comprise most of the arboviruses. Togaviruses are worldwide, and are common in the tropics, where they cause many severe diseases such as encephalitides of man and other vertebrates. The togaviruses are comprised of the alphaviruses (group A arboviruses) such as Sindbis, Semliki Forest and various equine encephalitis viruses, and the flavoviruses (group B arboviruses), which include dengue, Japanese B encephalitis and yellow fever viruses. There are also various smaller groups of which the bunyavirus group is the largest. All replicate both in vertebrates and in their invertebrate vectors. Most are transmitted by mosquitoes, but some of the flavoviruses of cold regions are transmitted by ticks. All have isometric particles 50 to 100 nm in diameter, which have a lipid-containing outer layer that is acquired when the nucleocapsid buds through a cellular membrane (Fig. 16.5b) and in this respect they resemble carrot mottle virus (Chapter 2; Fig. 5.14h).

The picornaviruses [R/1: 2.7/25: S/S: V,I/0] have small icosahedral ether-resistant particles, each containing a single infective RNA molecule and sixty subunits, each of four proteins, which are derived by the post translational cleavage of a single larger protein. The group contains the enteroviruses, such as poliovirus and encephalomyocarditis virus, which are primarily viruses of the gut and have acid resistant particles, and also the rhinoviruses (common cold viruses), foot and mouth disease virus and vesicular exanthema virus, all of which have particles that are sensitive to low pH. Poliovirus no longer causes widespread epidemics of human infantile paralysis in developed countries, having been controlled by vaccines, at first of inactivated virus, but more recently of live attenuated virus.

The orthomyxoviruses or influenza viruses [R/1: Σ5/1: S/E: V/0] cause widespread epidemics of respiratory disease in mammals and birds. Major epidemics of 'flu in man may perhaps result from pseudo-recombinants formed between human strains and strains from other animals that carry novel antigenic determinants. The virus particles are pleomorphic, about 100 nm in diameter, and contain the filamentous nucleocapsid. The genome is divided into about eight parts, and on entering a susceptible cell is transcribed by a replicase, which is carried in the particle, into complementary messenger RNA.

By contrast the paramyxoviruses [R/1: 7/1: S/E: V/0] have their genome in one piece, even though it too is the complement of the messenger RNA. Although the particles of the paramyxoviruses (Fig. 16.5d) are superficially similar to those of the orthomyxoviruses, and the two groups have a similar ecology, they are quite unrelated. Paramyxoviruses cause many economically important diseases, including mumps, measles, dog distemper, rinderpest and Newcastle disease.

Finally there are the rhabdoviruses [R/1: 4/2:U/E:V,I,S/0,Ap,Au,Di], a worldwide group of viruses of plants, insects and vertebrates, even fish. It includes rabies, vesicular stomatitis (Fig. 16.5e, f) and Drosophila sigma viruses, and many others. All have similar complex bacilliform particles with a helically constructed tubular core and lipid-containing envelope.

16.4.3 *Viruses of other animals*

No viruses of Protozoa have been well characterized, though several have been reported. Virus-like agents have been found in parasitic protozoa such as *Entamoeba hystolytica, Naegleria gruberi* and *Plasmodium* (Gibbs, 1973b), and also in *Aphelidium,* a phycomycetaceous protozoon, found parasitizing *Scenedesmus armatus* (Schnepf, Soeder and Hegewald, 1970). Furthermore, phages or defective phages are possibly involved in the complex interaction between the ciliate *Paramecium aurelia* and the kappa killer factor that is associated with its endosymbiotic gram-negative bacteria (Beale, Jurand and Preer, 1969; Preer *et al*, 1971).

Plant parasitic nematodes transmit nepoviruses

and tobraviruses to plants in nature, although they are not hosts of the viruses (Chapters 2 and 13). Nematodes have also been implicated in the spread of some animal viruses, including swine influenza virus (an orthomyxovirus) and lymphocytic choriomeningitis virus (an arenavirus) (Hooper, 1973; Harrison, 1973; Taylor and Cadman, 1969). In addition, virus-like particles 25 nm in diameter are described in the intestinal epithelial cells of *Dolichodorus heterocephalus* (Zuckerman, Himmelhoch and Kisiel, 1973).

Finally, an irido-like virus has been found in association with a lethal and common cancer-like disease of octopi in the Bay of Naples (Rungger *et al*, 1971).

16.5 Origins of viruses

Although we have no direct knowledge of the origins of viruses, various hypotheses have been advanced, and some of the clues that have been obtained are worth discussing. There are three main hypotheses to explain the origin of viruses: (1) viruses are descended from primitive pre-cellular forms of life, (2) viruses are degenerate microorganisms, and (3) viruses are derivatives of cell components that have escaped from normal cellular control processes.

The first general point is that viruses use the same types of molecules as other forms of life, and furthermore use these molecules in much the same way as do organisms; even the ability to transcribe RNA sequences is now realized not to be confined to viruses. This biochemical unity suggests that viruses and organisms have common origins. Indeed, for their replication viruses totally depend on living organisms for energy, ribosomes and many enzymes. They must, therefore, have originated from a non-virus independent stage, presumably from an organism of some sort, and hypothesis (1) seems ruled out. Thus the virus genome is adapted to an intracellular environment; it does not contain genes concerned with anabolic metabolism but relies on its host for the products of such genes, and instead has special genes that sequestrate host metabolism and others that ensure spread of the virus genome from host to host.

The idea that viruses are degenerate pathogenic microorganisms (Green, 1935; Laidlaw, 1938) is also unattractive because with one exception virus particles lack features suggestive of cellular organization. The exception is the poxvirus group, consisting of viruses whose particles are large, complex in structure and that contain many different protein species (Joklik, 1974). In structure, particles of other viruses resemble more closely various cellular components that consist of protein subunits put together in regularly repeating arrangements by a self-assembly process; and the genetic plasticity of viruses relies in part on interactions between virus genome and host genome. We therefore conclude that viruses most probably are derived from cell components (Bawden and Pirie, 1953) but cannot specify which.

A classification of viruses, made using all their available properties, shows that most viruses fall into well-defined groups (§12.4). Members of these groups usually share very many of their properties, such as the type of particle they produce, their biochemical life cycle and their mode of transmission, and they differ most often only in host range and pathogenicity. The differences between virus groups are usually clear cut, and there is little evidence that the different major groups may have a common ancestor. Thus viruses are probably polyphyletic in origin, and we must search for almost as many origins as there are groups of viruses. As already pointed out, viruses are probably not so much a family, more a way of life.

Viruses seem to have more genetic plasticity than organisms, perhaps because their genomes are smaller and their replication rate greater. In Chapter 12 we discussed genetical variation among plant viruses, and there are many further examples with other viruses. For example, Mills, Kramer and Spiegelman (1973), using the replicase of ribophage Qβ, found that when various selection pressures were applied during a series of *in vitro* subcultures, a replicating RNA molecule only 218 nucleotides long was selected from the genome of the phage. Furthermore the genomes of a bacterium, its plasmids and its virulent and lysogenic DNA phages can be regarded as one large fluctuating gene pool that is able to react in a variety of ways to its environment. There is also much genetical plasticity among the animal viruses; complementation, recombination, pseudo-recombination and satellitism are common, and hybrids can be formed between viruses that seem totally unrelated, such as SV40 (a papovavirus) and adenovirus (Fenner *et al*, 1974; Lebowitz *et al*, 1974). One consequence of this genetic flexibility is that the genes of one particular virus may have not come from one particular free-living progenitor, but may be polyphyletic in origin themselves. It is also easy to see that virus genome size could increase by gene duplication, followed by mutation of one copy to assume a new function; and that, conversely, genome size could decrease when a single gene product takes on an additional function (as in the ribophages; Chapter 11), so rendering redundant a second gene that formerly performed this function. Thus there is a variety of mechanisms that are undoubtedly important for virus evolution even if not involved in the origin of the viruses.

Various kinds of extra-chromosomal gene movement, not involving virus infection, are now recognized, especially in bacteria where they are most easily studied (Hayes, 1969; Meynell, 1972), but also in eukaryotes (Ledoux, 1971). Thus it is possible that viruses may originate on rare occasions when gene movement has brought together particular combinations of genes that can free themselves of repression by other nearby genes and can then sequestrate the products of other genes, and obtain safe transport to other uninfected susceptible 'hosts'; possibly transmission to another 'host' is the crucial step that removes 'host' repression, and perhaps it is no coincidence that members of known virus groups almost always have the same mode of transmission. Some of the stages of this hypothetical process have been found among the plasmids and episomes of bacteria (Meynell, 1972), and indeed plasmids may be thought of as viruses that are particularly benign and often helpful to their hosts.

There are several possible ways of obtaining clues to the origins of particular virus groups or their genes. For example, it is interesting to note that rod-shaped viruses have been found only in lower plants and higher plants, suggesting that the examples infecting angiosperms may have an ancient ancestry. By contrast, insects seem to be the common factor in the host ranges of different rhabdoviruses and, possibly, reoviruses, and more primitive hosts are not known, suggesting that these virus groups may have a more recent origin. However, the most useful clues seem likely to come from detailed studies on the composition, structure and behaviour of the various virus macromolecules. Thus the nucleotide sequence of the genome of a virus may provide some clues to its origin. Several investigators have shown that the nearest-neighbour nucleotide composition (§5.2) of a nucleic acid is characteristic of the virus or organism from which it came and that related organisms have genomes with closely similar nearest-neighbour patterns (Subak-Sharpe *et al*, 1966; Subak-Sharpe, 1969; Russell *et al*, 1973). Subak-Sharpe and his colleagues found that the nearest-neighbour patterns of the genomes of viruses with small particles usually resemble closely that of their host, whereas the pattern of viruses with large particles was in most instances quite different from their host's. Subak-Sharpe concluded from this that the small viruses probably originated from their host whereas the large viruses came from elsewhere. This conclusion can be questioned because all the small viruses had patterns resembling their hosts, which are closely related, yet other properties suggest that the viruses are unrelated to one another. An alternative explanation is that the similarity in pattern between host and small virus is the consequence of some selective force that does not affect the large viruses (e.g. the pattern of the small virus genomes but not that of the large ones may be selected by the host's transcription apparatus). If this alternative is correct then nearest-neighbour patterns of the large viruses are more likely to contain clues to their ancestry than those of the small viruses.

However, it is worth remembering that many selective forces act on the nucleotide sequence of a virus genome; not only selection through ability to code for a useful protein, but also through ability to interact correctly with other parts of the genome and form base paired loops (Ball, 1972) that may control transcription, translation and particle assembly.

When discussing the value of protein and nucleic acid sequence data, Lanni (1964) predicted that 'Proteins may well have a better memory of distant ancestry', and it has been repeatedly shown that the amino-acid sequence of a protein is conserved during evolution, and that the extent of the sequence similarity of homologous proteins in different organisms is correlated with the relatedness of the organisms (Dayhoff, 1972). Information obtained from the closely related ribophages, f2, MS2 and R17, has confirmed the relatively greater evolutionary conservation of protein-related structure. Thus Robertson and Jeppesen (1972) found that the comparable sequences of the four untranslated regions of these ribophage genomes (sequences to which ribosomes or replicase bind) are almost identical, and the amino-acid sequences of the three coat proteins differ only slightly, whereas the cistrons coding for these coat proteins differ more. The coat-protein sequence and the protein-recognition sites of the genome have been much more rigorously conserved than the protein-specifying sequences, which constitute most of the genome. These observations suggest that it may be fruitful to hunt for clues to virus origins among the proteins. A preliminary comparison (Gibbs, unpublished) of ten virus coat protein sequences and about one hundred other proteins has shown several interesting sequence similarities of which the most definite is a similarity between the sequences of Qβ phage coat protein and the acyl carrier protein from the fatty-acid synthetase enzyme of *E. coli* (Dayhoff, 1972).

It would therefore seem that instead of simplifying ideas on the probable origins and evolutionary trends of viruses, modern studies are revealing new mechanisms of evolution and are broadening the range of possibilities.

Chapter 17

Plant pathogens confused with viruses

17.1 The range of pathogens

Several plant diseases that have long been thought to be caused by viruses are now known to be associated with pathogens of other kinds. Some are bacteria; for example, the so-called 'beet latent virus' is the bacterium *Pseudomonas aptata* (Yarwood *et al*, 1961). Plants affected by clover club leaf disease contain a rickettsia-like leafhopper-transmitted agent which is passed transovarially to the progeny of the leafhoppers and is sensitive to penicillin (Windsor and Black, 1972). Leafhopper-transmitted agents resembling rickettsias also occur in the xylem of plants with phony disease of peach, and Pierce's disease of grapevine (Goheen, Nyland and Lowe, 1973; Hopkins, French and Mollenhauer, 1973). However the biggest single group of plant pathogens confused with viruses are the mycoplasma-like organisms (MLOs) that are transmitted by leafhoppers and are associated with diseases of the witches' broom and yellows types (Doi *et al*, 1967; Maramorosch, Granados and Hirumi, 1970; Whitcomb and Davis, 1970b). It is this group of organisms that we will mainly discuss in this chapter, although we will also deal briefly with the rickettsia-like pathogens.

Plants infected with MLOs contain foreign cells less than 1 μm in diameter. Typically, these MLOs are transmissible by grafting but not by inoculation of sap, and several of them have leafhoppers or psyllids as vectors. It is therefore not surprising that they were for long confused with viruses, and indeed their study has produced many ideas that have proved valuable in plant virology. Although their taxonomic position is still controversial (Whitcomb, 1973), as also is that of the rickettsia-like agents, the plant MLOs are most like members of the Mycoplasmatales (Class Mollicutes) and we will now briefly discuss organisms in this group.

17.2 Characteristics of the Mycoplasmatales

Mycoplasmas are common in man and other vertebrates, and although one species, *Acholeplasma laidlawii*, was originally found in sewage, it was not rigorously proved to multiply there. Some species cause disease, such as *Mycoplasma pneumoniae* and *M. suipneumoniae*, which produce pneumonia in humans and pigs respectively. Other mycoplasmas are associated with particular diseases of vertebrates but are not proved to cause them; and several seem to be commensal inhabitants of the respiratory or urogenital tracts and do not produce any obvious symptoms.

The main characteristics of organisms in the Mycoplasmatales are:

(a) Their cells are pleomorphic, but mostly 0.1 to 1.0 μm in diameter, and will pass a filter with 450 nm diameter pores.

(b) Their cells are bounded only by a three-layered unit membrane about 10 nm thick.

(c) Their cells contain ribosomes of the bacterial type.

(d) Their cells contain both DNA and RNA; the DNA is in double-stranded circular molecules of molecular weight 4×10^8 to 1×10^9 and has a guanine + cytosine content of 23 to 41%.

(e) They probably multiply by binary fission.

(f) They can be cultured on artificial media; on agar media they typically produce minute colonies each of which is the shape of a fried egg.

(g) Typically they are resistant to penicillin and sensitive to tetracyclines. When cultured in penicillin-free media they do not revert to bacteria (in contrast to many L-forms of bacteria).

(h) They are themselves subject to virus infection (Gourlay, 1972).

For references to further information on these organisms, see section 2 of the appendix.

17.3 The plant-infecting MLOs

17.3.1 *Characteristics*

The first convincing evidence of MLOs in plants came from Japanese work; mycoplasma-like cells were found by electron microscopy in ultrathin sections of the phloem sieve tubes of plants affected by aster yellows, mulberry dwarf and several other plant diseases (Doi *et al*, 1967), and treatment of diseased mulberry plants with tetracycline caused remission of symptoms of dwarf disease (Ishiie *et al*, 1967). These plant MLOs resemble species in the Mycoplasmatales in points (a), (b), (c) and (g) in §17.2 but they differ in living intracellularly in plants, whereas mycoplasmas are usually found extracellularly in vertebrates. Some similarities and differences between the Mycoplasmatales, plant MLOs and plant viruses are listed in Table 17.1.

The plant MLOs share some properties with other organisms that are smaller than typical bacteria; these include Chlamydiae and L-forms of bacteria (Davis and Whitcomb, 1971). However, they differ from the Chlamydiae in having a less substantial outer cell wall, and in not being sensitive to penicillin. The MLOs differ from L-forms of bacteria in being pathogenic and in not reverting to bacteria, but this is a less satisfactory distinction because not all L-forms that have arisen from bacteria can also revert.

The commonest evidence cited for the presence of MLOs in plants is from electron microscopy of sections of plant tissue, and more than forty have been detected in this way; some are listed in Table 17.2. The cells of these pathogens closely resemble those of animal-infecting mycoplasmas. They are mostly 0.1 to 1.0 μm in diameter, they are bounded by a unit membrane usually about 10 nm thick, and their ribosomes are similar in size to those of bacteria (Fig. 17.1d, e). Typically these cells are rounded, but some may be elongated or shaped like a dumb-bell.

Different features have been emphasized in describing the form of the cells of different MLOs, but as the cells are pleiomorphic, and may have been exposed to different environmental conditions and different methods of preparation for electron microscopy, it is usually difficult to be confident that there are consistent morphological differences between the cells of the different agents. However, two organisms that seem morphologically different from several of the other MLOs are those occurring in maize with stunt disease and citrus with stubborn disease. Their cells are often helical in form (Cole *et al*,

Table 17.1 Comparison of some properties of plant viruses, animal mycoplasmas (Mycoplasmatales) and plant MLOs

Property	Plant viruses	Plant MLOs	Animal mycoplasmas
Size and shape of particles or cells	Definite shape; either rounded (diameter 17 to 100 nm) or elongated (12 × 2000 nm to 100 × 300 nm)	Pleiomorphic; rounded (usually 100 to 1000 nm diameter) or elongate	Pleiomorphic; rounded (usually 100 to 1000 nm diameter) or elongate
Outer membrane of particles or cells	Usually none, but lipid-containing envelope in particles of some viruses	Unit membrane	Unit membrane
Nucleic acid in particles or cells	Either DNA or RNA	Probably DNA and RNA	DNA and RNA
Genome size	0.4×10^6 daltons	1000×10^6 daltons in one instance	400 to 1000×10^6 daltons
Ribosomes	None	Plentiful, of bacterial size	Plentiful, of bacterial size
Replication	Particles disassemble during infection, virus components are synthesized, and progeny particles assemble	Probably by binary fission	By binary fission
Location in host	Intracellular	Intracellular	Extracellular
Cultivation on artificial media	Not cultured	'Fried egg' colonies produced in a few instances	Produce 'fried egg' colonies
Sensitivity to tetracyclines	Resistant	Sensitive	Sensitive
Natural method of spread	Mostly by vectors	By arthropod vectors	Mostly by contaminating the environment (droplets, etc.)

Table 17.2 Some plant MLOs and rickettsia-like organisms, and their vectors

Associated disease	Vector[1]
MLOs	
Aster yellows	*Macrosteles fascifrons*
Citrus greening	*Diaphorina citri*
Citrus stubborn	Not known
Clover phyllody	*Euscelis plebejus*
Corn stunt	*Dalbulus elimatus*
Mulberry dwarf	*Hishimonus sellatus*
Papaya bunchy top	*Empoasca papayae*
Pea 618M	*Acyrthosiphon pisum*
Peach western X	*Colladonus montanus*
Pear decline	*Psylla pyricola*
Phormium yellow leaf	*Oliarus atkinsoni*
Potato stolbur	*Hyalesthes obsoletus*
Rice white leaf	*Epitettix hiroglyphicus*
Rubus stunt	*Macropsis fuscula*
Sandal spike	*Jassus indicus*
Rickettsia-like agents	
Clover club leaf	*Agalliopsis novella*
Peach phony	*Homalodisca triquetra*
Pierce's grapevine and alfalfa dwarf	*Draeculacephala minerva*

[1] Many of the agents have more than one vector species.

1973; Fig. 17.2), with a width of about 250 nm and length up to 12 μm, and they are motile; they have been christened 'spiroplasmas' (Davis and Worley, 1973), and the citrus stubborn agent is characterized in detail under the name *Spiroplasma citri* (Saglio *et al*, 1973).

Few of the plant MLOs have yet been cultured on artificial media although some have been maintained *in vitro* for some weeks and then returned to plants. Lin, Lee and Chiu (1970) reported that they had cultured in liquid medium a MLO from sugarcane with white leaf disease; it produced 'fried egg' colonies on agar medium and reproduced white leaf disease when transmitted back to sugarcane by pricking the plants with contaminated pins. Similarly, a MLO from clover with phyllody disease was maintained in a semi-synthetic medium and returned to clover by leafhoppers which had fed on the medium through a membrane or were injected with it (Faivre-Amiot *et al*, 1970), and analogous results were reported with corn stunt MLO and its vector *Dalbulus elimatus* (Chen and Granados, 1970). However, the most detailed evidence for culture *in vitro* has been obtained with *Spiroplasma citri* (Fudl-Allah *et al*, 1971; Saglio *et al*, 1971, 1972, 1973). Using a medium containing 0.1% glucose, 0.1% fructose, 0.5% tryptone, 3% sucrose, 2%

PPLO broth, 10% of 25% fresh yeast extract and 20% horse serum, *S. citri* can be consistently obtained from trees with stubborn disease but not from healthy trees. It grows readily at 27 to 33°C and pH 7.5, produces 'fried egg' colonies on solid medium, is gram positive and has a DNA genome of 10^9 daltons that contains 26% of guanine + cytosine; *S. citri* is serologically unrelated to all tested isolates (over 100) of species in the Mycoplasmatales (Saglio *et al*, 1972; Fudl-Allah and Calavan, 1973) but is apparently distantly related to the corn stunt organism. Moreover it induced disease symptoms when transmitted to clover plants (Daniels *et al*, 1973) and citrus. An obvious and important task is now to culture a wide range of the plant MLOs and to test both their pathogenicity and their relatedness.

17.3.2 *Behaviour in plants*

The plant diseases associated with MLOs include several of considerable economic importance, including potato purple top and stolbur, mulberry dwarf, pear decline, sandal spike and papaya bunchy top, to name a few examples. Typically, the MLOs invade plants systemically, and their host range is usually, but not always, quite wide. Thus the MLO associated with aster yellows can infect many plant species, whereas that associated with rubus stunt is apparently confined to a few.

Some of the symptoms produced by the infected plants are very characteristic and, in combination, are rarely caused by other pathogens or conditions. These symptoms include witches' broom, which consists of a mass of spindly shoots produced when buds develop prematurely, following the loss of either apical dominance or dormancy (e.g. hair sprout of potato tubers; Fig. 17.1c). In some instances flowers remain green (*virescence*) or are transformed into leafy structures (*phyllody*) (Fig. 17.1b). Other symptoms include stunting of the shoots, and yellowing or reddening of the leaves, which give the name 'yellows-type diseases'; virus infection can also cause yellowing, reddening and stunting, though it is not known to cause witches' broom, or virescence. The symptoms of infection suggest that the MLOs derange the hormone metabolism of plants, but the details are not clear, although it is known that gibberellic acid reverses the stunting produced in maize by corn stunt MLO (Maramorosch, 1957). Elongated virus-like particles are reported to occur near the bodies of some MLOs in plant cells (Allen, 1972), and particles typical of a tailed bacteriophage occur in cultures of *Spiroplasma citri* (Cole *et al*, 1973), but the role of these particles in plant disease, if any, is unknown.

Electron microscopy indicates that the cells of most plant MLOs are restricted to the phloem, particularly the sieve tubes (Fig. 17.1e), although phloem parenchyma cells are also sometimes invaded; the pea 618M agent however is claimed to occur mainly in leaf parenchyma cells (Hampton *et al*, 1969). Although plant MLOs have cells that are larger than the phloem sieve pores, they probably pass through the sieve pores by

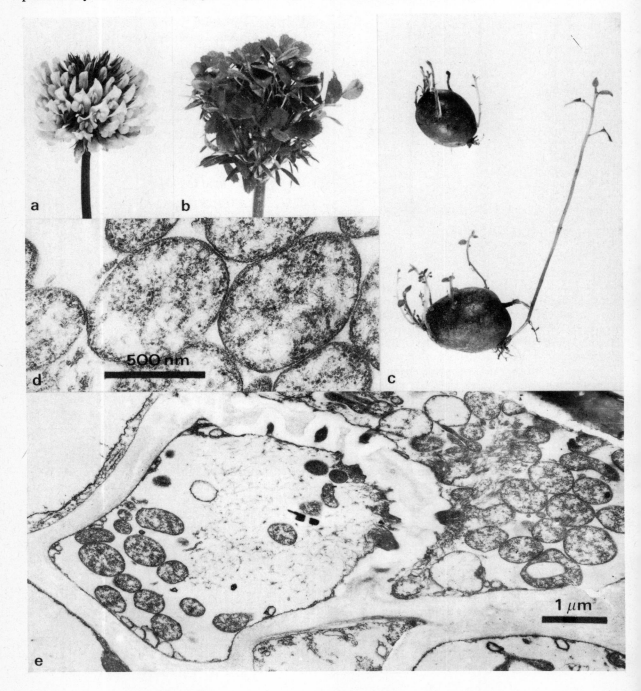

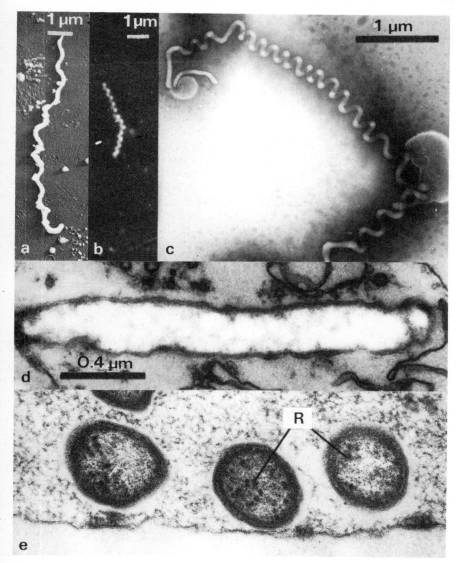

Fig. 17.2a-c Cells from *in vitro* cultures of *Spiroplasma citri*. (Courtesy R. M. Cole; Cole *et al* (1973).) **a,** Electron micrograph of fixed and shadowed cell. **b,** Fixed cell viewed by dark-field light microscopy. **c,** Electron micrograph of cell mounted in 6% ammonium molybdate **d** and **e,** Electron micrographs of transverse (R) and longitudinal sections of rickettsia-like cells in the phloem of periwinkle **(d),** and crimson clover **(e)** plants infected with the clover club leaf agent. Note the undulating cell wall in the longitudinal section. (Courtesy L. M. Black; Windsor and Black, 1973.)

temporarily distorting their cell shape. Not surprisingly, plant MLOs are readily transmitted by grafting and, interestingly, there is evidence that infection of a plant with one strain of a MLO

Fig. 17.1a Normal flower head of white clover (*Trifolium repens*). **b,** Flower head transformed into leafy structures, from white clover plant infected with clover phyllody MLO. **c,** Hair sprout, a symptom shown by potato (*Solanum tuberosum*) tubers infected by potato witches' broom MLO. (From Harrison, 1971.) **d,** Electron micrograph of MLO cells in sieve tube of potato infected with the potato witches' broom agent. **e,** As **(d)**, general view of two sieve tubes separated by a sieve plate. (From Harrison and Roberts, 1969.)

protects it from infection with related strains (Kunkel, 1955; Freitag, 1964). Like the particles of some viruses, the cells of several MLOs can be eliminated from plants, or greatly diminished in concentration, by growing the plants at 37 to 40°C (Kunkel, 1941). Conversely, mulberry dwarf MLO may be unable to survive in plant shoots at low temperatures; in summer in Japan it invades the shoots but in winter it is confined to the roots (Tahama, 1968).

17.3.3 *Behaviour in vectors*

Leafhoppers have long been known to transmit

the causal agents of diseases since found to be associated with MLOs (Table 17.2). More recently other MLOs, such as those of citrus greening and of pear decline, were found to have psyllid vectors, and the agent causing yellow leaf of *Phormium tenax* is transmitted by the plant hopper *Oliarus atkinsoni*. There is an unconfirmed report that the pea 618M agent is transmitted by the aphid *Acyrthosiphon pisum* (Hampton *et al*, 1969). Some MLOs, such as those of rubus stunt and of citrus greening seem to be transmitted by only one or a few vector species whereas others, such as Californian aster yellows MLO, have many vector species.

Several MLOs are known to multiply in their insect vectors and all probably do so. Aster yellows MLO can be acquired from plants by the leafhopper *Macrosteles fascifrons* in 30 minutes and infective leafhoppers can probably inoculate it to healthy plants in a similar period. However the probability of transmission is greatly increased by increasing these access periods to a day or more, and there is also a latent period of at least 14 days (at 20°C) between the start of acquisition feeding and becoming able to inoculate the MLO (Kunkel, 1954). This is the period when the MLO is multiplying in the insect (Maramorosch, 1952) and moving from the alimentary canal to the salivary glands and saliva. The length of the latent period depends on the temperature and is increased by decreasing the amount of inoculum acquired by feeding or by injection. During the latent period, the titre of the MLO in the vector increases logarithmically and transmission usually begins when a plateau value is reached (Whitcomb, Jensen and Richardson, 1966).

Examination of sections of leafhopper tissue in electron microscopes has revealed MLO cells and provided some details of events in the vector during the latent period of the aster yellows, clover phyllody and peach western X MLOs (Hirumi and Maramorosch, 1969; Sinha and Paliwal, 1970; Nasu *et al*, 1970). About two weeks after acquisition by feeding, MLO cells are found in intestinal epithelium cells but later they are found in other tissues, including the salivary glands. For example, the peach western X MLO was found in the salivary glands, brain and fat body of *Colladonus montanus* 27 days after acquisition, and in various other tissues after 50 days. When the MLO was injected into the haemocoele instead of being acquired by feeding, its cells were found in the brain after as little as 20 days and in the salivary glands after 22 days (Nasu, Jensen and Richardson, 1970). For an insect to become infective after feeding on an infected plant, the MLO seems to need first to multiply in intestinal cells, then to pass through the haemocoele and finally to multiply in salivary gland cells before passing into the saliva and becoming transmissible to healthy plants. MLOs seem usually to infect their vectors systemically and, once infective, the vectors usually remain so for much or all of their life. Not all MLOs that multiply in a leafhopper, however, may be transmitted. For example, aster yellows MLO survives and possibly multiplies to a limited extent in the leafhopper *Agallia quadripunctata* when acquired by feeding, but it is not transmitted (Sinha and Chiykowski, 1967). Transovarial transmission of MLOs in leafhopper vectors seems rare although it is not uncommon with viruses.

Not surprisingly, effects of MLOs on leafhoppers have been found in addition to those observed by electron microscopy. For example, aster yellows MLO induces nuclear changes in the fat body cells of *Macrosteles fascifrons* (Littau and Maramorosch, 1960). Peach western X MLO has effects on this and several other tissues of *Colladonus montanus*, causing the accumulation of basophilic material and production of enlarged or multinucleate cells (Whitcomb, Jensen and Richardson, 1968); moreover the life of *C. montanus* is shortened (Jensen, 1959).

There is some evidence with insects, as with plants, that infection by one MLO strain may protect from infection by a second; *Macrosteles fascifrons* carrying one strain of aster yellows MLO were unable to acquire and transmit a second strain (Kunkel, 1957). However, with other strains the situation was more complicated (Freitag, 1967). Another interesting effect of a MLO on a leafhopper is that of aster yellows MLO on *Dalbulus maidis*, which does not transmit it; *D. maidis*, which cannot survive on healthy asters, can survive on those infected with aster yellows MLO and, once infected with this MLO, becomes able to survive on healthy asters (Maramorosch, 1958).

17.3.4 *Ecology and control*

The plant diseases associated with MLOs are most prevalent in areas where the climate is hot for at least a part of the year, probably because the leafhopper vectors thrive in these conditions. Many MLOs and their vectors have wide plant-host ranges, which include species growing on uncultivated land and weeds of cultivated land, as well as crop plants. For example, Kunkel (1931) transmitted aster yellows MLO by leaf-hoppers to 120 plant species, including many that occur outdoors in the United States. Indeed many MLOs seem to live in wild plants and only spread into crop plants when circumstances are favourable. For example, in Czechoslovakia, stolbur MLO has often been found in *Convolvulus arvensis* (Valenta, Musil and Misiga, 1961) and other weed species, and spreads from these to potato and tomato; similarly, in Scotland potato witches' broom MLO spreads from wild plants to potato but not from one potato plant to another.

Some MLOs seem to undergo large cyclical fluctuations in prevalence. Thus stolbur was a common and serious disease of potato in central European countries from 1947 to 1953 but by 1960 had become very scarce, only to increase in prevalence again in 1963 to 1964 (Kozljarova and Bojnansky, 1969). This change paralleled a similar change in the prevalence of the leafhopper *Hyalesthes obsoletus*, the main vector of stolbur MLO. Similarly, pear decline disease appeared and greatly increased in prevalence in many parts of the north-western United States after 1947, when the eradication programme for *Psylla pyricola* ended, and the prevalence of this vector greatly increased as a result (Kaloostian, Oldfield and Jones, 1968).

Because many MLOs spread mainly to crop plants from sources outside the crop it is unlikely that treating crops with insecticide will be generally useful in keeping them healthy. The use of disease-escaping cultivars and the manipulation of cultural practices seem more promising. Thus the incidence of purple top disease in potatoes (probably caused by a strain of aster yellows MLO) in the coastal valleys of Peru is decreased by delaying planting until after most leafhopper vectors have migrated from their weed hosts.

The similarity of plant-infecting MLOs to animal-infecting mycoplasmas suggests the possibility of chemotherapy. In early work, Stoddard (1947) found that soaking infected peach buds in solutions of compounds such as 8-hydroxy-quinoline sulphate before grafting them to healthy stock seedlings decreased the proportion of plants that developed peach X disease. More recently, Ishiie *et al* (1967) showed that antibiotics can be used therapeutically. They found that tetracycline and chlortetracycline suppressed the symptoms of dwarf disease in mulberry plants,

whereas penicillin or kanamycin did not. Work with aster yellows, corn stunt, peach western X and other MLOs has shown similar effects, and confirmed that the treated plants contain fewer mycoplasma-like cells and are poorer sources of inoculum for vectors. Treating vector insects that have acquired MLOs or are about to do so also decreases transmission to healthy plants (Freitag and Smith, 1969; Whitcomb and Davis, 1970a). Unfortunately these antibiotic treatments often do not eliminate MLOs from plant or vector and, unless the antibiotic treatment is repeated at regular intervals, the plants again develop disease symptoms, and the vectors resume transmitting the MLOs to healthy plants. The use of antibiotics on crops will therefore depend on finding some way of prolonging their effect.

Heat therapy can also be used to free infected plants or infective vectors from MLOs. Kunkel (1937, 1941) freed periwinkles and aster leaf-hoppers from aster yellows MLO by keeping them at 37 to 40°C for long periods. This might be a useful method of obtaining a nucleus of healthy material from vegetatively propagated cultivars that are totally infected. Fortunately, however, there seems to have been no report of a totally infected cultivar and the technique is mainly of academic interest, although it has lately been extensively applied to virus-infected plants (see Chapter 15).

17.4 The rickettsia-like pathogens

Evidence that some well-known plant diseases are associated with and probably caused by rickettsia-like organisms is very recent and largely based on examination of tissue sections in the electron microscope. Cell-like structures have been found in and between the xylem vessels of grapevine with Pierce's disease but not in healthy plants; they are about 0.4 μm in diameter and up to 3 μm long, and have substantial walls with regular variations in thickness (Goheen, Nyland and Lowe, 1973). Similar cells were found in the xylem of roots of peach plants with phony disease but not in healthy peach (Hopkins, French and Mollenhauer, 1973). The rickettsia-like cells in the phloem tissue of clover with club leaf disease seem to be a little smaller and to have a different wall structure (Fig. 17.2); they are also known to be sensitive to penicillin (Windsor and Black, 1972, 1973). All these organisms differ from MLOs in having a thick outer wall and, in the one instance tested, in being sensitive to penicillin, but they have not been cultured on artificial media.

The rickettsia-like organisms are not transmitted by inoculation of sap, but have insect vectors. The peach phony disease organism is transmitted by leafhoppers (sharpshooters; Turner, 1952) and the Pierce's disease agent by both sharpshooters and spittlebugs (Frazier and Freitag, 1946; Severin, 1950); these feed primarily on the xylem. The clover club leaf organism has the leafhopper *Agalliopsis novella* as a vector, and is passed transovarially to progeny leafhoppers. Black (1950) kept infective *A. novella* on immune plants, recorded twenty-one successive generations of transovarial passage, and thereby proved that the clover club leaf organism multiplies in its vector. This is a further point of resemblance to rickettsias, which are widespread in arthropods.

All three plant-infecting organisms become systemic in plants, although Pierce's disease and peach phony organisms are both confined to woody tissue, and peach phony agent is more uniformly distributed in peach roots than shoots (Hutchins *et al*, 1953). The two xylem-limited organisms are only transmitted by grafting when the diseased scion material includes woody tissue.

The symptoms of peach phony disease include premature growth in spring, and stunting, but most of the many species that can be infected with Pierce's disease organism are symptomless. However infected lucerne is stunted and the shoots of infected grapevines wilt and die back, probably as the result of water shortage. In California, the Pierce's disease agent was found in thirty-six naturally infected species in eighteen plant families, and spreads from these to grapevine, especially during periods when the rainfall is above average for two or more years (Winkler, 1949; Freitag, 1951). There seems to be little spread from grapevine to grapevine.

Peach phony disease is economically important in the south-eastern and central United States, and Pierce's disease can be a serious problem in California. Control measures include hot water treatment of planting material (for instance 3 hours at 45°C; Hutchins and Rue, 1939; Goheen, Nyland and Lowe, 1973), which cures infected plants and should ensure that the organisms are not introduced into new areas. Also, lucerne crops should not be grown near grapevines.

APPENDIX

Further reading and information.

Section 1 General texts, journals, reviews etc.
Section 2 Reviews on special topics, listed under the chapter to which they are relevant
Section 3 Some additional information for those wishing to use plant viruses for class exercises; sources and storage of material

(N.B. the references given here are independent of those in the main reference list, and are given in full the first time they are cited.)

Section 1 General texts, journals, reviews, etc.

BAWDEN, F. C. (1956). *Plant viruses and virus diseases*, 3rd edition. Waltham, Mass.: Chronica Botanica.
BAWDEN, F. C. (1964). *Plant viruses and virus diseases*, 4th edition. New York: Ronald Press Co.
C.M.I./A.A.B. *Descriptions of plant viruses* (Nos. 1–80, A. J. GIBBS, B. D. HARRISON and A. F. MURANT (eds.); 80– , B. D. HARRISON and A. F. MURANT (eds.). Kew: Commonwealth Mycological Institute and Association of Applied Biologists.
CORBETT, M. K. and SISLER, H. D. (eds.) (1964). *Plant virology*. Gainesville: University of Florida Press.
GIBBS, A. J. (ed.) (1973). *Viruses and invertebrates*. Amsterdam: North-Holland/Elsevier.
KADO, C. I. and AGRAWAL, H. O. (eds.) (1972). *Principles and techniques in plant virology*. New York: Van Nostrand Reinhold Co.
KNIGHT, C. A. (1974). *Molecular virology*. New York: McGraw Hill.
MARAMOROSCH, K. (ed.) (1969). *Viruses, vectors and vegetation*. New York: Interscience Publishers.
MARAMOROSCH, K. and KOPROWSKI, H. (eds.) (1967–71). *Methods in virology*, Vols. 1–3 (1967), 4 (1968), 5 (1971). New York: Academic Press.
MARTYN, E. B. (1968). *Plant virus names*. Phytopathological papers No. 9. Kew: Commonwealth Mycological Institute.
MARTYN, E. B. (1971). *Ibid*. Supplement 1.
MATTHEWS, R. E. F. (1970). *Plant virology*. New York: Academic Press.
NOORDAM, D. (1973). *Identification of plant viruses. Methods and experiments*. Wageningen: Pudoc.
SMITH, K. M. (1972). *A textbook of plant virus diseases*, 3rd edition. London: Longman.

Many of the papers reporting research on plant viruses are published in the following journals:

Annals of Applied Biology. London: Association of Applied Biologists.

Journal of General Virology. London: Cambridge University Press.
Netherlands Journal of Plant Pathology. Wageningen: Pudoc.
Phytopathologische Zeitschrift. Berlin: Paul Parey.
Phytopathology. St Paul, Minn.: American Phytopathological Society.
Virology. New York: Academic Press.

Papers dealing with plant viruses are abstracted in:

Review of Plant Pathology. Kew: Commonwealth Mycological Institute.
Virology Abstracts. London: Information Retrieval.

In addition to books of special reviews, there are reviews on plant virus topics in:

Advances in Virus Research. New York: Academic Press.
Annual Review of Phytopathology. Palo Alto: Annual Reviews Inc.

Section 2 Reviews on special topics listed under the chapter to which they are relevant

Chapter 1 *The history and scope of plant virology*
HOLMES, F. O. (1968). Trends in the development of plant virology. *Ann. Rev. Phytopath*. **6**, 41.

Chapter 2 *Some plant viruses and their names*
BELLETT, A. J. D., FENNER, F. and GIBBS, A. J. (1973). The viruses. In GIBBS (1973), p. 41.
BRANDES, J. and BERCKS, R. (1965). Gross morphology and serology as a basis for classification of elongated plant viruses. *Adv. Virus Res*. **11**, 1.
C.M.I./A.A.B. *Descriptions of Plant Viruses* (1970–).
DIENER, T. O. (1972). Viroids. *Adv. Virus Res*. **17**, 295.
EDWARDSON, J. R. (1974a). Some properties of the potato virus Y-group. *Florida Agricultural Experiment Stations Monograph Series* **4**.
EDWARDSON, J. R. (1974b). Host ranges of viruses in the PVY-group. *Florida Agricultural Experiment Stations Monograph Series* **5**.
GIBBS, A. J. (1969). Plant virus classification. *Adv. Virus Res*. **14**, 263.
GIBBS, A. J. and HARRISON, B. D. (1968). Realistic approach to virus classification and nomenclature. *Nature, Lond*. **218**, 927.
GIBBS, A. J., HARRISON, B. D., WATSON, D. H. and WILDY, P. (1966). What's in a virus name? *Nature, Lond*. **209**, 450.
HARRISON, B. D., FINCH, J. T., GIBBS, A. J., HOLLINGS, M., SHEPHERD, R. J., VALENTA, V. and WETTER, C. (1971).

Sixteen groups of plant viruses. *Virology* **45**, 356.

LANE, L. C. (1974). The bromoviruses. *Adv. Virus Res.* **19**, 151.

MARTYN (1968; 1971).

SMITH (1972).

Chapter 3 *Effects of viruses on plants*

BALD, J. G. (1966). Cytology of plant virus infections. *Adv. Virus Res.* **12**, 103.

BALD, J. G. (1966). Some cell reactions during plant virus synthesis seen under the light microscope. In BEEMSTER, A. B. R. and DIJKSTRA, J. (eds.) (1966). *Viruses of plants.* Amsterdam: North-Holland, p. 66.

BOS, L. (1970). *Symptoms of virus diseases in plants,* 2nd edition. Wageningen: Pudoc.

DIENER, T. O. (1963). Physiology of virus-infected plants. *Ann. Rev. Phytopath.* **1**, 197.

ESAU, K. (1967). Anatomy of plant virus infections. *Ann. Rev. Phytopath.* **5**, 45.

ESAU, K. (1968). *Viruses in plant hosts.* Madison: University of Wisconsin Press.

HOLMES, F. O. (1964). Symptomatology of viral diseases in plants. In CORBETT and SISLER (1964), p. 17.

MATSUI, C. and YAMAGUCHI, A. (1966). Some aspects of plant viruses *in situ. Adv. Virus Res.* **12**, 127.

MATTHEWS, R. E. F. (1973). Induction of disease by viruses with special reference to turnip yellow mosaic virus. *Ann. Rev. Phytopath.* **11**, 147.

RUBIO-HUERTOS, M. (1972). Inclusion bodies. In KADO and AGRAWAL (1972), p. 62.

SCHLEGEL, D. E., SMITH, S. H. and DE ZOETEN, G. A. (1967). Sites of virus synthesis within cells. *Ann. Rev. Phytopath.* **5**, 223.

Chapter 4 *Experimental transmission*

(N.B. references on the experimental culture and manipulation of vectors given under Chapter 13)

BENNETT, C. W. (1967). Plant viruses: transmission by dodder. In MARAMOROSCH and KOPROWSKI (1967), Vol. 1, p. 393.

FULTON, R. W. (1964). Transmission of plant viruses by grafting, dodder, seed and mechanical inoculation. In CORBETT and SISLER (1964), p. 39.

FULTON, R. W. (1966). Mechanical transmission of viruses of woody plants. *Ann. Rev. Phytopath.* **4**, 79.

GARNER, R. J. (1958). *The grafter's handbook.* London: Faber and Faber.

KADO, C. I. (1972). Mechanical and biological inoculation principles. In KADO and AGRAWAL (1972), p. 3.

KASSANIS, B. (1967). Plant tissue culture. In MARAMOROSCH and KOPROWSKI (1967), Vol. 1, p. 537.

YARWOOD, C. E. and FULTON, R. W. (1967). Mechanical transmission of plant viruses. In MARAMOROSCH and KOPROWSKI (1967), Vol. 1, p. 237.

Chapter 5 *The composition and structure of the particles of plant viruses*

BANCROFT, J. B. (1970). The self-assembly of spherical plant viruses. *Adv. Virus Res.* **16**, 99.

BRUENING, G. E. (1972). Virus degradation and nucleic acid isolation. In KADO and AGRAWAL (1972), p. 444.

CASPAR, D. L. D. (1964). Structure and function of regular virus particles. In CORBETT and SISLER (1964), p. 267.

CASPAR, D. L. D. and KLUG, A. (1962). Physical principles in the construction of regular viruses. *Cold Spring Harbor Symp. Quant. Biol.* **27**, 1.

DAYHOFF, M. O. (1972). *Atlas of protein sequence and structure 1972. Vol. 5.* Washington: National Biomedical Research Foundation.

EISERLING, F. A. and DICKSON, R. C. (1972). Assembly of viruses. *Ann. Rev. Biochem.* **41**, 467.

FINCH, J. T. and HOLMES, K. C. (1967). Structural studies of viruses. In MARAMOROSCH and KOPROWSKI (1967), Vol. 3, p. 351.

FRAENKEL-CONRAT, H. (1969). *The chemistry and biology of viruses.* New York: Academic Press.

GIBBS, A. J. and SKEHEL, J. J. (1973). Viral RNA. In STEWART, P. R. and LETHAM, D. S. (eds.) (1973). *The ribonucleic acids.* Berlin: Springer, p. 207.

KAPER, J. M. (1968). The small RNA viruses of plants, animals and bacteria. A. Physical properties. In FRAENKEL-CONRAT, H. (ed.) (1968). *Molecular basis of virology.* A.C.S. Monograph 164. New York: Reinhold, p. 1.

KUSHNER, D. J. (1969). Self-assembly of biological structure. *Bact. Rev.* **33**, 302.

LEBERMAN, R. (1968). The disaggregation and assembly of simple viruses. In CRAWFORD, L. V. and STOKER, M. G. P. (eds.) (1968). *The molecular biology of viruses.* London: Cambridge University Press, p. 183.

RALPH, R. K. and BERGQUIST, P. L. (1967). Separation of viruses into components. In MARAMOROSCH and KOPROWSKI (1967), Vol. 2, p. 463.

TSUGITA, A. and HIRASHIMA, A. (1972). Isolation and properties of virus proteins. In KADO and AGRAWAL (1972), p. 413.

Chapter 6 *The purification of virus particles, and some properties of purified preparations*

BRAKKE, M. K. (1960). Density gradient centrifugation and its application to plant viruses. *Adv. Virus Res.* **7**, 193.

FRANCKI, R. I. B. (1972). Purification of viruses. In KADO and AGRAWAL (1972), p. 295.

SCHUMAKER, V. and REES, A. (1972). Preparative centrifugation in virus research. In KADO and AGRAWAL (1972), p. 336.

STEERE, R. L. (1959). The purification of plant viruses. *Adv. Virus Res.* **6**, 1.

Chapter 7 *Infectivity assay*

BRAKKE, M. K. (1970). Systemic infection for the assay of plant viruses. *Ann. Rev. Phytopath.* **8**, 61.

GIBBS, A. J. and GOWER, J. C. (1960). The use of a multiple transfer method in plant virus transmission studies—some statistical points arising in the analysis of results. *Ann. Appl. Biol.* **48**, 75.

KLECZKOWSKI, A. (1968). Experimental design and statistical methods of assay. In MARAMOROSCH and KOPROWSKI (1968), Vol. 4, p. 17.

ROBERTS, D. A. (1964). Local lesion assay of plant viruses. In CORBETT and SISLER (1964), p. 194.

Chapter 8 *Serological methods*

BERCKS, R., KOENIG, R. and QUERFURTH, G. (1972). Plant virus serology. In KADO and AGRAWAL (1972), p. 466.

MATTHEWS, R. E. F. (1957). *Plant virus serology.* London: Cambridge University Press.

MATTHEWS, R. E. F. (1970). *Plant virology.* New York and London: Academic Press.

NOORDAM, D. (1973). *Identification of plant viruses. Methods and experiments.* Wageningen: Pudoc.

VAN DER VEKEN, J. A., VAN SLOGTEREN, D. H. M. and VAN DER WANT, J. P. H. (1962). Immunological methods. In LINSKENS, H. F. and TRACEY, M. V. (eds.) (1962). *Modern methods of plant analysis.* Vol. 5. Berlin: Springer, p. 422.

VAN REGENMORTEL, M. H. V. (1966). Plant virus serology. *Adv. Virus Res.* **12**, 207.

WETTER, C. (1965). Serology in virus disease diagnosis. *Ann. Rev. Phytopath.* **3**, 19.

Chapter 9 *Physical and chemical methods of assay and analysis*

COLOWICK, S. P. and KAPLAN, N. O. (eds.) (1955–). *Methods in enzymology* Vol. 1– . New York and London: Academic Press.

DAVIDSON, J. N. (1972). *The biochemistry of the nucleic acids,* 7th Edition. London: Chapman and Hall.

DAWSON, R. M. C., ELLIOTT, D. C., ELLIOTT, W. H. and JONES, K. M. (1969). *Data for biochemical research.* Oxford: Clarendon Press.

FINCH, J. T. and HOLMES, K. C. (1967). In MARAMOROSCH and KOPROWSKI (1967) Vol. 3, p. 351.

HALL, C. E. (1962). Electron microscopy: principles and application to virus research. In CORBETT and SISLER (1964), p. 253.

HOLMES, K. C. and BLOW, D. M. (1965). The use of X-ray diffraction in the study of protein and nucleic acid structure. *Meths. Biochem. Analy.* **13**, 113.

KAY, D. (ed.) (1965). *Techniques for electron microscopy.* Oxford: Blackwell Scientific Publications.

LIN, T. H. and MAES, R. F. (1967). Methods of degrading nucleic acids and separating the components. In MARAMOROSCH and KOPROWSKI (1967) Vol. 2, p. 547.

MARKHAM, R. (1962). The analytical centrifuge as a tool for the investigation of plant viruses. *Adv. Virus Res.* **9**, 241.

MARKHAM, R. (1968). The optical diffractometer. In MARAMOROSCH and KOPROWSKI (1968), Vol. 4, p. 503.

MILNE, R. G. (1972). Electron microscopy of viruses. In KADO and AGRAWAL (1972), p. 76.

NOORDAM (1973).

SCHACHMAN, H. K. (1959). *Ultracentrifugation in biochemistry.* New York: Academic Press.

SCHACHMAN, H. K. and WILLIAMS, R. C. (1959). The physical properties of infective particles. In BURNET, F. M. and STANLEY, W. M. (eds.) (1959). *The viruses.* Vol. 1. New York: Academic Press, p. 223.

TRAUTMAN, R. and HAMILTON, M. G. (1972). Analytical centrifugation. In KADO and AGRAWAL (1972), p. 491.

WANG, C. H. and WILLIS, D. L. (1965). *Radiotracer methodology in biological science.* New York: Prentice-Hall.

Chapter 10 *The effects of inactivators on virus particles*

BAWDEN, F. C. (1954). Inhibitors and plant viruses. *Adv. Virus Res.* **2**, 31.

BOS, L., HAGEDORN, D. J. and QUANTZ, L. (1960). Suggested procedures for international identification of legume viruses. *T. Pl. Ziekten* **66**, 328.

CHESSIN, M. (1972). Effect of radiation on viruses. In KADO and AGRAWAL (1972), p. 531.

GINOZA, W. (1968). Inactivation of viruses by ionizing radiation and by heat. In MARAMOROSCH and KOPROWSKI (1968), Vol. 4, p. 139.

MATTHEWS, R. E. F. (1960). Virus inactivation *in vitro* and *in vivo.* In HORSFALL, J. G. and DIMOND, A. E. (eds.) (1960). *Plant pathology, an advanced treatise.* Vol. 2. New York: Academic Press, p. 461.

NOORDAM (1973).

PRICE, W. C. (1963). Inactivation and denaturation of plant viruses. *Adv. Virus Res.* **10**, 171.

Chapter 11 *Behaviour of viruses in plants*
(N.B. see also reviews listed under Chapter 3)

GIBBS, A. J. and SKEHEL, J. J. (1973). In STEWART and LETHAM (1973), p. 207.

HAMILTON, R. I. (1974). Replication of plant viruses. *Ann. Rev. Phytopath.* **12**, 223.

JASPARS, E. M. J. (1974). Plant viruses with a multipartite genome. *Adv. Virus Res.* **19**, 37.

KASSANIS, B. (1963). Interactions of viruses in plants. *Adv. Virus Res.* **10**, 219.

LOEBENSTEIN, G. (1972). Inhibition, interference and acquired resistance during infection. In KADO and AGRAWAL (1972), p. 32.

LOEBENSTEIN, G. (1972). Localization and induced resistance in virus infected plants. *Ann. Rev. Phytopath.* **10**, 177.

MUNDRY, K. W. (1963). Plant virus-host cell relations. *Ann. Rev. Phytopath.* **1**, 173.

SIEGEL, A. and ZAITLIN, M. (1964). Infection process in plant virus diseases. *Ann. Rev. Phytopath.* **2**, 179.

SUGIYAMA, T., KORANT, B. D. and LONBERG-HOLM, K. K. (1972). RNA virus gene expression and its control. *Ann. Rev. Microbiol.* **26**, 467.

ZAITLIN, M. and BEACHY, R. N. (1974). The use of protoplasts and separated cells in plant virus research. *Adv. Virus Res.* **19**, 1.

Chapter 12 *Variation, strains and classification*
(N.B. references on plant virus classification under Chapter 2)

BERGAN, T. (1971). Survey of numerical techniques for grouping. *Bact. Rev.* **35**, 379.

BOS, L., HAGEDORN, D. J. and QUANTZ, L. (1960). *Tijdschr. Pl Ziekt.* **66**, 328.

GOWER, J. C. (1969). A survey of numerical methods useful in taxonomy. *Acarologia* **11**, 357.

HENNIG, B. and WITTMAN, H. G. (1972). Tobacco mosaic virus: mutants and strains. In KADO and AGRAWAL (1972), p. 546.

LANNI, F. (1964). Viruses and molecular taxonomy. In CORBETT and SISLER (1964), p. 386.

NOORDAM (1973).

PRICE, W. C. (1964). Strains, mutation, acquired immunity, and interference. In CORBETT and SISLER (1964), p. 93.

ROCHOW, W. F. (1972). The role of mixed infections in the transmission of plant viruses by aphids. *Ann. Rev. Phytopath.* **10**, 101.

SNEATH, P. H. A. and SOKAL, R. R. (1973). *Numerical taxonomy. The principles and practice of numerical classification.* San Francisco: Freeman.

VAN KAMMEN, A. (1972). Plant viruses with a divided genome. *Ann. Rev. Phytopath.* **10**, 125.

Chapter 13 *Transmission by vectors and in other natural ways*

There are many excellent reviews, too numerous to list, in:

CORBETT and SISLER (1964).

GIBBS (1973).

KADO and AGRAWAL (1972).

MARAMOROSCH, K. (ed.) (1969). *Viruses, vectors and vegetation*. New York: Interscience.

MARAMOROSCH and KOPROWSKI (1967–71).

also:

BENNETT, C. W. (1969). Seed transmission of plant viruses. *Adv. Virus Res.* **14**, 221.

BLACK, L. M. (1969). Insect tissue cultures as tools in plant virus research. *Ann. Rev. Phytopath.* **7**, 73.

CARTER, W. (1973). *Insects in relation to plant disease*, 2nd edition. New York: Wiley-Interscience.

GROGAN, R. G. and CAMPBELL, R. N. (1966). Fungi as vectors and hosts of viruses. *Ann. Rev. Phytopath.* **4**, 29.

KENNEDY, J. S., DAY, M. F. and EASTOP, V. F. (1962). *A conspectus of aphids as vectors of plant viruses*. London: Commonwealth Institute of Entomology.

OLDFIELD, G. N. (1970). Mite transmission of plant viruses. *Ann. Rev. Ent.* **15**, 343.

ROCHOW, W. F. (1970). Barley yellow dwarf virus: phenotypic mixing and vector specificity. *Science* **167**, 875.

WALTERS, H. J. (1969). Beetle transmission of plant viruses. *Adv. Virus Res.* **15**, 339.

WATSON, M. A. and PLUMB, R. T. (1972). Transmission of plant pathogenic viruses by aphids. *Ann. Rev. Ent.* **17**, 425.

Chapter 14 *Virus ecology*

BENNETT, C. W. (1967). Epidemiology of leafhopper-transmitted viruses. *Ann. Rev. Phytopath.* **5**, 87.

BROADBENT, L. (1964). Control of plant virus diseases. In CORBETT and SISLER (1964), p. 330.

CADMAN, C. H. (1963). Biology of soil-borne viruses. *Ann. Rev. Phytopath.* **1**, 143.

CARTER, W. (1973).

DUFFUS, J. E. (1971). Role of weeds in the incidence of virus diseases. *Ann. Rev. Phytopath.* **9**, 319.

GREGORY, P. H. (1968). Interpreting plant disease dispersal gradients. *Ann. Rev. Phytopath.* **6**, 189.

THRESH, J. M. (1974). Temporal patterns of virus spread. *Ann. Rev. Phytopath.* **12**, 111.

WALLACE, H. R. (1963). *The biology of plant parasitic nematodes*. London: Edward Arnold.

Chapter 15 *Ways of preventing crop losses*
(N.B. also see reviews under Chapter 14).

BROADBENT, L. (1964). Control of plant virus diseases. In CORBETT and SISLER (1964), p. 330.

HEATHCOTE, G. D. (1973). Control of viruses spread by invertebrates to plants. In GIBBS (1973), p. 587.

HOLLINGS, M. (1965). Disease control through virus-free stock. *Ann. Rev. Phytopath.* **3**, 367.

HOLMES, F. O. (1965). Genetics of pathogenicity in viruses and of resistance in host plants. *Adv. Virus Res.* **11**, 139.

NYLAND, G. and GOHEEN, A. C. (1969). Heat therapy of virus diseases of perennial plants. *Ann. Rev. Phytopath.* **7**, 331.

TODD, J. M. (1961). The incidence and control of aphid-borne potato virus diseases in Scotland. *Eur. Potato J.* **4**, 316.

Chapter 16 *Viruses of organisms other than higher plants. Origins of viruses*

ANDREWES, C. H. (1967). *The natural history of viruses*. New York: Norton.

ANDREWES, C. H. and PEREIRA, H. G. (1972). *Viruses of vertebrates*, 3rd Edition. London: Baillière Tindall.

BELLETT, A. J. D., FENNER, F. and GIBBS, A. J. (1973). The viruses. In GIBBS (1973), p. 41.

BRADLEY, D. (1967). Ultrastructure of bacteriophages and bacteriocins. *Bact. Rev.* **31**, 320.

BROWN, R. M. (1972). Algal viruses. *Adv. Virus Res.* **17**, 243.

DAYHOFF, M. O. (1972). *Atlas of protein sequence and structure 1972*, Vol. 5. Washington: Nat. Biomed. Res. Found.

FENNER, F., MCAUSLAN, B. R., MIMS, C. A., SAMBROOK, J. and WHITE, D. O. (1974). *The biology of animal viruses*. New York: Academic Press.

HAYES, W. (1969). *The genetics of bacteria and their viruses*, 2nd edition. Oxford and Edinburgh: Blackwell.

HINDLEY, J. (1973). Structure and strategy in phage RNA. *Prog. Biophys. mol. Biol.* **26**, 269.

HOLLINGS, M. and STONE, O. M. (1971). Viruses that infect fungi. *Ann. Rev. Phytopath.* **9**, 93.

JOKLIK, W. K. (1974). Evolution in viruses. In CARLILE, M. J. and SKEHEL, J. J. (eds.) (1974). *Evolution in the microbial world*. London: Cambridge University Press, p. 293.

MEYNELL, G. G. (1972). *Bacterial plasmids. Conjugation, colicinogeny and transmissible drug-resistance*. London: Macmillan.

Chapter 17 *Plant pathogens confused with viruses*

DAVIS, R. E. and WHITCOMB, R. F. (1971). Mycoplasmas, rickettsiae, and chlamydiae: possible relation to yellows diseases and other disorders of plants and insects. *Ann. Rev. Phytopath.* **9**, 119.

ELLIOTT, K. and BIRCH, J. (eds.) (1972). *Pathogenic mycoplasmas*. Amsterdam: Associated Scientific Publishers.

HAYFLICK, L. (ed.) (1967). Biology of the mycoplasmas. *Ann. N.Y. Acad. Sci.* **143**, 522.

HAYFLICK, L. (ed.) (1969). *The Mycoplasmatales and the L-phase of bacteria*. New York: Appleton-Century-Crofts.

MARAMOROSCH, K., GRANADOS, R. R. and HIRUMI, H. (1970). Mycoplasma diseases of plants and insects. *Adv. Virus Res.* **16**, 135.

WHITCOMB, R. F. (1973). Diversity of procaryotic plant pathogens. *Proc. N. Central Branch Entomol. Soc. America* **28**, 38.

Section 3 Some additional information for those wishing to use plant viruses for class exercises; sources and storage of material

Many interesting and instructive experiments can be done using plant viruses; for suggestions see Noordam

(1973), and the *Source book of laboratory exercises of plant pathogens* (1967; American Phytopathological Society, San Francisco: Freeman).

Plants for inoculation are best kept in a glasshouse that can be heated, or shaded and cooled, to maintain temperatures between 15 and 25°C, and that also has sufficient moist evaporative surfaces (e.g. soil, sand, peat or coke) to keep the air humid. It is important to inspect plants regularly for insects and, if necessary, to spray or fumigate them with insecticides, and it is some help to have glasshouse apertures covered with insect proof gauze. Plants should be grown in soil or potting compost that has been sterilized or fumigated.

Plant pathologists and other biologists working in research laboratories or agricultural advisory services often have inocula of viruses or know where to obtain them; in the U.S.A. inocula can be bought from the American Type Culture Collection (12301 Parklawn Drive, Rockville, Maryland 20852, U.S.A.). Alternatively, viruses may be got from naturally infected plants. In most parts of the world it is easy to obtain:

1. *Tobacco mosaic virus* from tobacco. A sample of tobacco from cigarettes or, better still, cigars from less well-known tobacco-growing regions of the world is ground in a little water and inoculated to young leaves of tobacco (*Nicotiana tabacum*), tomato (*Lycopersicon esculentum*) or petunia (*Petunia hybrida*). If the tobacco contained TMV, these plants will develop necrotic or chlorotic lesions in 3 to 6 days and, depending upon variety, may develop a systemic distorting mosaic. Tomato strains of TMV may be obtained in the same way from extracts of commercially grown tomato fruit, especially those from glasshouse crops. The ribgrass and cucumber strains of TMV may also be obtained in many parts of the world from *Plantago lanceolata* and commercially grown cucumbers (*Cucumis sativus*) respectively.

2. *Tobacco necrosis virus* and its fungus vector from the roots of plants that have been kept for some time in a heated humid glasshouse, when unsterilized soil is used. Samples of soil and roots are moistened and put in humid containers with seeds of mung bean (*Phaseolus aureus*; obtained from distributors of Chinese food) or seedlings of most legumes or solanaceous plants. The bait plants are incubated with the source for two days and mung bean roots may develop necrotic lesions. Extracts of roots are inoculated to leaves of French beans (*Phaseolus vulgaris*) which may develop the brown necrotic local lesions of tobacco necrosis virus in 2 to 4 days.

3. *White clover mosaic virus* (a potexvirus) which is common in white clover (*Trifolium repens*) plants, especially in mown lawns. Sap from white clover leaves is inoculated to the young primary leaves of French beans or peas, and if the sap contains the virus the leaves develop chlorotic lesions with necrotic flecking of the veins, and the tip leaves develop severe chlorosis, necrosis and stunting.

4. *Alfalfa mosaic virus* which is common in old crops of alfalfa (*Medicago sativa*) and sometimes in ornamental lupins (*Lupinus* spp.). When inoculated to tobacco plants, sap from infected plants produces local and systemic necrotic ringspots and chlorosis, and most strains of the virus produce brown necrotic local lesions in French bean leaves.

5. *Broad bean wilt virus*, which is common in nasturtium (*Tropaeolum majus*) in which it causes chlorotic mottling and ringspots. This virus has a wide host range, including broad bean (*Vicia faba*), and is readily transmitted by sap inoculation, or by aphids, in the non-persistent manner.

6. *Pea mosaic virus, bean yellow mosaic virus* and various other *potyviruses*. These are common in garden and field crops and cause obvious mosaic symptoms. They are readily transmitted by inoculating sap of diseased plants to healthy plants of the same species; many have limited host ranges, and some of these are seed transmitted, such as bean common mosaic virus and lettuce mosaic virus, which are found in some varieties of *Phaseolus vulgaris* and lettuce (*Lactuca sativa*) respectively. Some potyviruses have wider host ranges, and these include bean yellow mosaic virus, which is common in ornamental gladioli. All are transmitted by aphids in the non-persistent manner.

7. *Barley yellow dwarf virus*, which is common in festucoid grasses and cereals. It may be obtained by collecting aphids from grasses and confining them on young seedlings of oats (*Avena sativa*). Infected oats develop chlorotic terminal leaves and reddened older leaves. Aphids that survive on the oats may be cultured, and virus-free clones obtained from them by collecting larval aphids as they are produced, and putting them on virus-free oat seedlings.

Using the viruses described above various exercises are possible. Estimating the incidence of virus infection can illustrate the principles of sampling and pooling of samples. Simplest for this type of exercise is estimation of the incidence of white clover mosaic in lawns and pastures containing, for example, different white clover populations, or ones that have been maintained in different ways. It is also interesting to estimate the incidence of TMV in tomato fruit from different sources, or in cigars, cigarettes and snuff from various parts of the world.

The particles of TMV are easily purified by various simple methods that require, at most, a simple bench centrifuge. The particles are heat resistant (75°C for 5 min precipitates host proteins), and precipitate at their isoelectric point or in one-third saturated ammonium sulphate solution. The concentration of particles in suspensions may be roughly assayed during purification merely by looking for streaming birefringence while swirling the suspensions between two sheets of Polaroid. TMV is of great value to laboratory biologists; the composition and physical properties of the particles and their constituent macromolecules are well defined, and they are readily labelled with radioactive compounds; hence they can be used as standards in various experiments.

Alfalfa mosaic virus, broad bean wilt virus and the potyviruses are transmitted by aphids in the non-persistent manner, whereas barley yellow dwarf virus is transmitted in the persistent manner. Alfalfa mosaic and broad bean wilt viruses have much wider host ranges than the potyviruses, and give a variety of symptoms in different species.

Tobacco necrosis and alfalfa mosaic viruses give easily counted local lesions on the primary leaves of

French bean, but have particles with contrasting biophysical behaviour; they differ in stability and composition, and alfalfa mosaic but not tobacco necrosis virus has a multipartite genome.

All the viruses mentioned above, except for barley yellow dwarf virus, are readily stored, for long periods, as frozen or freeze-dried sap or leaves (Hollings and Stone, 1970; McKinney and Silber, 1968; Noordam, 1973). One of the simplest ways to freeze-dry leaves, is to put them in a sealed container with a desiccant such as silica gel or calcium chloride, and put the container in a freezer.

REFERENCES

Italic figures in parentheses after each reference indicate the chapter in which the reference appears.

AAIJ, C. and BORST, P. (1972). *Biochim. biophys. Acta* **269**, 192. (*9*)

A'BROOK, J. (1964). *Ann. appl. Biol.* **54**, 199. (*14*)

A'BROOK, J. (1968). *Ann. appl. Biol.* **61**, 289. (*14*)

A'BROOK, J. (1973a). *Ann. appl. Biol.* **74**, 279. (*14*)

A'BROOK, J. (1973b). *Ann. appl. Biol.* **74**, 263. (*14*)

ABU SALIH, H. S., MURANT, A. F. and DAFT, M. J. (1968a). *J. gen. Virol.* **3**, 299. (*8*)

ABU SALIH, H. S., MURANT, A. F. and DAFT, M. J. (1968b). *J. gen. Virol.* **2**, 155. (*8*)

ACKERMANN, H. W. and EISENSTARK, A. (1974). *Intervirology* **3**, 201. (*16*)

ADANSON, M. (1763). *Familles des plantes* Vol. 1, Paris: Vincent. (*12*)

AHMED, M. E., BLACK, L. M., PERKINS, E. G., WALKER, B. L. and KUMMEROW, F. A. (1964). *Biochem. biophys. Res. Commun.* **17**, 103. (*5, 9*)

AINSWORTH, G. C. (1935). *Ann. appl. Biol.* **22**, 55 (*12*)

AINSWORTH, G. C. (1937). *Ann. appl. Biol.* **24**, 545. (*12*)

ALBOUY, J., KUSIAK, C., LAPIERRE, H., LAQUERRIERE, F. and MAURY, Y. (1970). *Ann. Phytopathol.* **2**, 607. (*11*)

ALDRICH, D. T. A., GIBBS, A. J. and TAYLOR, L. R. (1965). *Pl. Path.* **14**, 11. (*15*)

ALLEN, T. C. (1972). *Virology* **47**, 491. (*17*)

AMMAN, J., DELIUS, H. and HOFSCHNEIDER, P. H. (1964). *J. mol. Biol.* **10**, 55. (*11*)

AMOS, P., HATTON, R. G., KNIGHT, R. C. and MASSEE, A. M. (1927). *Rep. E. Malling Res. Sta. Suppl.* **11**, 126. (*13*)

ANDERER, F. A. (1959). *Z. Naturforsch.* **14b**, 642. (*5*)

ANDERER, F. A. and SCHLUMBERGER, H. D. (1966). *Biochim. biophys. Acta* **115**, 222. (*8*)

ANDERSON, N. G. (1968). *Anal Biochem.* **23**, 72. (*9*)

ANDREWES, C. H. and PEREIRA, H. G. (1972). *Viruses of vertebrates*, 3rd edition. London: Ballière Tindall. (*16*)

ANDREWES, C. H. and SNEATH, P. H. A. (1958). *Nature, Lond.* **182**, 12. (*12*)

ANNAND, P. C., CHAMBERLAIN, J. C., HENDERSON, C. F. and WATERS, H. A. (1932). *U.S.D.A. Circular* **244**, 1. (*14*)

AOKI, S. and TAKEBE, I. (1969). *Virology* **39**, 439. (*4, 11*)

ASSINK, A. M., SWAANS, H. and VAN KAMMEN, A. (1973). *Virology* **53**, 384. (*11*)

ASTELL, C., SILVERSTEIN, S. C., LEVIN, D. H. and ACS, G. (1972). *Virology* **48**, 648 (*11*)

ATABEKOV, J. (1971). *Acta Phytopathol. Acad. Sci. Hung.* **6**, 57. (*11*)

ATABEKOV, J. G., SCHASKOLSKAYA, N. D., ATABEKOVA, T. I. and SACHAROVSKAYA, G. A. (1970). *Virology* **41**, 397. (*11*)

AYALA, A. and ALLEN, M. W. (1968). *J. Agr. Univ. Puerto Rico* **52**, 101. (*13*)

BABOS, P. (1971). *Virology* **43**, 597. (*11*)

BABOS, P. and KASSANIS, B. (1963). *Virology* **20**, 490. (*10*)

BADAMI, R. S. (1959). *Ann. appl. Biol.* **47**, 78. (*11*)

BAKER, P. E. (1960). *Ann. appl. Biol.* **48**, 2. (*14*)

BALD, J. G. (1966). In BEEMSTER and DIJKSTRA (1966), p. 66. (*3, 11*)

BALD, J. G. and SAMUEL, G. (1931). *Aust. Council Sci. Ind. Res. Bull.* No. 54. (*13*)

BALD, J. G. and TINSLEY, T. W. (1967). *Virology* **32**, 328. (*12*)

BALL, E. M. and BRAKKE, M. K. (1968). *Virology* **36**, 152. (*8*)

BALL, E. M. and BRAKKE, M. K. (1969). *Virology* **39**, 746. (*8*)

BALL, L. A. (1972). *J. theoret. Biol.* **36**, 313. (*16*)

BALL, L. A., MINSON, A. C. and SHIH, D. S. (1973). *Nature New Biology* **246**, 206. (*9*)

BANCROFT, J. B. (1970). *Adv. Virus Res.* **16**, 99. (*5*)

BANCROFT, J. B. (1972). *J. gen. Virol.* **14**, 223. (*12*)

BANCROFT, J. B. and FLACK, I. H. (1972). *J. gen. Virol.* **15**, 247 (*6*)

BANCROFT, J. B. and LANE, L. C. (1973). *J. gen. Virol.* **19**, 381. (*12*)

BANCROFT, J. B., MCLEAN, G. D., REES, M. W. and SHORT, M. N. (1971). *Virology* **45**, 707. (*11*)

BANG, F. B. (1971). *Infect. Immun.* **3**, 617. (*16*)

BANKS, G. T., BUCK, K. W., CHAIN, E. B., DARBYSHIRE, J. E. and HIMMELWEIT, F. (1969a). *Nature, Lond.* **222**, 89. (*16*)

BANKS, G. T., BUCK, K. W., CHAIN, E. B., DARBYSHIRE, J. E. and HIMMELWEIT, F. (1969b). *Nature, Lond.* **223**, 155. (*16*)

BANKS, G. T., BUCK, K. W., CHAIN, E. B., HIMMELWEIT, F., MARKS, J. E., TYLER, J. M., HOLLINGS, M., LAST, F. T. and STONE, O. M. (1968). *Nature, Lond.* **218**, 542. (*16*)

BANTTARI, E. E. and ZEYEN, R. J. (1969). *Phytopathology* **59**, 183. (*6*)

BARNETT, O. W. and MURANT, A. F. (1970). *Ann. appl. Biol.* **65**, 435. (*6*)

BARTELS, R. (1963–4). *Phytopath. Z.* **49**, 257. (*11*)

BASSI, M. and FAVALI, M. A. (1972). *J. gen. Virol.* **16**, 153. (*11*)

BAWDEN, F. C. (1935). *Br. J. exp. Pathol.* **16**, 435. (*10*)

BAWDEN, F. C. (1941). *Plant viruses and virus diseases*, 1st edition. Waltham, Mass.: Chronica Botanica. (*12*)

BAWDEN, F. C. (1950). *Plant viruses and virus diseases*, 3rd edition. Waltham, Mass.: Chronica Botanica. (*1, 15*)

BAWDEN, F. C. (1954). *Adv. Virus Res.* **2**, 31. (*10*)

BAWDEN, F. C. and FREEMAN, G. G. (1952). *J. gen. Microbiol.* **7**, 154. (*11*)

BAWDEN, F. C. and HARRISON, B. D. (1955). *J. gen. Microbiol.* **13**, 494. (*11*)

BAWDEN, F. C. and KASSANIS, B. (1945). *Ann. appl. Biol.* **32**, 52. (*11*)

BAWDEN, F. C. and KASSANIS, B. (1950). *Ann. appl. Biol.* **37**, 215. (*3, 11*)

BAWDEN, F. C. and KASSANIS, B. (1951). *Ann. appl. Biol.* **38**, 402. (*11*)

BAWDEN, F. C. and KASSANIS, B. (1954). *J. gen. Microbiol.* **10**, 160. (*11*)

BAWDEN, F. C. and KASSANIS, B. (1965). *Rep. Rothamsted Exp. Sta. 1964*, p. 282. (*15*)

BAWDEN, F. C. and KLECZKOWSKI, A. (1955). *J. gen. Microbiol.* **13**, 370. (*10, 11*)

BAWDEN, F. C. and KLECZKOWSKI, A. (1959). *Nature, Lond.* **183**, 503. (*10*)

BAWDEN, F. C. and PIRIE, N. W. (1937). *Proc. Roy. Soc.* B **123**, 274. (*1, 6, 9*)

BAWDEN, F. C. and PIRIE, N. W. (1938). *Br. J. exp. Pathol.* **19**, 251. (*1, 6*)

BAWDEN, F. C. and PIRIE, N. W. (1940). *Biochem. J.* **34**, 1278. (*5*)

BAWDEN, F. C. and PIRIE, N. W. (1950). *J. gen. Microbiol.* **4**, 482. (*10*)

BAWDEN, F. C. and PIRIE, N. W. (1953). In FILDES and VAN HEYNINGEN (1953), p. 21. (*16*)

BAWDEN, F. C. and PIRIE, N. W. (1959). *J. gen. Microbiol.* **21**, 438. (*6*)

BAWDEN, F. C., PIRIE, N. W., BERNAL, J. D. and FANKUCHEN, I. (1936). *Nature, Lond.* **138**, 1051. (*5*)

BAWDEN, F. C. and ROBERTS, F. M. (1948). *Ann. appl. Biol.* **35**, 418. (*4*)

BAWDEN, F. C. and SINHA, R. C. (1961). *Virology* **14**, 198. (*11*)

BEACHY, R. N. and MURAKISHI, H. H. (1973). *Virology* **55**, 320. (*11*)

BEALE, G. H., JURAND, A. and PREER, J. R. (1969). *J. Cell Sci.* **5**, 65. (*16*)

BEDBROOK, J. R. and MATTHEWS, R. E. F. (1973). *Virology* **53**, 84. (*3*)

BEEMSTER, A. B. R. and DIJKSTRA, J. (eds.) (1966). *Viruses of Plants.* Amsterdam: North-Holland.

BEER, S. V. and KOSUGE, T. (1968). *Phytopathology* **58**, 1042. (*9*)

BEIJERINCK, M. W. (1898). *Verhand. Kon Akad. Weten. Amsterdam* **6**, 3. (*Phytopathological Classic 7*, 33). (*1, 6*)

BELLETT, A. J. D. (1967a). *J. gen. Virol.* **1**, 583. (*12*)

BELLETT, A. J. D. (1967b). *J. Virol.* **1**, 245. (*12*)

BELLETT, A. J. D., FENNER, F. and GIBBS, A. J. (1973). In GIBBS (1973a), p. 41. (*16*)

BELLMAN, S. H., BENDER, R., GORDON, R. and ROWE, J. E. (1971). *J. theoret. Biol.* **32**, 205. (*9*)

BENDA, G. T. A. (1956). *Virology* **2**, 820. (*11*)

BENDA, G. T. A. and NAYLOR, A. W. (1958). *Am. J. Bot.* **45**, 33. (*11*)

BENDER, S. (1952). *J. chem. Educ.* **20**, 15. (*9*)

BENNETT, C. W. (1934). *J. agr. Res.* **48**, 665. (*11*)

BENNETT, C. W. (1940). *Phytopathology* **30**, 2. (*4*)

BENNETT, C. W. (1944). *Phytopathology* **34**, 77. (*4*)

BENNETT, C. W. (1955). *Phytopathology* **45**, 531. (*11*)

BENNETT, C. W. (1956). *Ann. Rev. Pl. Physiol.* **7**, 143. (*11*)

BENNETT, C. W. and COSTA, A. S. (1949). *J. agr. Res.* **78**, 207. (*4*)

BENNETT, C. W. and WALLACE, H. E. (1938). *J. agr. Res.* **56**, 31. (*13*)

BERCKS, R. (1967). *Phytopath. Z.* **58**, 1. (*8*)

BERCKS, R. and BRANDES, J. (1961). *Phytopath. Z.* **42**, 45. (*6*)

BERCKS, R. and QUERFURTH, G. (1971a). *J. gen. Virol.* **12**, 25. (*8*)

BERCKS, R. and QUERFURTH, G. (1971b). *Phytopath. Z.* **72**, 354. (*8, 12*)

BERNAL, J. D. and FANKUCHEN, I. (1937). *Nature, Lond.* **139**, 923. (*1*)

BERNAL, J. D. and FANKUCHEN, I. (1941). *J. gen. Physiol.* **25**, 147. (*5*)

BEST, R. J. (1939). *Aust. J. exp. Biol. Med. Sci.* **17**, 1. (*10*)

BEST, R. J. (1966). *Enzymologia* **31**, 333. (*10*)

BETTO, E., BASSI, M., FAVALI, M. A. and CONTI, G. G. (1972). *Phytopath. Z.* **75**, 193. (*11*)

BEVAN, E. A., HERRING, A. J. and MITCHELL, D. J. (1973). *Nature, Lond.* **245**, 81. (*16*)

BEWLEY, W. F. (1923). *Diseases of glasshouse plants.* London: Benn. (*12*)

BISHOP, D. H. L. and ROY, P. (1972). *J. Virol.* **10**, 234. (*11*)

BLACK, D. R. and KNIGHT, C. A. (1970). *J. Virol.* **6**, 194. (*5, 9*)

BLACK, L. M. (1950). *Nature, Lond.* **166**, 852. (*17*)

BLACK, L. M. (1953). *Phytopathology* **43**, 466. (*12*)

BLACK, L. M. (1969). *Ann. Rev. Phytopathol.* **7**, 73. (*4, 11*)

BLACK, L. M. and BRAKKE, M. K. (1952). *Phytopathology* **42**, 269. (*1, 13*)

BLACK, L. M., SMITH, K. M., HILLS, G. J. and MARKHAM, R. (1965). *Virology* **27**, 446. (*6*)

BLACK, L. M., WOLCYRZ, S. and WHITCOMB, R. F. (1958). *7th Int. Cong. Microbiol. Abstracts*, No. 14a. (*13*)

BLATTNÝ, C. and KRÁLÍK, O. (1968). *Ceska Mykol.* **22**, 161. (*16*)

BOARDMAN, N. K., FRANCKI, R. I. B. and WILDMAN, S. G. (1965). *Biochemistry* **4**, 872. (*6*)

BOARDMAN, N. K., FRANCKI, R. I. B. and WILDMAN, S. G. (1966). *J. mol. Biol.* **17**, 470. (*6*)

BOEDTKER, H. (1968). *J. mol. Biol.* **35**, 61 (*9*)

BOEDTKER, H. (1971). *Biochim. biophys. Acta* **240**, 448. (*9*)

BOL, J. F. and VAN VLOTEN-DOTING, L. (1973). *Virology* **51**, 102. (*7, 11*)

BOL, J. F., VAN VLOTEN-DOTING, L. and JASPARS, E. M. J. (1971). *Virology* **46**, 73. (*6*)

BONAMI, J. R. and VAGO, C. (1971). *Experientia* **27**, 1363. (*16*)

BORDER, D. J. (1972). *Nature, New Biology* **236**, 87. (*16*)

BORDET, J. and GENGOU, O. (1901). *Ann. Inst. Pasteur* **15**, 289. (*8*)

BOS, L., HAGEDORN, D. J. and QUANTZ, L. (1960). *Tijdschr. Pl. Ziekt.* **66**, 328. (*10, 12*)

BOVÉ, C., MOCQUOT, B. and BOVÉ, J. M. (1972). *Symp. Biol. Hung.* **13**, 43. (*11*)

BOVÉ, J. M. (1972). *Proc. 2nd. Int. Cong. Virology, Budapest* (ed. by J. L. Melnick). Basel: S. Karger, p. 223. (*11*)

BOZARTH, R. F., WOOD, H. A. and MANDLEBROT, A. (1971). *Virology* **45**, 516. (*16*)

BRADLEY, D. E. (1966). *J. gen. Microbiol.* **44**, 383. (*9*)

BRADLEY, D. E. (1967). *Bact. Rev.* **31**, 230. (*16*)

BRADLEY, R. H. E. (1963). *Can. J. Microbiol.* **9**, 369 (*15*)

BRADLEY, R. H. E. (1964). In CORBETT and SISLER (1964), p. 148. (*13*)

BRADLEY, R. H. E. and COUSIN, M. T. (1969). *Virology* **39**, 338. (*13*)

BRADLEY, R. H. E. and GANONG, R. Y. (1955). *Can. J. Microbiol.* **1**, 775. (*13*)

BRADLEY, R. H. E., MOORE, C. A. and POND, D. D. (1966). *Nature, Lond.* **209**, 1370. (*15*)

BRAKKE, M. K. (1955). *Arch. Biochem. Biophys.* **55**, 175. (*6*)

BRAKKE, M. K. (1958). *Virology* **6**, 96. (*9*)

BRAKKE, M. K. (1959). *Virology* **9**, 506. (*6*)

BRAKKE, M. K. (1960). *Adv. Virus Res.* **7**, 193. (*6*)

BRAKKE, M. K. (1964). *Arch. Biochem. Biophys.* **107**, 388. (*6*)

BRAKKE, M. K. (1970). *Ann. Rev. Phytopath.* **8**, 61. (*7*)

BRAKKE, M. K. and VAN PELT, N. (1970a). *Anal. Biochem.* **38**, 56. (*6, 9*)

BRAKKE, M. K. and VAN PELT, N. (1970b). *Virology* **42**, 699. (*9*)

BRAKKE, M. K., VATTER, A. E. and BLACK, L. M. (1954). *Brookhaven Symp. Biol.* **6**, 137. (*13*)

BRAMHALL, S., NOACK, N., WU, M. and LOEWENBERG, J. R. (1969). *Anal. Biochem.* **31**, 146. (*9*)

BRANDES, J. (1964). *Mitt. Biol. Bundesanstalt Land-Forstwirtsch. Berlin-Dahlem* **110**, 1. (*6*)

BRANDES, J. and CHESSIN, M. (1965).*Virology* **25**, 673. (*6*)

BRANDES, J. and WETTER, C. (1959). *Virology* **8**, 99. (*1, 2, 9, 12*)

BRANTS, D. H. (1964). *Virology* **23**, 588. (*11*)

BRANTS, D. H. (1971). *Neth. J. Pl. Path.* **77**, 175. (*13, 16*)

BRENNER, S. and HORNE, R. W. (1959). *Biochim. biophys. Acta* **34**, 103. (*9*)

BROADBENT, L. (1957). *Investigation of virus diseases of brassica crops.* London: Cambridge University Press. (*14*)

BROADBENT, L. (1962). *Ann. appl. Biol.* **50**, 461. (*12*)

BROADBENT, L. (1965). *Ann. appl. Biol.* **56**, 177. (*13, 14*)

BROADBENT, L., BURT, P. E. and HEATHCOTE, G. D. (1958). *Proc. Third Conf. Potato Virus Diseases, Lisse-Wageningen, 1957*, p. 91. (*15*)

BROADBENT, L. and FLETCHER, J. T. (1966). *Ann. appl. Biol.* **57**, 113. (*15*)

BROWN, R. M. (1972). *Adv. Virus Res.* **17**, 243. (*16*)

BUCK, K. W. and KEMPSON-JONES, G. F. (1970). *Nature, Lond.* **225**, 945. (*16*)

BUCK, K. W. and KEMPSON-JONES, G. F. (1973). *J. gen. Virol.* **18**, 223. (*16*)

BURNET, F. M. and STANLEY, W. M. (eds.) (1959). *The viruses*, Vols. 1–3. New York and London: Academic Press.

BURTON, K. (1956). *Biochem. J.* **62**, 315. (*9*)

BUSSEY, H. (1972). *Nature, New Biology* **235**, 73. (*16*)

BUSSEY, H. and SHERMAN, D. (1973). *Biochim. biophys. Acta* **298**, 868. (*16*)

BUTLER, P. J. G. and KLUG, A. (1971). *Nature, New Biology* **229**, 47. (*5*)

C.M.I./A.A.B. *Descriptions of Plant Viruses*, 1970– . Kew: Commonwealth Mycological Institute. (*2*)

CADMAN, C. H. (1959). *J. gen. Microbiol.* **20**, 113. (*10*)

CADMAN, C. H. (1962). *Nature, Lond.* **193**, 49. (*11, 12*)

CADMAN, C. H. (1965). *Pl. Dis. Reptr.* **49**, 230. (*13*)

CADMAN, C. H., DIAS, H. F. and HARRISON, B. D. (1960). *Nature, Lond.* **187**, 577. (*4*)

CADMAN, C. H. and HARRISON, B. D. (1959). *Ann. appl. Biol.* **47**, 542. (*11, 12*)

CAMERINI-OTERO, R. D. and FRANKLIN, R. M. (1972). *Virology* **49**, 385. (*16*)

CAMPBELL, R. N. (1962). *Nature, Lond.* **195**, 675. (*13*)

CAMPBELL, R. N. and GROGAN, R. G. (1964). *Phytopathology* **54**, 681. (*13*)

CAPOOR, S. P. (1949). *Ann. appl. Biol.* **36**, 307. (*11*)

CAPOOR, S. P. (1950). *Curr. Sci. (India)* **19**, 22. (*12*)

CARLILE, M. J. and SKEHEL, J. J. (eds.) (1974). *Evolution in the microbial world.* London: Cambridge University Press.

CARPENTER, J. M. and KLECZKOWSKI, A. (1969). *Virology* **39**, 542. (*10*)

CARTER, W. (1927). *J. agr. Res.* **34**, 449. (*4*)

CARTER, W. (1973). *Insects in relation to plant disease*, 2nd edition. New York: Wiley-Interscience. (*14*)

CASPAR, D. L. D. (1956). *Nature, Lond.* **177**, 476. (*5*)

CASPAR, D. L. D. (1964). In CORBETT and SISLER (1964) p. 267. (*5*)

CASPAR, D. L. D. and KLUG, A. (1962). *Cold Spring Harbor Symp. Quant. Biol.* **27**, 1. (*5*)

CATHERALL, P. L. (1966). *Ann. appl. Biol.* **57**, 155. (*15*)

CATHERALL, P. L. (1970). *Pl. Path.* **19**, 101. (*13*)

CATHERALL, P. L. and GRIFFITHS, E. (1966). *Ann. appl. Biol.* **57**, 149. (*15*)

CECH, M., KRÁLÍK, O. and BLATTNÝ, C. (1961). *Phytopathology* **51**, 183. (*16*)

CERIOTTI, G. (1955). *J. Biol. Chem.* **214**, 59. (*9*)

CHALCROFT, J. and MATTHEWS, R. E. F. (1967). *Virology* **33**, 167. (*3*)

CHARLES, H. P. and KNIGHT, B. C. J. G. (eds.) (1970). *Organization and control in prokaryotic and eukaryotic cells.* London: Cambridge University Press.

CHEN, T. A. and GRANADOS, R. R. (1970). *Science* **167**, 1633. (*17*)

CHEO, P. C. (1955). *Phytopathology* **45**, 17. (*13*)

CHEO, P. C. (1972). *Pl. Dis. Reptr.* **56**, 1010. (*16*)

CHESSIN, M. (1958). *Proc. 3rd Conf. Potato Virus Diseases, Lisse-Wageningen, 1957*, p. 80. (*3*)

CHIU, R. J. and BLACK, L. M. (1967). *Nature, Lond.* **215**, 1076. (*4*)

CHIU, R. J. and BLACK, L. M. (1969). *Virology* **37**, 667. (*7, 8, 13*)

CHIU, R. J., LIU, H. Y., MACLEOD, R. and BLACK, L. M. (1970). *Virology* **40**, 387. (*4, 8*)

CHIU, R. J., REDDY, D. V. R. and BLACK, L. M. (1966). *Virology* **30**, 562. (*13*)

CHRISTIE, R. G. (1967). *Virology* **31**, 268. (*3*)

CLARK, M. F. (1968). *J. gen. Virol.* **3**, 427. (*6*)

CLARK, M. F. and LISTER, R. M. (1971). *Virology* **43**, 338. (*6*)

CLARKE, J. T. (1964). *Ann. N.Y. Acad. Sci.* **121**, 428. (*9*)

CLARK-WALKER, G. D. and PRIMROSE, S. B. (1971). *J. gen. Virol.* **11**, 139. (*16*)

CLOSE, R. C. (1964). *Ann. appl. Biol.* **53**, 151. (*11*)

COCKBAIN, A. J. (1971). *Proc. 6th Br. Insectic. Fungic. Conf.*, 302. (*13*)

COCKBAIN, A. J. and COSTA, C. L. (1973). *Ann. appl. Biol.* **73**, 167. (*14*)

COCKBAIN, A. J., GIBBS, A. J. and HEATHCOTE, G. D. (1963). *Ann. appl. Biol.* **52**, 133. (*14*)

COCKERHAM, G. (1943). *Ann. appl. Biol.* **30**, 338. (*15*)

COCKERHAM, G. (1955). *Proc. 2nd Conf. Potato Virus Diseases, Lisse-Wageningen, 1954*, p. 89. (*15*)

COCKERHAM, G. (1970). *Heredity* **25**, 309. (*15*)

COCKING, E. C. and POJNAR, E. (1969). *J. gen. Virol.* **4**, 305. (*11*)

COHEN, S. and HARPAZ, I. (1964). *Ent. exp. appl.* **7**, 155. (*13*)

COLE, R. M., TULLY, J. G., POPKIN, T. J. and BOVÉ, J. M. (1973). *J. Bact.* **115**, 367. (*17*)

COLOWICK, S. P. and KAPLAN, N. O. (eds.) (1955–   ). *Methods in enzymology*, Vol. 1– . New York and London: Academic Press.

COMMERFORD, S. (1971). *Biochemistry* **10**, 1993. (*9*)

COMMONER, B. and MERCER, F. L. (1952). *Arch. Biochem. Biophys.* **35**, 278. (*11*)

COOPER, J. I. and HARRISON, B. D. (1973a). *Rep. Scott. Hort. Res. Inst. 1972*, 65. (*15*)

COOPER, J. I. and HARRISON, B. D. (1973b). *Ann. appl. Biol.* **73**, 53. (*14, 15*)

COOPER, J. I. and MAYO, M. A. (1972). *J. gen. Virol.* **16**, 285. (*6, 9*)

COOPER, J. I. and THOMAS, P. R. (1971). *Ann. appl. Biol.* **69**, 23. (*15*)

COOPER, P. D. (1969). In LEVY (1969), p. 177. (*12*)

CORBETT, M. K. and SISLER, H. D. (eds.) (1964). *Plant virology*. Gainesville: University of Florida Press.

COSTA, A. S. (1969). In MARAMOROSCH (1969), p. 95. (*13*)

COSTA, A. S. and CARVALHO, A. M. B. (1960). *Bragantia* **19**, 27. (*13*)

COSTA, A. S., DE SILVA, D. M. and DUFFUS, J. E. (1958). *Virology* **5**, 145. (*13*)

COSTA, A. S., DUFFUS, J. E. and BARDIN, R. (1959). *J. Am. Soc. Sugar Beet Tech.* **10**, 371. (*13*)

COWAN, K. M. and GRAVES, J. H. (1968). *Virology* **34**, 544. (*8, 9*)

COX, R. A., KANAGALINGAM, K. and SUTHERLAND, E. S. (1970). *Biochem. J.* **120**, 549. (*16*)

COX, R. A., KANAGALINGAM, K. and SUTHERLAND, E. S. (1971). *Biochem. J.* **125**, 655. (*16*)

CRICK, F. H. C. and WATSON, J. D. (1953a). *Nature, Lond.* **171**, 737. (*5*)

CRICK, F. H. C. and WATSON, J. D. (1953b). *Nature, Lond.* **171**, 964. (*5*)

CRICK, F. H. C. and WATSON, J. D. (1956). *Nature, Lond.* **177**, 473. (*5*)

CROWLEY, N. C. (1959). *Virology* **8**, 116. (*13*)

CROWLEY, N. C. and HANSON, J. B. (1960). *Virology* **12**, 603. (*11*)

CROWTHER, R. A. and KLUG, A. (1971). *J. theoret. Biol.* **32**, 199. (*9*)

CROWTHER, R. A., GEELEN, J. L. M. C. and MELLEMA, J. E. (1974). *Virology* **57**, 20. (*5*)

DADD, R. H. and KRIEGER, D. L. (1967). *J. econ. Ent.* **60**, 1512. (*4*)

DADD, R. H. and MITTLER, T. E. (1966). *Experientia* **22**, 832. (*4*)

DALE, W. T. (1953). *Imp. Coll. Trop. Agr. Rep. 1953*, p. 130. (*14*)

DALE, W. T. (1962). In WILLS (1962), p. 286. (*13*)

DAMIRDAGH, I. S. and SHEPHERD, R. J. (1970). *Phytopathology* **60**, 132. (*6*)

DANEHOLT, B., EDSTROM, J. E., EGYHAZI, E., LAMBERT, B. and RINGBORG, U. (1969). *Chromosoma (Berlin)* **28**, 379. (*9*)

DANIELS, M. J., MARKHAM, P. G., MEDDINS, B. M., PLASKITT, A. K., TOWNSEND, R. and BAR-JOSEPH, M. (1973). *Nature, Lond.* **244**, 523. (*17*)

DAS, S. and RASKI, D. J. (1968). *Nematologica* **14**, 55. (*13*)

DAVIDSON, J. N. (1972). *The biochemistry of the nucleic acids*. London: Methuen (*5*)

DAVIDSON, J. N. and WAYMOUTH, C. (1944). *Biochem. J.* **38**, 379. (*9*)

DAVIES, J. W. and KAESBERG, P. (1974). *J. gen. Virol.* **25**, 11. (*11*)

DAVIES, J. W., KAESBERG, P. and DIENER, T. O. (1974). *Virology* **61**, 281. (*12*)

DAVIS, R. E. and WHITCOMB, R. F. (1971). *Ann. Rev. Phytopathol.* **9**, 119. (*17*)

DAVIS, R. E. and WORLEY, J. F. (1973). *Phytopathology* **63**, 403. (*17*)

DAVISON, P. F. (1968). *Science* **161**, 906. (*9*)

DAWSON, R. M. C., ELLIOTT, D. C., ELLIOTT, W. H. and JONES, K. M. (1969). *Data for biochemical research*. Oxford: Clarendon Press. (*9*)

DAYHOFF, M. O. (1972). *Atlas of protein sequence and structure 1972*, Vol. 5. Washington: Nat. Biomed. Res. Found. (*12, 16*)

DE BOKX, J. A. (1967). *Eur. Potato J.* **10**, 221. (*3*)

DE BOKX, J. A. (ed.) (1972). *Viruses of potatoes and seed-production*. Wageningen: Pudoc.

DESJARDINS, P. R., LATTERELL, R. L. and MITCHELL, J. E. (1954). *Phytopathology* **44**, 86. (*13*)

DE SEQUEIRA, O. A. and LISTER, R. M. (1969). *Phytopathology* **59**, 1740. (*6*)

DE ZOETEN, G. A. and SCHLEGEL, D. E. (1967). *Virology* **32**, 416. (*11*)

D'HERELLE, F. (1917). *C.r. hebd. Séanc. Acad. Sci. Paris* **165**, 373. (*16*)

DIAS, H. F. (1970). *Virology* **40**, 828. (*13*)

DIAS, H. F. and HARRISON, B. D. (1963). *Ann. appl. Biol.* **51**, 97. (*8*)

DIELEMAN-VAN ZAAYEN, A. (1968). *Champignoncultuur* **12**, 192. (*16*)

DIELEMAN-VAN ZAAYEN, A. (1970). *M.G.A. Bull.* **244**, 158. (*16*)

DIELEMAN-VAN ZAAYEN, A. (1971). *Neth. J. Agr. Sci.* **19**, 154. (*16*)

DIELEMAN-VAN ZAAYEN, A. (1972). *Virology* **47**, 94. (*16*)

DIELEMAN-VAN ZAAYEN, A. and IGESZ, O. (1969). *Virology* **39**, 147. (*16*)

DIELEMAN-VAN ZAAYEN, A., IGESZ, O. and FINCH, J. T. (1970). *Virology* **42**, 534. (*16*)

DIELEMAN-VAN ZAAYEN, A. and TEMMINCK, J. H. M. (1968). *Neth. J. Pl. Pathol.* **74**, 48. (*16*)

DIENER, T. O. (1972). *Adv. Virus Res.* **17**, 295. (*1, 2*)

DIENER, T. O. and SCHNEIDER, I. R. (1968). *Arch. Biochem. Biophys.* **124**, 401. (*5*)

DIJKSTRA, J. (1966). In BEEMSTER and DIJKSTRA (1966) p. 19. (*11*)

DINGJAN-VERSTEEGH, A., VAN VLOTEN-DOTING, L. and JASPARS, E. M. J. (1972). *Virology* **49**, 716. (*12*)

DOBROV, E. N., KUST, S. V., TIKCHONENKO, T. I. (1972). *J. gen. Virol.* **16**, 161. (*10*)

DOI, Y., TERANAKA, M., YORA, K. and ASUYAMA, H. (1967). *Ann. Phytopathol. Soc. Japan* **33**, 259. (*1, 17*)

DONCASTER, J. P. and GREGORY, P. H. (1948). *Agr. Res. Counc. Rep.* **7**, London: HMSO. (*14*)

DOOLITTLE, S. P. (1916). *Phytopathology* **6**, 145. (*13*)

DOTY, P. (1961). *Harvey Lectures* **55**, 103. (*9*)

DOTY, P., FRESCO, J. R., HASELKORN, R. and LITT, M. (1959). *Proc. natn. Acad. Sci. U.S.A.* **45**, 482. (*5*)

DOWNER, D. N., ROGERS, H. W. and RANDALL, C. C. (1973). *Virology* **52**, 13. (*9*)

DUBOS, R. J. (1958). In POLLARD (1959), p. 291. (*1*)

DUFFUS, J. E. (1963). *Virology* **21**, 194. (*13*)

DUFFUS, J. E. and GOLD, A. H. (1969). *Virology* **37**, 150. (*8*)

DUNN, D. B. and HITCHBORN, J. H. (1965). *Virology* **25**, 171. (*6*)

DUNNING, R. A. and WINDER, G. H. (1965). *Pl. Pathol.* **14**, 30. (*15*)

DURHAM, A. C. H., FINCH, J. T. and KLUG, A. (1971). *Nature, New Biology* **229**, 37. (*5*)

DURHAM, A. C. H. and KLUG, A. (1971). *Nature, New Biology* **229**, 42. (*5*)

DWURAZNA, M. M. and WEINTRAUB, M. (1969). *Can. J. Bot.* **47**, 731. (*3*)

EDWARDSON, J. R. (1966). *Am. J. Bot.* **53**, 359. (*3*)

EDWARDSON, J. R., PURCIFULL, D. E. and CHRISTIE, R. G. (1968). *Virology* **34**, 250. (*3*)

EFRON, D. and MARCUS, A. (1973). *Virology* **53**, 343. (*11*)

ELLIOTT, A., LOWY, J. and SQUIRE, J. M. (1968). *Nature, Lond.* **219**, 1224. (*9*)

ELLIOTT, K. and BIRCH, J. (eds.) (1972). *Pathogenic mycoplasmas.* Amsterdam: Associated Scientific Publishers.

EL MANNA, M. M. and BRUENING, G. (1973). *Virology* **56**, 198. (*5*)

EL NAGAR, S. and MURANT, A. F. (1973). *Rep. Scott. Hort. Res. Inst. for 1972*, p. 66. (*13*)

ENGLANDER, S. W. and EPSTEIN, H. T. (1957). *Arch. Biochem. Biophys.* **68**, 144. (*9*)

EPSTEIN, H. T. (1953). *Nature, Lond.* **171**, 394. (*10*)

ESAU, K. (1941). *Hilgardia* **13**, 437. (*3*)

ESAU, K. and CRONSHAW, J. (1967). *J. Cell Biol.* **33**, 665. (*3*)

ESAU, K., CRONSHAW, J. and HOEFERT, L. L. (1966). *Proc. natn. Acad. Sci. U.S.A.* **55**, 486. (*3*)

EUCHIDA, T., PAPPENHEIMER, A. M. and GREANY, R. (1973). *J. biol. Chem.* **248**, 3838. (*16*)

EVANS, A. C. (1954). *Nature, Lond.* **173**, 1242. (*15*)

FAIVRE-AMIOT, A., MOREAU, J. P., COUSIN, M. T. and STARON, T. (1970). *Ann. Phytopathol.* **2**, 251. (*17*)

FAVALI, M. A., BASSI, M. and APPIANO, A. (1974). *J. gen. Virol.* **24**, 563. (*11*)

FAVALI, M. A., BASSI, M. and CONTI, G. G. (1973). *Virology* **53**, 115. (*11*)

FENNER, F. (1970). *Ann. Rev. Microbiol.* **24**, 297. (*12*)

FENNER, F., MCAUSLAN, B. R., MIMS, C. A., SAMBROOK, J. and WHITE, D. O. (1974). *The biology of animal viruses.* New York: Academic Press. (*16*)

FILDES, P. and VAN HEYNINGEN, W. E. (eds.) (1953). *The nature of virus multiplication.* London: Cambridge University Press.

FINCH, J. T. (1972). *J. mol. Biol.* **66**, 291. (*5*)

FINCH, J. T. and HOLMES, K. C. (1967). In MARAMOROSCH and KOPROWSKI (1967) Vol. 3, p. 351. (*5, 9*)

FINNEY, D. J. (1964). *Statistical method in biological assay*, 2nd edition. London: Griffin. (*7*)

FISHER, R. A. and YATES, F. (1963). *Statistical tables.* London: Oliver and Boyd. (*7*)

FORD, R. E. and ROSS, A. F. (1962). *Phytopathology* **52**, 71. (*11*)

FRAENKEL-CONRAT, H. (1956). *J. Am. Chem. Soc.* **78**, 882. (*1*)

FRAENKEL-CONRAT, H. (ed.) (1968a). *Molecular basis of virology.* A.C.S. Monograph 164. New York: Reinhold Book Corp.

FRAENKEL-CONRAT, H. (1968b). In FRAENKEL-CONRAT (1968a) p. 134. (*5*)

FRAENKEL-CONRAT, H. (1969). *The chemistry and biology of viruses.* New York and London: Academic Press. (*5*)

FRAENKEL-CONRAT, H. and WILLIAMS, R. C. (1955). *Proc. natn. Acad. Sci. U.S.A.* **41**, 690. (*5, 10*)

FRANCKI, R. I. B. (1968). *Virology* **34**, 694. (*10*)

FRANCKI, R. I. B. and MATTHEWS, R. E. F. (1962a). *Virology* **17**, 22. (*11*)

FRANCKI, R. I. B. and MATTHEWS, R. E. F. (1962b). *Virology* **17**, 367. (*11, 15*)

FRANCKI, R. I. B. and RANDLES, J. W. (1972). *Virology* **47**, 270. (*5, 9*)

FRANCKI, R. I. B. and RANDLES, J. W. (1973). *Virology* **54**, 349. (*11*)

FRANCKI, R. I. B., RANDLES, J. W., CHAMBERS, T. C. and WILSON, S. B. (1966). *Virology* **28**, 729. (*8*)

FRANCKI, R. I. B., ZAITLIN, M. and GRIVELL, C. J. (1971). *Aust. J. Biol. Sci.* **24**, 815. (*12*)

FRANKLIN, R. E. (1955). *Nature, Lond.* **175**, 379. (*5*)

FRANKLIN, R. E., KLUG, A. and HOLMES, K. C. (1956). Ciba Symposium on *The nature of viruses*, p. 39. (*5*)

FRANKLIN, R. M. (1966). *Proc. natn. Acad. Sci. U.S.A.* **55**, 1504. (*11*)

FRAZIER, N. W. and FREITAG, J. H. (1946). *Phytopathology* **36**, 634. (*17*)

FREIFELDER, D. (1970). *J. mol. Biol.* **54**, 567. (*9*)

FREITAG, J. H. (1936). *Hilgardia* **10**, 305. (*13*)

FREITAG, J. H. (1951). *Phytopathology* **41**, 920. (*17*)

FREITAG, J. H. (1956). *Phytopathology* **46**, 73. (*13*)

FREITAG, J. H. (1964). *Virology* **24**, 401. (*17*)

FREITAG, J. H. (1967). *Phytopathology* **57**, 1016. (*17*)

FREITAG, J. H. and SMITH, S. H. (1969). *Phytopathology* **59**, 1820. (*17*)

FRITSCH, C., STUSSI, C., WITZ, J. and HIRTH, L. (1973). *Virology* **56**, 33. (*5*)

FROST, R. R., HARRISON, B. D. and WOODS, R. D. (1967). *J. gen. Virol.* **1**, 57. (*12*)

FUDL-ALLAH, A. E. A. and CALAVAN, E. C. (1973). *Phytopathology* **63**, 256. (*17*)

FUDL-ALLAH, A. E. A., CALAVAN, E. C. and IGWEGBE, E. C. K. (1971). *Phytopathology* **61**, 1321. (*17*)

FUKUSHI, T. (1933). *J. Fac. Agr. Hokkaido Imp. Univ.* **37**, 41. (*13*)

FUKUSHI, T. (1940). *J. Fac. Agr. Hokkaido Imp. Univ.* **45**, 83. (*13*)

FUKUSHI, T. (1965). *Conf. on relationships between arthropods and plant-pathogenic viruses, Tokyo, 1965*, 1. (*1*)

FULTON, J. P. (1962). *Phytopathology* **52**, 375. (*13*)

FULTON, J. P. (1969). *Phytopathology* **59**, 236. (*16*)

FULTON, R. W. (1950). *Phytopathology* **40**, 219. (*12*)

FULTON, R. W. (1968). *TagBer. dt. Akad. LandwWiss. Berl.* **97**, 123. (*2*)

FURUMOTO, W. A. and MICKEY, R. (1967a). *Virology* **32**, 216. (*7*)

FURUMOTO, W. A. and MICKEY, R. (1967b). *Virology* **32**, 224. (*7*)

GALTON, F. (1878). *Nature, Lond.* **18**, 97. (*9*)

GAMEZ, R. and BLACK, L. M. (1967). *Nature, Lond.* **215**, 173. (*7*)

GAMEZ, R. and BLACK, L. M. (1968). *Virology* **34**, 444. (*7, 13*)

GANDY, D. G. (1960). *Ann. appl. Biol.* **48**, 427. (*16*)

GANDY, D. G. (1962). *Mushr. Sci.* **5**, 468. (*16*)

GANDY, D. G. and HOLLINGS, M. (1962). *Rep. Glasshouse Crops Res. Inst. 1961*, p. 103. (*16*)

GARNER, R. J. (1958). *The grafter's handbook.* London: Faber and Faber. (*4*)

GARNSEY, S. M. and JONES, J. W. (1967). *Pl. Dis. Reptr* **51**, 410. (*13*)

GARRETT, R. G. (1973). In GIBBS (1973a), p. 476. (*13*)

GEELEN, J. L. M. C., REZELMAN, G. and VAN KAMMEN, A. (1973). *Virology* **51**, 279. (*10*)

GENDRON, Y. and KASSANIS, B. (1954). *Ann. appl. Biol.* **41**, 183. (*4*)

GEORGE, J. A. and DAVIDSON, T. R. (1963). *Can. J. Plant Sci.* **43**, 276. (*13, 14*)

GHABRIAL, S. A. and LISTER, R. M. (1973). *Virology* **51**, 485. (*5*)

GIBBS, A. J. (1969). *Adv. Virus Res.* **14**, 263. (*2, 11, 12*)

GIBBS, A. J. (ed.) (1973a). *Viruses and invertebrates.* Amsterdam: North-Holland Elsevier. (*16*)

GIBBS, A. J. (1973b). In GIBBS (1973a), p. 526. (*16*)

GIBBS, A. J., DALE, M. B., KINNS, H. R. and MACKENZIE, H. G. (1971). *Syst. Zool.* **20**, 417. (*12*)

GIBBS, A. J. and GOWER, J. C. (1960). *Ann. appl. Biol.* **48**, 75. (*7*)

GIBBS, A. J. and HARRISON, B. D. (1968). *Nature, Lond.* **218**, 927. (*2, 12*)

GIBBS, A. J., HARRISON, B. D., WATSON, D. H. and WILDY, P. (1966a). *Nature, Lond.* **209**, 450. (*2*)

GIBBS, A. J., HECHT-POINAR, E., WOODS, R. D. and MCKEE, R. K. (1966b). *J. gen. Microbiol.* **44**, 177. (*8*)

GIBBS, A. J. and MACDONALD, P. W. (1974). *Intervirology* **4**, 52. (*11*)

GIBBS, A. J. and MACINTYRE, G. A. (1969). *J. gen. Virol.* **5**, 379. (*12*)

GIBBS, A. J. and MACINTYRE, G. A. (1970a). *J. gen. Virol.* **9**, 51. (*5*)

GIBBS, A. J. and MACINTYRE, G. A. (1970b). *Europ. J. Biochem.* **16**, 1. (*12*)

GIBBS, A. J., NIXON, H. L. and WOODS, R. D. (1965). *Virology* **19**, 441. (*6*)

GIBBS, A. J. and SKEHEL, J. J. (1973). In STEWART and LETHAM (1973), p. 207. (*11*)

GIBBS, A. J., SKOTNICKI, A. H., GARDINER, J. E., WALKER, E. S. and HOLLINGS, M. (1975). *Virology* **64**, 571. (*16*)

GIERER, A. and MUNDRY, K. W. (1958). *Nature, Lond.* **182**, 1457. (*1*)

GIERER, A. and SCHRAMM, G. (1956). *Nature, Lond.* **177**, 702. (*1, 5, 6*)

GILLIES, S., BULLIVANT, S. and BELLAMY, A. R. (1972). *Science* **174**, 694. (*11*)

GINOZA, W. (1968). In MARAMOROSCH and KOPROWSKI (1968), Vol. 4, p. 139 (*10*)

GINOZA, W., HOELLE, C. J., VESSEY, K. B. and CARMACK, C. (1964). *Nature, Lond.* **203**, 606. (*10*)

GOHEEN, A. C., NYLAND, G. and LOWE, S. K. (1973). *Phytopathology* **63**, 341. (*17*)

GOMATOS, P. J. and TAMM, I. (1963). *Proc. natn. Acad. Sci. U.S.A.* **50**, 878. (*5*)

GOMATOS, P. J., TAMM, I., DALES, S. and FRANKLIN, R. M. (1962). *Virology* **17**, 441. (*9*)

GOODMAN, P. J., WATSON, M. A. and HILL, A. R. C. (1965). *Ann. appl. Biol.* **56**, 65. (*3*)

GORDON, R., BENDER, R. and HERMAN, G. T. (1970). *J. theoret. Biol.* **29**, 471. (*9*)

GORDON, M. P. and STAEHELIN, M. (1959). *Biochim. biophys. Acta* **36**, 351. (*10*)

GOUDIE, R. B., HORNE, C. H. W. and WILKINSON, P. C. (1966). *Lancet* **2**, 1224. (*8*)

GOURLAY, R. N. (1971). *J. gen. Virol.* **12**, 65. (*16*)

GOURLAY, R. N. (1972). In ELLIOTT and BIRCH (1972), p. 145. (*17*)

GOURLAY, R. N., BRUCE, J. and GARWES, D. J. (1971). *Nature, New Biology* **229**, 118. (*16*)

GOVIER, D. A. (1958). *Scottish Pl. Breed. Sta. Rec.* p. 77. (*8*)

GOVIER, D. A. and KASSANIS, B. (1974). *Virology* **57**, 285. (*13*)

GOVIER, D. A. and KLECZKOWSKI, A. (1970). *Photochem. Photobiol.* **12**, 345. (*10*)

GOVIER, D. A. and WOODS, R. D. (1971). *J. gen. Virol.* **13**, 127. (*5*)

GOWER, J. C. (1966). *Biometrika* **53**, 325. (*12*)

GRACE, T. D. C. (1962). *Nature, Lond.* **195**, 788. (*4*)

GRACE, T. D. C. (1973). In GIBBS (1973a), p. 321. (*4, 13*)

GREEN, R. G. (1935). *Science* **82**, 443. (*16*)

GREGORY, P. H. (1948). *Ann. appl. Biol.* **35**, 412. (*14*)

GREGORY, P. H. (1968). *Ann. Rev. Phytopath.* **6**, 189. (*14*)

GREGORY, P. H. and READ, D. R. (1949). *Ann. appl. Biol.* **36**, 475. (*14*)

GUILLEY, M. H., STUSSI, C. and HIRTH, L. (1971). *C.r. hebd. Séanc. Acad. Sci. Paris Sér. D.* **272**, 1181. (*5*)

HADIDI, A. and FRAENKEL-CONRAT, H. (1973). *Virology* **52**, 363. (*9, 11*)

HADIDI, A., HARIHARASUBRAMANIAN, V. and FRAENKEL-CONRAT, H. (1973). *Intervirology* **1**, 201. (*11*)

HALL, C. E. (1955). *J. biophys. biochem. Cytol.* **1**, 1 (*9*)

HAMILTON, R. I. (1965). *Phytopathology* **55**, 798. (*13*)

HAMPTON, R. E. and FULTON, R. W. (1961). *Virology* **13**, 44. (*10, 11*)

HAMPTON, R. O., STEVENS, J. O. and ALLEN, T. C. (1969). *Pl. Dis. Reptr.* **53**, 499. (*17*)

HANSEN, H. P. (1966). *Kgl. Vet.-og. Land-bohøjskole Års. 1966*, p. 191. (*12*)

HARIHARASUBRAMANIAN, V., HADIDI, A., SINGER, B. and FRAENKEL-CONRAT, H. (1973). *Virology* **54**, 190. (*11*)

HARIHARASUBRAMANIAN, V. and ZAITLIN, M. (1968). *Virology* **36**, 521. (*8*)

HARPAZ, I. (1961). *F.A.O. Plant Protect. Bull.* **9**, 144. (*15*)

HARPAZ, I. and COHEN, S. (1965). *Phytopath. Z.* **54**, 240. (*13*)

HARRAP, K. A. (1973). In GIBBS (1973a) p. 271. (*16*)

HARRIS, J. I. and HINDLEY, J. (1965). *J. mol. Biol.* **13**, 894. (*5*)

HARRIS, J. I. and KNIGHT, C. A. (1955). *J. biol. Chem.* **214**, 215. (*10*)

HARRISON, B. D. (1956a). *Ann. appl. Biol.* **44**, 215. (*11*)

HARRISON, B. D. (1956b). *J. gen. Microbiol.* **15**, 210. (*11*)

HARRISON, B. D. (1958). *J. gen. Microbiol.* **18**, 450. (*11, 12*)

HARRISON, B. D. (1964a). *Virology* **22**, 544. (*13, 14*)

HARRISON, B. D. (1964b). *Virology* **24**, 228. (*16*)

HARRISON, B. D. (1966). *Ann. appl. Biol.* **57**, 121. (*15*)

HARRISON, B. D. (1971). In WESTERN (1971), p. 123. (*17*)

HARRISON, B. D. (1973). In GIBBS (1973a), p. 512. (*13, 16*)

HARRISON, B. D. and CADMAN, C. H. (1959). *Nature, Lond.* **184**, 1624. (*14*)

HARRISON, B. D. and CROCKATT, A. A. (1971). *J. gen. Virol.* **12**, 183. (*11*)

HARRISON, B. D. and CROWLEY, N. C. (1965). *Virology* **26**, 297. (*10, 11*)

HARRISON, B. D., FINCH, J. T., GIBBS, A. J., HOLLINGS, M., SHEPHERD, R. J., VALENTA, V. and WETTER, C. (1971). *Virology* **45**, 356. (*1, 2, 12*)

HARRISON, B. D. and JONES, R. A. C. (1971a). *Ann. appl. Biol.* **67**, 377. (*11*)

HARRISON, B. D. and JONES, R. A. C. (1971b). *Ann. appl. Biol.* **68**, 281. (*3, 11*)

HARRISON, B. D. and KLUG, A. (1966). *Virology* **30**, 738. (*9*)

HARRISON, B. D., MOWAT, W. P. and TAYLOR, C. E. (1961). *Virology* **14**, 480. (*13*)

HARRISON, B. D., MURANT, A. F. and MAYO, M. A. (1972). *J. gen. Virol.* **16**, 339. (*12*)

HARRISON, B. D., MURANT, A. F., MAYO, M. A. and ROBERTS, I. M. (1974). *J. gen. Virol.* **22**, 233. (*12, 13*)

HARRISON, B. D. and NIXON, H. L. (1959a). *J. gen. Microbiol.* **21**, 569. (*9*)

HARRISON, B. D. and NIXON, H. L. (1959b). *J. gen. Microbiol.* **21**, 591. (*6*)

HARRISON, B. D., PEACHEY, J. E. and WINSLOW, R. D. (1963). *Ann. appl. Biol.* **52**, 243. (*15*)

HARRISON, B. D. and ROBERTS, I. M. (1969). *Ann. appl. Biol.* **63**, 347. (*17*)

HARRISON, B. D., ROBERTSON, W. M. and TAYLOR, C. E. (1974). *J. Nematol.* **6**, 155. (*13*)

HARRISON, B. D., STEFANAC, Z. and ROBERTS, I. M. (1970). *J. gen. Virol.* **6**, 127. (*3, 9, 11*)

HARRISON, B. D. and WINSLOW, R. D. (1961). *Ann. appl. Biol.* **49**, 621. (*14*)

HARRISON, B. D. and WOODS, R. D. (1966). *Virology* **28**, 610. (*6*)

HARTMAN, F. W., HORSFALL, F. L. and KIDD, J. E. (eds.) (1954). *The dynamics of virus and rickettsial infections.* New York: McGraw-Hill.

HARTREE, E. F. (1972). *Anal. Biochem.* **48**, 422. (*9*)

HARVEY, J. D. (1973). *Virology* **56**, 365. (*9*)

HATTA, T., BULLIVANT, S. and MATTHEWS, R. E. F. (1973). *J. gen. Virol.* **20**, 37. (*11*)

HAY, J. and SUBAK-SHARPE, H. (1968). *J. gen. Virol.* **2**, 469. (*5*)

HAYASHI, T., MACHIDA, H., ABE, T. and KIHO, Y. (1969). *Jap. J. Microbiol.* **13**, 386. (*11*)

HAYES, W. (1969). *The genetics of bacteria and their viruses,* 2nd edition. Oxford and Edinburgh: Blackwell. (*12, 16*)

HEATHCOTE, G. D. (1957). *Ann. appl. Biol.* **45**, 133. (*14*)

HEATHCOTE, G. D. (1958). *Pl. Pathol.* **7**, 32. (*14*)

HEBERT, T. T. (1963). *Phytopathology* **53**, 362. (*6*)

HELMS, K. and MACINTYRE, G. A. (1967). *Virology* **32**, 489. (*4*)

HEDRICK, J. L. and SMITH, A. J. (1968). *Arch. Biochem. Biophys.* **126**, 155. (*9*)

HENNIG, B. and WITTMANN, H. G. (1972). In KADO and AGRAWAL (1972), p. 546. (*12*)

HEROLD, F., BERGOLD, G. H. and WEIBEL, J. (1960). *Virology* **12**, 335. (*5*)

HERRING, A. J. and BEVAN, E. A. (1974). *J. gen. Virol.* **22**, 387. (*16*)

HERSHEY, A. D. (ed.) (1971). *The bacteriophage lambda.* New York: Cold Spring Harbor Laboratory. (*16*)

HERSHEY, A. D. and CHASE, M. (1952). *J. gen. Physiol.* **36**, 39. (*1*)

HEWITT, W. B., RASKI, D. J. and GOHEEN, A. C. (1958). *Phytopathology* **48**, 586. (*1, 13*)

HIDAKA, Z. and TAGAWA, A. (1962). *Ann. Phytopathol. Soc. Japan* **27**, 77. (*13*)

HIEBERT, E. and MCDONALD, J. G. (1973). *Virology* **56**, 349. (*5, 11*)

HIEBERT, E., PURCIFULL, D. E., CHRISTIE, R. G. and CHRISTIE, S. R. (1971). *Virology* **43**, 638. (*3, 5*)

HILLE RIS LAMBERS, D. (1955). *Ann. appl. Biol.* **42**, 355. (*15*)

HILLS, G. J. and CAMPBELL, R. N. (1968). *J. Ultrastruct. Res.* **24**, 134. (*5*)

HINDLEY, J. (1973). *Prog. Biophys. mol. Biol.* **26**, 269. (*11, 16*)

HIRUMI, H. and MARAMOROSCH, K. (1969). *J. Virol.* **3**, 82. (*17*)

HITCHBORN, J. H. and HILLS, G. J. (1965). *Virology* **26**, 756. (*6*)

HITCHBORN, J. H. and HILLS, G. J. (1968). *Virology* **35**, 50. (*3*)

HOLDEN, M. and PIRIE, N. W. (1955). *Biochem. J.* **60**, 53. (*9*)

HOLDEN, M. and TRACEY, M. V. (1948). *Biochem. J.* **43**, 151. (*3*)

HOLLINGS, M. (1962). *Nature, Lond.* **196**, 962. (*16*)

HOLLINGS, M. (1965). *Ann. Rev. Phytopathol.* **3**, 367. (*15*)

HOLLINGS, M. (1974). *Acta Horticulturae* **36**, 23. (*12*)

HOLLINGS, M. and STONE, O. M. (1962). *Nature, Lond.* **184**, 607. (*8*)

HOLLINGS, M. and STONE, O. M. (1964). *Ann. appl. Biol.* **53**, 103. (*15*)

HOLLINGS, M. and STONE, O. M. (1969). *Sci. Prog., Oxf.* **57**, 371. (*16*)

HOLLINGS, M. and STONE, O. M. (1970). *Ann. appl. Biol.* **65**, 411. (*6, 10*)

HOLLINGS, M. and STONE, O. M. (1971). *Ann. Rev. Phytopathol.* **9**, 93. (*16*)

HOLLINGS, M. and STONE, O. M. (1973). *Ann. appl. Biol.* **74**, 333. (*10*)

HOLMES, F. O. (1928). *Bot. Gaz.* **86**, 66. (*7*)

HOLMES, F. O. (1929). *Bot. Gaz.* **87**, 39. (*1, 7*)

HOLMES, F. O. (1932). *Contribs. Boyce Thompson Inst.* **4**, 297. (*7*)

HOLMES, F. O. (1939). *Handbook of phytopathogenic viruses.* Minneapolis: Burgess. (*2, 12*)

HOLMES, F. O. (1941). *Phytopathology* **31**, 1089. (*12*)

HOLMES, F. O. (1956). *Virology* **2**, 611. (*11*)

HOLMES, K. C. and BLOW, D. M. (1965). *Meth. biochem. Analysis* **13**, 113. (*9*)

HOOKER, W. J. and FRONEK, F. R. (1960). *Proc. 4th Conf. Potato Virus Diseases, Braunschweig,* p. 76. (*14*)

HOOPER, D. J. (1973). In GIBBS (1973a), p. 212. (*16*)

HOPKINS, D. L., FRENCH, W. J. and MOLLENHAUER, H. H. (1973). *Phytopathology* **63**, 443. (*17*)

HORST, J., FRAENKEL-CONRAT, H. and MANDELES, S. (1971). *Biochemistry* **10**, 4748. (*5*)

HSU, H. T. and BLACK, L. M. (1973). *Virology* **52**, 284. (*7*)

HULL. R. (1954). *Agriculture* **61**, 205. (*14*)

HULL, R. (1965). *Ann. appl. Biol.* **56**, 345. (*15*)

HULL, R. (1968). *Virology* **35**, 333. (*16*)

HULL, R. and HEATHCOTE, G. D. (1967). *Ann. appl. Biol.* **60**, 469. (*1*)

HULL, R. and LANE, L. C. (1973). *Virology* **55**, 1. (*6, 9*)

HULL, R. and SELMAN, I. W. (1965). *Ann. appl. Biol.* **55**, 39. (*14*)

HUMMELER, K. (1971). In MARAMOROSCH and KURSTAK (1971), p. 361. (*11*)

HUNTER, D. E., DARLING, H. M. and BEALE, W. L. (1969). *Am. Potato J.* **46**, 247. (*13*)

HUNTER, J. A., CHAMBERLAIN, E. E. and ATKINSON, J. D. (1958). *N.Z. J. Agric. Res.* **1**, 80. (*13*)

HUTCHINS, L. M., COCHRAN, L. C., TURNER, W. F. and WEINBERGER, J. H. (1953). *Phytopathology* **43**, 691. (*17*)

HUTCHINS, L. M. and RUE, J. L. (1939). *Phytopathology* **29**, 12. (*17*)

HYDE, B. B., HODGE, A. J., KAHN, A. and BIRNSTIEL, M. L. (1963). *J. Ultrastruct. Res.* **9**, 248. (*6*)

IKEGAMI, M. and FRANCKI, R. I. B. (1973). *Virology* **56**, 404. (*8*)

INCARDONA, N. L. and KAESBERG, P. (1964). *Biophys. J.* **4**, 11. (*5*)

ISHIIE, T., DOI, Y., YORA, K. and ASUYAMA, H. (1967). *Ann. Phytopathol. Soc. Japan* **33**, 267. (*17*)

IVANOWSKI, D. (1892). *St Petersb. Acad. Imp. Sci. Bull.* **37**, 67. (*Phytopathological Classic* **7**, 27). (*1*)

JACKSON, A. O. and BRAKKE, M. K. (1973). *Virology* **55**, 483. (*6*)

JACKSON, A. O., MITCHELL, D. M. and SIEGEL, A. (1971). *Virology* **45**, 182. (*11*)

JACKSON, A. O., ZAITLIN, M., SIEGEL, A. and FRANCKI, R. I. B. (1972). *Virology* **48**, 655. (*4, 11*)

JASPARS, E. M. J. (1974). *Adv. Virus Res.* **19**, 37. (*11*)

JENSEN, D. D. (1959). *Virology* **8**, 164. (*17*)

JENSEN, D. D. and GOLD, H. A. (1951). *Phytopathology* **41**, 648. (*12*)

JENSEN, J. H. (1933). *Phytopathology* **23**, 964. (*12*)

JOCKUSCH, H. (1966). *Z. Vererbungsl.* **98**, 344. (*12*)

JOCKUSCH, H. (1968). *Virology* **35**, 94. (*12*)

JOHNSON, C. G. (1950). *Ann. appl. Biol.* **37**, 80. (*14*)

JOHNSON, J. (1927). *Wisconsin Agr. Exp. Sta. Res. Bull.* **76**, (*2, 12*)

JOHNSON, J. (1947). *Phytopathology* **37**, 822. (*12*)

JOHNSON, M. W., WAGNER, G. W. and BANCROFT, J. B. (1973). *J. gen. Virol.* **19**, 263. (*5*)

JOKLIK, W. K. (1974). In CARLILE and SKEHEL (1974), p. 293. (*16*)

JOKLIK, W. K., SKEHEL, J. J. and ZWEERINK, H. J. (1970). *Cold Spring Harbor Symp. Quant. Biol.* **35**, 791. (*11*)

JONES, I. M. and REICHMANN, M. E. (1973). *Virology* **52**, 49. (*11*)

JONES, R. A. C. and HARRISON, B. D. (1968). *Rep. Scott. Hort. Res. Inst., 1967*, p. 60. (*14, 15*)

JONES, R. A. C. and HARRISON, B. D. (1969). *Ann. appl. Biol.* **63**, 1. (*13, 14*)

JONES, R. A. C. and HARRISON, B. D. (1972). *Ann. appl. Biol.* **71**, 47. (*14*)

KABAT, E. A. and MAYER, M. M. (1961). *Experimental immunochemistry* 2nd edition. Illinois: Thomas. (*8*)

KADO, C. I. and AGRAWAL, H. O. (eds.) (1972). *Principles and techniques in plant virology.* New York: Van Nostrand Reinhold.

KADO, C. I. and KNIGHT, C. A. (1966). *Proc. natn. Acad. Sci. U.S.A.* **55**, 1276. (*5, 10, 12*)

KADO, C. I. and KNIGHT, C. A. (1968). *J. mol. Biol.* **36**, 15. (*5*)

KADO, C. I., VAN REGENMORTEL, M. H. V. and KNIGHT, C. A. (1968). *Virology* **34**, 17. (*12*)

KALLENBACH, N. R. (1968). *J. mol. Biol.* **37**, 445. (*9*)

KALMAKOFF, J., LEWANDOWSKI, L. J. and BLACK, D. R. (1969). *J. Virol.* **4**, 851. (*5*)

KALMUS, H. and KASSANIS, B. (1945). *Ann. appl. Biol.* **32**, 230. (*4*)

KALOOSTIAN, G. H., OLDFIELD, G. N. and JONES, L. S. (1968). *Phytopathology* **58**, 1236. (*17*)

KAMEI, T., RUBIO-HUERTOS, M. and MATSUI, C. (1969). *Virology* **37**, 507. (*11*)

KAPER, J. M. (1968). In FRAENKEL-CONRAT (1968a), p. 1. (*5, 9*)

KAPER, J. M. and HALPERIN, J. E. (1965). *Biochemistry* **4**, 2434. (*5*)

KAPER, J. M. and HOUWING, C. (1962). *Arch. Biochem. Biophys.* **97**, 449. (*5*)

KAPER, J. M. and WATERWORTH, H. E. (1973). *Virology* **51**, 183. (*9*)

KARLING, J. S. (1968). *The Plasmodiophorales*, 2nd edition. New York: Hafner Publishing Co. (*13*)

KASSANIS, B. (1952). *Ann. appl. Biol.* **39**, 358. (*4, 11*)

KASSANIS, B. (1954). *Ann. appl. Biol.* **41**, 420. (*15*)

KASSANIS, B. (1957a). *Virology* **4**, 187. (*12*)

KASSANIS, B. (1957b). *Ann. appl. Biol.* **45**, 422. (*15*)

KASSANIS, B. (1960). *Virology* **10**, 353. (*6, 11*)

KASSANIS, B. (1961). *Virology* **13**, 93. (*13*)

KASSANIS, B. (1962). *J. gen. Microbiol.* **27**, 477. (*11*)

KASSANIS, B. (1967). In MARAMOROSCH and KOPROWSKI (1967), Vol. 3, p. 537. (*4*)

KASSANIS, B. and BASTOW, C. (1971a). *J. gen. Virol.* **11**, 157. (*11*)

KASSANIS, B. and BASTOW, C. (1971b). *J. gen Virol.* **11**, 171. (*12*)

KASSANIS, B. and GOVIER, D. A. (1971a). *J. gen. Virol.* **10**, 99. (*13*)

KASSANIS, B. and GOVIER, D. A. (1971b). *J. gen. Virol.* **13**, 221. (*13*)

KASSANIS, B. and KLECZKOWSKI, A. (1948). *J. gen. Microbiol.* **2**, 143. (*10*)

KASSANIS, B. and KLECZKOWSKI, A. (1965). *Photochem. Photobiol.* **4**, 209. (*10*)

KASSANIS, B. and MACFARLANE, I. (1964). *J. gen. Microbiol.* **36**, 79. (*13*)

KASSANIS, B. and MACFARLANE, I. (1965). *Virology* **26**, 603. (*4, 13*)

KASSANIS, B. and MACFARLANE, I. (1968). *J. gen. Virol.* **3**, 227. (*13*)

KASSANIS, B. and NIXON, H. L. (1961). *J. gen. Microbiol.* **25**, 459. (*11*)

KASSANIS, B. and PHILLIPS, M. P. (1970). *J. gen. Virol.* **9**, 119. (*8, 11*)

KASSANIS, B. and VARMA, A. (1967). *Ann. appl. Biol.* **59**, 447. (*15*)

KASSANIS, B. and WELKIE, G. W. (1963). *Virology* **21**, 540. (*11, 12*)

KASSANIS, B. and WHITE, R. F. (1972). *J. gen. Virol.* **16**, 177. (*11*)

KASSANIS, B., WOODS, R. D. and WHITE, R. F. (1972). *J. gen. Virol.* **14**, 123. (*12*)

KAUSCHE, G. A., PFANKUCH, E. and RUSKA, H. (1939). *Naturwissenschaften* **27**, 292. (*1*)

KAZAMA, F. Y. and SCHORNSTEIN, K. L. (1973). *Virology* **52**, 478. (*16*)

KENDREW, J. C. (1962). *Brookhaven Symp. Biol.* **15**, 216. (*5*)

KENNEDY, J. S., DAY, M. F. and EASTOP, V. F. (1962). *A conspectus of aphids as vectors of plant viruses.* London: Commonwealth Institute of Entomology. (*14*)

KIELLAND-BRANDT, M. C. and NILSSON-TILGREN, T. (1973). *Mol. gen. Genet.* **121**, 229. (*11*)

KIHO, Y. (1970). *Jap. J. Microbiol.* **14**, 291. (*11*)

KIHO, Y. (1972a). *Jap. J. Microbiol.* **16**, 152. (*11*)

KIHO, Y. (1972b). *Jap. J. Microbiol.* **16**, 259. (*11*)

KIHO, Y., MACHIDA, H. and OSHIMA, N. (1972). *Jap. J. Microbiol.* **16**, 451. (*11*)

KING, J. and LAEMMLI, U. K. (1973). *J. mol. Biol.* **75**, 315. (*16*)

KING, J. and MYKOLAJEWYCZ, N. (1973). *J. mol. Biol.* **75**, 339. (*16*)

KISIMOTO, R. (1973). In GIBBS (1973a), p. 137. (*13*)

KITAJIMA, E. W., MULLER, G. W., COSTA, A. S. and YUKI, W. (1972). *Virology* **50**, 254. (*13*)

KITANO, T., HARUNA, I. and WATANABE, I. (1961). *Virology* **15**, 503. (*16*)

KLECZKOWSKI, A. (1949). *Ann. appl. Biol.* **36**, 139. (*7*)

KLECZKOWSKI, A. (1950). *J. gen. Microbiol.* **4**, 53. (*7*)

KLECZKOWSKI, A. (1955). *J. gen. Microbiol.* **13**, 91. (*7*)

KLECZKOWSKI, A. (1962). *Photochem. Photobiol.* **1**, 291. (*10*)

KLECZKOWSKI, A. (1966). In BEEMSTER and DIJKSTRA (1966), p. 196. (*8*)

KLECZKOWSKI, A. (1968). In MARAMOROSCH and KOPROWSKI (1968) Vol. 4, p. 615. (*7*)

KLEIN, W. H. and CLARK, J. M. (1973). *Biochemistry* **12**, 1528. (*11*)

KLEIN, W. H., NOLAN, C., LAZAR, J. M. and CLARK, J. M. (1972). *Biochemistry* **11**, 2009. (*9, 11, 12*)

KLEINSCHMIDT, A. K., LANG, D., JACHERTS, D. and ZAHN, R. K. (1962). *Biochim. biophys. Acta* **61**, 857. (*6*)

KLEINSCHMIDT, A. K. and ZAHN, R. K. (1959). *Z. Naturforsch.* **14b**, 770. (*9*)

KLEINSCHMIDT, W. J., CLINE, J. C. and MURPHY, E. B. (1964). *Proc. natn. Acad. Sci. U.S.A.* **52**, 741. (*16*)

KLEINSCHMIDT, W. J. and ELLIS, L. F. (1968). *Ciba Found. Symp.* No. 39. (*16*)

KLEINSCHMIDT, W. J., ELLIS, L. F., VAN FRANK, R. M. and MURPHY, E. B. (1968). *Nature, Lond.* **220**, 167. (*16*)

KLUG, A. and BERGER, J. F. (1964). *J. mol. Biol.* **10**, 565. (*9*)

KLUG, A. and DE ROSIER, D. J. (1966). *Nature, Lond.* **212**, 29. (*9*)

KLUG, A., FINCH, J. T. and FRANKLIN, R. E. (1957). *Biochim. biophys. Acta* **25**, 242. (*5*)

KLUG, A., LONGLEY, W. and LEBERMAN, R. (1966). *J. mol. Biol.* **15**, 315. (*6*)

KNIGHT, C. A. (1947). *J. biol. Chem.* **171**, 297. (*12*)

KNUDSON, D. L. (1973). *J. gen. Virol.* **20**, Suppl. 105. (*11*)

KNUDSON, D. L. and MACLEOD, R. (1972). *Virology* **47**, 285. (*5*)

KNUHTSEN, H., HIEBERT, E. and PURCIFULL, D. E. (1974). *Virology* **61**, 200. (*11*)

KODAMA, T. and BANCROFT, J. B. (1964). *Virology* **22**, 23. (*11*)

KOENIG, R. (1972). *Virology* **50**, 263. (*5, 9*)

KOENIG, R., STEGEMANN, H., FRANCKSEN, H. and PAUL, H. L. (1970). *Biochim. biophys. Acta* **207**, 184. (*5*)

KOLTIN, Y., BERICK, R., STANBERG, J. and BEN-SHAUL, Y. (1973). *Nature, New Biology* **241**, 108. (*16*)

KOZLJAROVA, O. and BOJNANSKY, V. (1969). *Proc. 6th Conf. Czechoslovak Plant Virologists, Olomouc, 1967*, p. 216. (*17*)

KUHN, C. W. (1965). *Virology* **25**, 9. (*11*)

KUMMERT, J. and SEMAL, J. (1970). *J. gen. Virol.* **16**, 11. (*11*)

KUNKEL, L. O. (1931). *Contrib. Boyce Thompson Inst.* **3**, 85. (*17*)

KUNKEL, L. O. (1936). *Phytopathology* **26**, 809. (*15*)

KUNKEL, L. O. (1937). *Am. J. Bot.* **24**, 316. (*17*)

KUNKEL, L. O. (1940). In MOULTON (1940), p. 22. (*12*)

KUNKEL, L. O. (1941). *Am. J. Bot.* **28**, 761. (*17*)

KUNKEL, L. O. (1954). In HARTMAN, HORSFALL and KIDD (1964), p. 150. (*17*)

KUNKEL, L. O. (1955). *Adv. Virus Res.* **3**, 251. (*17*)

KUNKEL, L. O. (1957). *Science* **126**, 3285. (*17*)

KURACHI, K., FUNATSU, G., FUNATSU, M. and HIDAKA, S. (1972). *Agr. Biol. Chem.* **36**, 1109. (*12*)

KURTZ-FRITSCH, C. and HIRTH, L. (1972). *Virology* **47**, 385. (*11*)

KUSHNER, D. J. (1969). *Bact. Rev.* **33**, 302. (*5*)

LAIDLAW, P. T. (1938). *Virus diseases and viruses.* London: Cambridge University Press. (*16*)

LAMPSON, G. P., TYTELL, A. A., FIELD, A. K., NEMES, M. M. and HILLEMAN, M. R. (1967). *Proc. natn. Acad. Sci. U.S.A.* **58**, 782. (*16*)

LANCE, G. N. and WILLIAMS, W. T. (1966). *Computer J.* **9**, 90. (*12*)

LANCE, G. N. and WILLIAMS, W. T. (1967). *Computer J.* **9**, 373. (*12*)

LANE, L. C. and KAESBERG, P. (1971). *Nature, New Biology* **232**, 40. (*12*)

LANNI, F. (1964). In CORBETT and SISLER (1964), p. 386. (*16*)

LAPIERRE, H., ASTIER-MANIFACIER, S. and CORNUET, P. (1971). *C.r. hebd. Séanc. Acad. Sci. Paris Sér. D* **273**, 992. (*16*)

LAPIERRE, H., FAIVRE-AMIOT, A., KUSIAK, C. and MOLIN, G. (1972). *C.r. hebd. Séanc. Acad. Sci. Paris Sér. D* **274**, 1867. (*16*)

LAPIERRE, H., LEMAIRE, J. M., JOUAN, B. and MOLIN, G. (1970). *C.r. hebd. Séanc. Acad. Sci. Paris Sér. D* **271**, 1833. (*16*)

LAUFFER, M. A. (1938). *J. biol. Chem.* **126**, 443. (*1*)

LAUFFER, M. A. and PRICE, W. C. (1940). *J. biol. Chem.* **133**, 1. (*10*)

LAUFFER, M. A. and STEVENS, C. L. (1968). *Adv. Virus Res.* **13**, 1. (*5*)

LEA, D. E. (1962). *Actions of radiation on living cells.* London: Cambridge University Press. (*10*)

LEBERMAN, R. (1966). *Virology* **30**, 341. (*6*)

LEBERMAN, R. (1968). *Symp. Soc. Gen. Microbiol.* **18**, 183. (*5*)

LEBOWITZ, P., KELLY, T. J., NATHANS, D., LEE, T. N. H. and LEWIS, A. M. (1974). *Proc. natn. Acad. Sci. U.S.A.* **71**, 441. (*16*)

LEDOUX, L. G. H. (ed.) (1971). *Informative molecules in biological systems.* Amsterdam: North-Holland. (*16*)

LEE, R. E. (1971). *J. Cell Sci.* **8**, 623. (*16*)

LEMKE, P. A., NASH, C. H. and PIEPER, S. W. (1973). *J. gen. Microbiol.* **76**, 265. (*16*)

LESNAW, J. and REICHMANN, M. E. (1970). *Proc. natn. Acad. Sci. U.S.A.* **66**, 140. (*5*)

LEVY, H. B. (ed.) (1969). *The biochemistry of viruses.* New York and London: Marcel Dekker.

LEWIS, T. (1966). *Ann. appl. Biol.* **58**, 371. (*14*)

LHOAS, P. (1971a). *Nature, Lond.* **230**, 248. (*16*)

LHOAS, P. (1971b). *J. gen. Virol.* **13**, 365. (*16*)

LHOAS, P. (1972). *Nature, New Biology* **236**, 86. (*16*)

LIN, S. C., LEE, C. S. and CHIU, R. J. (1970). *Phytopathology* **60**, 705. (*17*)

LINDEGREN, C. C., BANG, Y. N. and HIRANO, T. (1962). *N.Y. Acad. Sci. Ser. II* **24**, 540. (*16*)

LING, K. C. (1966). *Phytopathology* **56**, 1252. (*13*)

LISTER, R. M. (1964). *Ann. appl. Biol.* **54**, 167. (*8*)

LISTER, R. M. (1966). *Virology* **28**, 350. (*12*)

LISTER, R. M. (1968). *J. gen. Virol.* **2**, 43. (*12*)

LISTER, R. M., GHABRIAL, S. A. and SAKSENA, K. N. (1972). *Virology* **49**, 290. (*6*)

LISTER, R. M. and MURANT, A. F. (1967). *Ann. appl. Biol.* **59**, 49. (*13, 14*)

LISTER, R. M. and THRESH, J. M. (1955). *Nature, Lond.* **175**, 1047. (*12*)

LITTAU, V. C. and MARAMOROSCH, K. (1960). *Virology* **10**, 483. (*17*)

LLOYD, D. A. and MANDELES, S. (1970). *Biochemistry* **9**, 932. (*5*)

LOEBENSTEIN, G., DEUTSCH, M., FRANKEL, H. and SABAR, Z. (1966). *Phytopathology* **56**, 512. (*15*)

LOEFFLER, F. and FROSCH, P. (1898). *Zentralbl. Bakteriol. Parasitenk. Abt. 1*, **23**, 371. (*1*)

LOENING, U. E. (1968). *J. mol. Biol.* **38**, 355. (*6*)

LOVISOLO, O. and CONTI, M. (1966). *Atti della Accademia delle Scienze di Torino* **100**, 63. (*11*)

LOWE, H. J. B. and RUSSELL, G. E. (1969). *Ann. appl. Biol.* **63**, 337. (*15*)

LWOFF, A. (1957). *J. gen. Microbiol.* **17**, 239. (*1*)

LWOFF, A. and TOURNIER, P. (1966). *Ann. Rev. Microbiol.* **20**, 45. (*12*)

MACHIDA, H. and KIHO, Y. (1970). *Jap. J. Microbiol.* **14**, 441. (*11*)

MACKENZIE, D. R., ANDERSON, P. M. and WERNHAM, C. C. (1966). *Pl. Dis. Reptr.* **50**, 363. (*4*)

MANDEL, M. and MARMUR, J. (1968). *Methods in enzymology* **12B**, 195. (*9*)

MARAMOROSCH, K. (1952). *Phytopathology* **42**, 59. (*17*)

MARAMOROSCH, K. (1957). *Science* **126**, 651. (*3, 17*)

MARAMOROSCH, K. (1958). *Tijdschr. Pl. Ziekt.* **64**, 383. (*17*)

MARAMOROSCH, K. (ed.) (1962). *Biological transmission of disease agents.* New York: Academic Press.

MARAMOROSCH, K. (ed.) (1969). *Viruses, vectors and vegetation.* New York: Interscience.

MARAMOROSCH, K., GRANADOS, R. R. and HIRUMI, H. (1970). *Adv. Virus Res.* **16**, 135. (*17*)

MARAMOROSCH, K. and KOPROWSKI, H. (eds.) (1967–71). *Methods in virology.* Vol. 1–5. New York: Academic Press.

MARAMOROSCH, K. and KURSTAK, E. (eds.) (1971). *Comparative virology.* New York and London: Academic Press.

MARBROOK, J. and MATTHEWS, R. E. F. (1966). *Virology* **28**, 219. (*8*)

MARCUS, A. (1970). *J. biol. Chem.* **245**, 962. (*9, 11*)

MARCUS, A., EFRON, D. and WEEKS, D. P. (1973). *Methods in enzymology, nucleic acids and the wheat embryo protein synthesis. Part E.* (*11*)

MARCUS, A., LUGINBILL, B. and FEELEY, J. (1968). *Proc. natn. Acad. Sci. U.S.A.* **59**, 1243. (*9, 11*)

MARJANEN, L. A. and RYRIE, I. J. (1974). *Biochim. biophys. Acta* **371**, 442. (*9*)

MARKHAM, R. (1962). *Adv. Virus Res.* **9**, 241. (*9*)

MARKHAM, R. (1968). In MARAMOROSCH and KOPROWSKI (1968), Vol. 4, p. 503. (*9*)

MARKHAM, R., FREY, S. and HILLS, G. J. (1963). *Virology* **20**, 88. (*9*)

MARKHAM, R., HITCHBORN, J. H., HILLS, G. J. and FREY, S. (1964). *Virology* **22**, 342. (*9*)

MARKHAM, R. and SMITH, J. D. (1951). *Biochem. J.* **49**, 401. (*9*)

MARKHAM, R. and SMITH, J. D. (1952). *Biochem. J.* **52**, 552. (*9*)

MARKHAM, R. and SMITH, K. M. (1949). *Parasitology* **39**, 330. (*1, 6, 13*)

MARMUR, J. and DOTY, P. (1962). *J. mol. Biol.* **5**, 109. (*9*)

MARTYN, E. B. (ed.) (1968). *Plant virus names.* Phytopathological Papers, No. 9. Kew: Commonwealth Mycological Institute. (*1, 2, 12*)

MARTYN, E. B. (ed.) (1971). *Plant virus names.* Phytopathological Papers, No. 9, Supplement No. 1. Kew: Commonwealth Mycological Institute. (*2, 12*)

MATTHEWS, R. E. F. (1949). *Ann. appl. Biol.* **36**, 460. (*11, 12*)

MATTHEWS, R. E. F. (1953). *Ann. appl. Biol.* **40**, 377. (*4*)

MATTHEWS, R. E. F. (1956). *Biochim. biophys. Acta* **19**, 559. (*15*)

MATTHEWS, R. E. F. (1957). *Plant virus serology.* London: Cambridge University Press. (*8*)

MATTHEWS, R. E. F. (1966). *Virology* **30**, 82. (*5*)

MATTHEWS, R. E. F. (1970). *Plant virology.* New York and London: Academic Press. (*9*)

MATTOX, K. R., STEWART, K. D. and FLOYD, G. (1972). *Can. J. Microbiol.* **18**, 1620. (*16*)

MAY, J. T., GILLILAND, J. M. and SYMONS, R. H. (1970). *Virology* **41**, 653. (*9*)

MAY, J. T. and SYMONS, R. H. (1971). *Virology* **44**, 517. (*9, 11*)

MAYER, A. (1886). *Landw. Vers. Sta.* **32**, 451. (*Phytopathological Classic* **7**, 11). (*1*)

MAYO, M. A. and COOPER, J. I. (1973). *J. gen. Virol.* **18**, 281. (*10*)

MAYO, M. A., HARRISON, B. D., MURANT, A. F. and BARKER, H. (1973). *J. gen. Virol.* **19**, 155. (*10*)

MAYO, M. A., MURANT, A. F. and HARRISON, B. D. (1971). *J. gen. Virol.* **12**, 175. (*5*)

MAYOR, H. D. and HILL, N. O. (1961). *Virology* **14**, 264 (*9*)

MAYR, E. (1953). *Ann. N.Y. Acad. Sci.* **56**, 391. (*12*)

MCCARTHY, D., JARVIS, B. C. and THOMAS, B. J. (1970). *J. gen. Virol.* **9**, 9. (*11*)

MCCARTHY, D., LANDER, D. E., HAWKES, S. P. and KETTERIDGE, S. W. (1972). *J. gen. Virol.* **17**, 91. (*11*)

MCCARTY, K. S., STAFFORD, D. and BROWN, O. (1968). *Anal. Biochem.* **24**, 314. (*9*)

MCKINNEY, H. H. (1923). *J. agr. Res.* **23**, 771. (*14*)

MCKINNEY, H. H. (1929). *J. agr. Res.* **39**, 557. (*11*)

MCKINNEY, H. H. (1944). *J. Wash. Acad. Sci.* **34**, 139. (*2, 12*)

MCKINNEY, H. H. (1948). *Phytopathology* **38**, 1003. (*15*)

MCKINNEY, H. H. and SILBER, G. (1968). In MARAMOROSCH and KOPROWSKI (1968) Vol. 4, p. 491.

MCLEAN, G. D. and FRANCKI, R. I. B. (1967). *Virology* **31**, 585. (*6*)

MELLEMA, J. E. and AMOS, L. A. (1972). *J. mol. Biol.* **72**, 819. (*9*)

MERRETT, M. J. and BAYLEY, J. (1969). *Bot. Rev.* **35**, 372. (*3*)

MESSIEHA, M. (1969). *Phytopathology* **59**, 943. (*13*)

MEYNELL, G. G. (1972). *Bacterial plasmids: conjugation, colicinogeny and transmissible drug resistance.* London: Macmillan. (*16*)

MILES, P. W. (1959). *Nature, Lond.* **183**, 756. (*13*)

MILLER, G. L. (1959). *Anal. Chem.* **31**, 964. (*9*)

MILLS, D. R., KRAMER, F. R. and SPIEGELMAN, S. (1973). *Science* **180**, 916. (*16*)

MILNE, R. G. (1966). *Virology* **28**, 79. (*11*)

MILNE, R. G. (1967). *J. gen. Virol.* **1**, 403. (*6*)

MILNE, R. G. (1970). *J. gen. Virol.* **6**, 267. (*11*)

MILNE, R. G. (1972). In KADO and AGRAWAL (1972), p. 76. (*9*)

MILNE, R. G., CONTI, M. and LISA, V. (1973). *Virology* **53**, 130. (*5*)

MILNE, R. G., THOMPSON, G. W. and TAYLOR-ROBINSON, D. (1972). *Arch. ges. Virusforsch.* **37**, 378. (*16*)

MINK, G. I. and SAKSENA, K. N. (1971). *Virology* **45**, 755. (*10*)

MINSON, A. C. and DARBY, G. (1973a). *J. mol. Biol.* **77**, 337. (*5*)

MINSON, A. C. and DARBY, G. (1973b). *J. gen. Virol.* **19**, 253. (*9*)

MIURA, K. I., KIMURA, I. and SUZUKI, N. (1966). *Virology* **28**, 571. (*5*)

MOERICKE, V. (1949). *Anz. Schadlingsk.* **22**, 139. (*14*)

MOFFITT, E. M. and LISTER, R. M. (1973). *Virology* **52**, 301. (*16*)

MOHAMED, N. A., RANDLES, J. W. and FRANCKI, R. I. B. (1973). *Virology* **56**, 12. (*5, 9*)

MOREL, G. and MARTIN, C. (1952). *C.r. hebd. Séanc. Acad. Sci. Paris* **235**, 1324. (*15*)

MOREL, G. and MARTIN, C. (1955). *C.r. hebd. Séanc. Acad. Agr.* **41**, 472. (*15*)

MORGAN, J. F. (ed.) (1969). *Proceedings of the 8th Canadian cancer research conference.* Oxford: Pergamon Press.

MORRIS, T. J. and SEMANCIK, J. S. (1973). *Virology* **52**, 314. (*11*)

MOSCH, W. H. M., HUTTINGA, H. and RAST, A. T. B. (1973). *Neth. J. Pl. Pathol.* **79**, 104. (*12*)

MOTOYOSHI, F., BANCROFT, J. B. and WATTS, J. W. (1973). *J. gen. Virol.* **21**, 159. (*7*)

MOTOYOSHI, F., BANCROFT, J. B., WATTS, J. W. and BURGESS, J. (1973). *J. gen. Virol.* **20**, 177. (*7*)

MOULTON, F. R. ed. (1940). *The genetics of pathogenic organisms* A.A.A.S. Publication **12**.

MOUND, L. A. (1973). In GIBBS (1973a), p. 229. (*13*)

MULLIGAN, T. E. (1960). *Ann. appl. Biol.* **48**, 575. (*13*)

MUNDRY, K. W. (1957). *Z. ind. Abst. Vererbungsl.* **88**, 115. (*12*)

MUNDRY, K. W. (1959). *Virology* **9**, 722. (*12*)

MUNDRY, K. W. (1960). *Z. Vererbungsl.* **91**, 87. (*12*)

MUNDRY, K. W. (1963). *Ann. Rev. Phytopathol.* **1**, 173. (*9*)

MUNDRY, K. W. (1968). *Molec. gen. Genet.* **105**, 361. (*5*)

MUNDRY, K. W. and GIERER, A. (1958). *Z. ind. Abst. Vererbungsl.* **89**, 614. (*12*)

MUNOZ, J. (1957). *Proc. Soc. exp. Biol. Med.* **95**, 757. (*8*)

MURAKISHI, H. H., HARTMANN, J. X., BEACHY, R. N. and PELCHER, L. E. (1971). *Virology* **43**, 62. (*4, 11*)

MURAKISHI, H. H., HARTMANN, J. X., PELCHER, L. E. and BEACHY, R. N. (1970). *Virology* **41**, 365. (*11*)

MURANT, A. F. and GOOLD, R. A. (1968). *Ann. appl. Biol.* **62**, 123. (*13, 14*)

MURANT, A. F., GOOLD, R. A., ROBERTS, I. M. and CATHRO, J. (1969). *J. gen. Virol.* **4**, 329. (*5, 6, 11*)

MURANT, A. F. and LISTER, R. M. (1967). *Ann. appl. Biol.* **59**, 63. (*14*)

MURANT, A. F., MAYO, M. A., HARRISON, B. D. and GOOLD, R. A. (1972). *J. gen. Virol.* **16**, 327. (*6*)

MURANT, A. F., MAYO, M. A., HARRISON, B. D. and GOOLD, R. A. (1973). *J. gen. Virol.* **19**, 275. (*6*)

MURANT, A. F., ROBERTS, I. M. and GOOLD, R. A. (1973). *J. gen. Virol.* **21**, 269. (*11*)

MURANT, A. F. and TAYLOR, C. E. (1965). *Ann. appl.*

MURANT, A. F., TAYLOR, C. E. and CHAMBERS, J. (1968). *Ann. appl. Biol.* **61**, 175. (*15*)

MUSSELL, H. W., WOOD, H. A., ADLER, J. P. and BOZARTH, R. F. (1973). *2nd Int. Cong. Plant Pathol. Minneapolis.* Abstract no. 0908. (*16*)

NAGARAJ, A. N. (1962). *Virology* **18**, 329. (*8*)

NAGARAJ, A. N. (1965). *Virology* **25**, 133. (*8, 11*)

NASU, S. (1965). *Jap. J. appl. Ent. Zool.* **9**, 225. (*13*)

NASU, S., JENSEN, D. D. and RICHARDSON, J. (1970). *Virology* **41**, 583. (*17*)

NATHANS, D., OESCHGER, M. P., POLMAR, S. K. and EGGEN, K. (1969). *J. mol. Biol.* **39**, 279. (*11*)

NIBLETT, C. L. and SEMANCIK, J. S. (1969). *Virology* **38**, 685. (*5, 10*)

NIENHAUS, F. (1971). *Virology* **46**, 504. (*13, 16*)

NILSSON-TILGREN, T. (1970). *Mol. gen. Genet.* **109**, 246. (*11*)

NIXON, H. L. and FISHER, H. L. (1958). *Brit. J. appl. Phys.* **9**, 68. (*9*)

NOLL, H. (1967). *Nature, Lond.* **215**, 360. (*9*)

NOORDAM, D. (1973). *Identification of plant viruses. Methods and experiments.* Wageningen: Pudoc. (*6, 9, 10, 12*)

NOVACKY, A. and HAMPTON, R. E. (1968). *Phytopathology* **58**, 301. (*3*)

NOZU, Y. and OKADA, Y. (1970). *Virology* **40**, 1066. (*12*)

NYLAND, G. and GOHEEN, A. C. (1969). *Ann. Rev. Phytopathol.* **7**, 331. (*15*)

OAKLEY, C. L. and FULTHORPE, A. J. (1953). *J. path. Bact.* **65**, 49. (*8*)

ÖBERG, B. F. and SHATKIN, A. J. (1972). *Proc. natn. Acad. Sci. U.S.A.* **69**, 3589. (*11*)

OKADA, Y., OHASHI, Y., OHNO, T. and NOZU, Y. (1970). *Virology* **42**, 243. (*5*)

O'LOUGHLIN, G. T. (1969). Ph.D. Thesis. University of Melbourne. (*13*)

O'LOUGHLIN, G. T. and CHAMBERS, T. C. (1967). *Virology* **33**, 262. (*13*)

OORTWIJN BOTJES, J. G. (1920). Dissertation, Wageningen, The Netherlands. (*13*)

ORLOB, G. B. (1966). *Phytopath. Z.* **55**, 218. (*13*)

ORLOB, G. B. (1967). *Virology* **31**, 402. (*10*)

ORLOB, G. B. (1968). *Virology* **35**, 121. (*13*)

OTSUKI, Y., SHIMOMURA, T. and TAKEBE, I. (1972). *Virology* **50**, 45. (*11*)

OTSUKI, Y. and TAKEBE, I. (1969). *Virology* **38**, 497. (*8*)

OUCHTERLONY, O. (1948). *Acta Path. Microbiol. Scand.* **25**, 186. (*8*)

OUDIN, J. (1946). *C.r. hebd. Séanc. Acad. Sci. Paris* **222**, 115. (*8*)

OWENS, R. A., BRUENING, G. and SHEPHERD, R. J. (1973). *Virology* **56**, 390. (*10*)

PADAN, E. and SHILO, M. (1973). *Bact. Rev.* **37**, 343. (*16*)

PALIWAL, Y. C. and SINHA, R. C. (1970). *Virology* **42**, 668. (*13*)

PALIWAL, Y. C. and SLYKHUIS, J. T. (1967). *Virology* **32**, 344. (*13*)

PAUL, H. L. (1959). *Z. Naturforsch.* **14b**, 427. (*9*)

PAUL, H. L. and BUCHTA, U. (1971). *J. gen. Virol.* **11**, 11 (*9*)

PAUL, H. L. and QUANTZ, L. (1959). *Archiv. Mikrobiol.* **32**, 312. (*11*)

PAUL, H. L., WETTER, C., WITTMANN, H. G. and BRANDES, J. (1965). *Z. Vererbungsl.* **97**, 186. (*12*)

PAULING, L. and COREY, R. B. (1951). *Proc. natn. Acad. Sci. U.S.A.* **37**, 305. *(5)*

PEDEN, K. W. C., MAY, J. T. and SYMONS, R. H. (1972). *Virology* **47**, 498. *(11)*

PELHAM, J., FLETCHER, J. T. and HAWKINS, J. H. (1970). *Ann. appl. Biol.* **65**, 293. *(15)*

PERKINS, F. O. (1969). *J. invert. Pathol.* **13**, 199. *(16)*

PETERS, D. and BLACK, L. M. (1970). *Virology* **40**, 847. *(4, 13)*

PETERS, D. and KITAJIMA, E. W. (1970). *Virology* **41**, 135. *(5)*

PICKETT-HEAPS, J. D. (1972). *J. Phycol.* **8**, 44. *(16)*

PIERPOINT, W. S. and HARRISON, B. D. (1963). *J. gen. Microbiol.* **32**, 429. *(10)*

PIRIE, N. W. (1962). *Perspectives in Biology and Medicine* **5**, 446. *(1)*

PLUMB, R. T. (1971). *Proc. 6th Br. Insectic. Fungic. Conf. 1971*, p. 307. *(14)*

POLLARD, M. (ed.) (1959). *Perspectives in virology*. Vol. 1. New York: Wiley.

POLSON, A. (1958). *Biochim. biophys. Acta* **29**, 426. *(8)*

POLSON, A. (1961). *Biochim. biophys. Acta* **50**, 565. *(9)*

POLSON, A. and VAN REGENMORTEL, M. H. V. (1961). *Virology* **15**, 397. *(9)*

PORTER, C. A. and WEINSTEIN, L. H. (1960). *Contrib. Boyce Thompson Inst.* **20**, 307. *(3)*

POSNETTE, A. F. and TODD, J. M. (1955). *Ann. appl. Biol.* **43**, 433. *(15)*

POUND, G. S. (1949). *J. agr. Res.* **78**, 161. *(11)*

POUND, G. S. and WALKER, J. C. (1945). *J. agr. Res.* **71**, 471. *(11)*

POWELL, H. M., CULBERTSON, C. G., MCGUIRE, J. M., HOEHN, M. M. and BAKER, L. A. (1952). *Antibiot. Chemother.* **2**, 432. *(16)*

PREER, J. R., PREER, L. B., RUDMAN, B. and JURAND, A. (1971). *Mol. gen. Genet.* **111**, 202. *(16)*

PRICE, W. C. (1936a). *Phytopathology* **26**, 503. *(11)*

PRICE, W. C. (1936b). *Phytopathology* **26**, 665. *(11)*

PRICE, W. C. (1943). *Phytopathology* **33**, 586. *(11)*

PRICE, W. C. (1964). In CORBETT and SISLER (1964), p. 93. *(12)*

PRINGLE, C. R., SLADE, W. R., ELWORTHY, P. and O'SULLIVAN, M. (1970). *J. gen. Virol.* **6**, 213. *(12)*

PROESELER, G. (1966). *Phytopath. Z.* **56**, 213. *(13)*

PURCIFULL, D. E., HIEBERT, E. and MCDONALD, J. G. (1973). *Virology* **55**, 275. *(11)*

PURDY, H. A. (1929). *J. exp. Med.* **49**, 919. *(1)*

RAFF, M. C. (1973). *Nature, Lond.* **242**, 19. *(8)*

RAGETLI, H. W. J. and WEINTRAUB, M. (1964). *Science* **144**, 1023. *(8)*

RAGETLI, H. W. J. and WEINTRAUB, M. (1965). *Biochim. biophys. Acta* **111**, 522. *(8)*

RAIMONDO, L. M., LUNDH, N. P. and MARTINEZ, R. J. (1968). *J. Virol.* **2**, 256. *(16)*

RALPH, R. K. (1969). *Adv. Virus Res.* **15**, 61. *(11)*

RALPH, R. K., BULLIVANT, S. and WOJCIK, S. J. (1971). *Virology* **43**, 713. *(11, 13)*

RALPH, R. K. and WOJCIK, S. J. (1969). *Virology* **37**, 276. *(11)*

RANDLES, J. W. and FRANCKI, R. I. B. (1972). *Virology* **50**, 297. *(11)*

RAO, A. S. and BRAKKE, M. K. (1969). *Phytopathology* **59**, 581. *(13, 14)*

RAPPAPORT, I. and SIEGEL, A. (1955). *J. Immunol.* **74**, 106. *(8)*

RAPPAPORT, I. and WILDMAN, S. G. (1957). *Virology* **4**, 265. *(11)*

RAPPAPORT, I. and WU, J. H. (1962). *Virology* **17**, 411. *(11)*

RAST, A. T. B. (1972). *Neth. J. Pl. Pathol.* **78**, 110. *(15)*

RATTI, G. and BUCK, K. W. (1972). *J. gen. Virol.* **14**, 165. *(16)*

RAWLINSON, C. J., HORNBY, D., PEARSON, V. and CARPENTER, J. M. (1973). *Ann. appl. Biol.* **74**, 197. *(16)*

RAYMER, W. B. and DIENER, T. O. (1969). *Virology* **37**, 343. *(7)*

RAZIN, S. (1973). *Adv. Microbiol. Physiol.* **10**, 1. *(16)*

REDDY, D. V. R. and BLACK, L. M. (1966). *Virology* **30**, 551. *(7, 8, 13)*

REDDY, D. V. R. and BLACK, L. M. (1973a). *Virology* **54**, 557. *(6, 9, 11)*

REDDY, D. V. R. and BLACK, L. M. (1973b). *2nd Int. Cong. Plant Pathol. Minneapolis. Abstract no. 0259.* *(13)*

REDDY, D. V. R. and BLACK, L. M. (1974). *Virology* **61**, 458. *(12)*

REEDER, G. S., KNUDSON, D. L. and MACLEOD, R. (1972). *Virology* **50**, 301. *(11)*

REES, M. W. and SHORT, M. N. (1965). *Virology* **26**, 596. *(12)*

REESTMAN, A. J. (1972). In DE BOKX (1972), p. 152. *(15)*

REEVES, P. (1972). *Bacteriocins.* London: Chapman and Hall. *(16)*

REICHMANN, M. E. (1958). *Can. J. Chem.* **36**, 1603. *(9)*

REICHMANN, M. E. (1959). *Can. J. Chem.* **37**, 384. *(9)*

REICHMANN, M. E. (1965). *Virology* **24**, 166. *(9)*

REID, M. S. and MATTHEWS, R. E. F. (1966). *Virology* **28**, 563. *(3)*

REIGNDERF, L., SLOOF, P., SIVLAL, G. and BORST, P. (1973). *Biochim. biophys. Acta* **324**, 320. *(9)*

RENKONEN, O., KÄÄRÄINEN, L., SIMONS, K. and GAHMBERG, C. G. (1971). *Virology* **46**, 318. *(9)*

RICHARDSON, D. E. and DOLING, D. A. (1957). *Nature, Lond.* **180**, 866. *(14)*

ROBERTS, B. E. and PATERSON, B. M. (1973). *Proc. natn. Acad. Sci. U.S.A.* **70**, 2330. *(11, 12)*

ROBERTS, F. M. (1946). *Nature, Lond.* **158**, 663. *(13)*

ROBERTS, F. M. (1948). *Ann. appl. Biol.* **35**, 266. *(14)*

ROBERTS, I. M. and HARRISON, B. D. (1970). *J. gen. Virol.* **7**, 47. *(11)*

ROBERTSON, H. D., JEPPESEN, P. G. N. (1972). *J. mol. Biol.* **68**, 417. *(16)*

ROBINSON, D. J. (1973a). *J. gen. Virol.* **18**, 215. *(12)*

ROBINSON, D. J. (1973b). *J. gen. Virol.* **21**, 499. *(12)*

ROCHOW, W. F. (1956). *Phytopathology* **46**, 133. *(12)*

ROCHOW, W. F. (1959). *Phytopathology* **49**, 126. *(7)*

ROCHOW, W. F. (1969). *Phytopathology* **59**, 1580. *(13, 14)*

ROCHOW, W. F. (1970). *Science* **167**, 875. *(8, 12, 13)*

ROCHOW, W. F. and ROSS, A. F. (1955). *Virology* **1**, 10. *(11)*

ROIVAINEN, O. (1971). *Proc. 3rd Int. Cocoa Res. Conf. Ghana*, p. 518. *(13)*

ROMBAUTS, W. and FRAENKEL-CONRAT, H. (1968). *Biochemistry* **7**, 3334. *(12)*

ROMERO, J. (1973). *Virology* **55**, 224. *(11)*

ROSENBERG, H., DISKIN, B., ORON, L. and TRAUB, A. (1972). *Proc. natn. Acad. Sci. U.S.A.* **69**, 3815. *(11)*

ROSS, A. F. (1961a). *Virology* **14**, 329. *(11)*

ROSS, A. F. (1961b). *Virology* **14**, 340. *(11)*

ROSS, A. F. (1964). In CORBETT and SISLER (1964), p. 68. (*12*)

ROY, D., FRAENKEL-CONRAT, H., LESNAW, J. and REICHMANN, M. E. (1969). *Virology* **38**, 368. (*5*)

RUBIO-HUERTOS, M., CASTRO, S., FUJISAWA, I. and MATSUI, C. (1972). *J. gen. Virol.* **15**, 257. (*11*)

RUBIO-HUERTOS, M. and HIDALGO, F. G. (1964). *Virology* **24**, 84. (*3*)

RUNGGER, D., RASTELLI, M., BRAENDLE, E. and MALSBERGER, R. G. (1971). *J. invert. Pathol.* **17**, 72. (*16*)

RUSHIZKY, G. W. and KNIGHT, C. A. (1960). *Proc. natn. Acad. Sci. U.S.A.* **46**, 945. (*5, 12*)

RUSHIZKY, G. W., KNIGHT, C. A. and MCLAREN, A. D. (1960). *Virology* **12**, 32. (*10*)

RUSHIZKY, G. W., SOBER, H. A. and KNIGHT, C. A. (1962). *Biochim. biophys. Acta* **61**, 56. (*12*)

RUSSELL, G. E. (1966a). *Ann. appl. Biol.* **57**, 311. (*15*)

RUSSELL, G. E. (1966b). *Ann. appl. Biol.* **57**, 425. (*14*)

RUSSELL, G. E. (1968). *Brit. Sugar Beet Rev.* **37**, 77. (*14*)

RUSSELL, G. J., FOLLETT, E. A. C., SUBAK-SHARPE, J. H. and HARRISON, B. D. (1971). *J. gen. Virol.* **11**, 129. (*5, 6, 11*)

RUSSELL, G. J., MCGEOCH, D. J., ELTON, R. A. and SUBAK-SHARPE, J. H. (1973). *J. molec. Evol.* **2**, 277. (*16*)

SADASIVAN, T. S. (1940). *Ann. appl. Biol.* **27**, 359. (*11*)

SAGLIO, P., LAFLÈCHE, D., BONISSOL, C. and BOVÉ, J. M. (1971). *Physiol. Veg.* **9**, 569. (*17*)

SAGLIO, P., LAFLÈCHE, D., L'HOSPITAL, M., DUPONT, G. and BOVÉ, J. (1972). In ELLIOTT and BIRCH (1972), p. 187. (*17*)

SAGLIO, P., L'HOSPITAL, M., LAFLÈCHE, D., DUPONT, G., BOVÉ, J. M., TULLY, J. G. and FREUNDT, A. E. (1973). *Int. J. syst. Bact.* **23**, 191. (*17*)

SAKAI, F. and TAKEBE, I. (1972). *Mol. gen. Genet.* **118**, 93. (*11*)

SAKIMURA, K. (1962). In MARAMOROSCH (1962), p. 33. (*13*)

SALAMAN, R. N. (1933). *Nature, Lond.* **131**, 468. (*11*)

SALIVAR, W. O., TZAGOLOFF, H. and PRATT, D. (1964). *Virology* **24**, 359. (*9*)

SAMMONS, I. M. and CHESSIN, M. (1961). *Nature, Lond.* **191**, 517. (*12*)

SAMUEL, G. (1931). *Ann. appl. Biol.* **18**, 494. (*7*)

SAMUEL, G. (1934). *Ann. appl. Biol.* **21**, 90. (*11*)

SÄNGER, H. L. (1968). *Mol. gen. Genet.* **101**, 346. (*11, 12*)

SASTRY, K. S. and GORDON, M. P. (1966). *Biochim. biophys. Acta* **129**, 32. (*10*)

SATO, T., KYOGOKU, Y., HIGUCKI, S., MITSUI, Y., IITAKA, Y. and TSUBOI, M. (1966). *J. mol. Biol.* **16**, 180. (*5*)

SCHACHMAN, H. K. (1957). In COLOWICK and KAPLAN (1955–), Vol. 4, p. 34. (*9*)

SCHACHMAN, H. K. (1959). *Ultracentrifugation in biochemistry.* New York and London: Academic Press. (*9*)

SCHACHMAN, H. K. and WILLIAMS, R. C. (1959). In BURNET and STANLEY (1959), Vol. 1, p. 223. (*9*)

SCHISLER, L. C., SINDEN, J. W. and SIGEL, E. M. (1967). *Phytopathology* **57**, 519. (*16*)

SCHLESINGER, M. (1934). *Biochem. Z.* **237**, 306. (*1*)

SCHMELZER, K. (1956). *Phytopath. Z.* **28**, 1. (*4*)

SCHMELZER, K. (1957). *Phytopath. Z.* **30**, 281. (*3*)

SCHMUTTERER, H. and EHRHARDT, P. (1964). *Z. PflKrankh. PflSchutz.* **71**, 647. (*13*)

SCHNEIDER, I. R. (1971). *Virology* **45**, 108. (*6*)

SCHNEIDER, I. R. and WORLEY, J. F. (1959a). *Virology* **8**, 230. (*11*)

SCHNEIDER, I. R. and WORLEY, J. F. (1959b). *Virology* **8**, 243. (*11*)

SCHNEIDER, W. C. (1957). In COLOWICK and KAPLAN (1955–), Vol. 3, p. 680. (*9*)

SCHNEPF, E., SOEDER, C. J. and HEGEWALD, E. (1970). *Virology* **42**, 482. (*16*)

SCHRAMM, G. (1947). *Z. Naturforsch.* **2b**, 112 and 249. (*5*)

SCHULZ, J. T. (1963). *Pl. Dis. Reptr* **47**, 594. (*13*)

SCHUMAKER, V., and REES, A. (1972). In KADO and AGRAWAL (1972), p. 336. (*6*)

SCHÜMMELFEDER, N. (1958). *J. Histochem. Cytochem.* **6**, 392. (*9*)

SCHUSTER, H. and SCHRAMM, G. (1958). *Z. Naturforsch.* **13b**, 697. (*12*)

SCHUSTER, H. and WILHELM, R. C. (1963). *Biochim. biophys. Acta* **68**, 554. (*12*)

SCHUSTER, H. and WITTMANN, H. G. (1963). *Virology* **19**, 421. (*12*)

SCHWARZ, R. E. (1965). *S. Afr. J. agr. Sci.* **8**, 839. (*14*)

SCOTT, H. A. (1963). *Virology* **20**, 103. (*6*)

SCOTT, H. A. (1968). *Virology* **34**, 79. (*8*)

SEHGAL. O. P., JONG-HO, J., BHALLA, R. B., MEEI MEEI SONG and KRAUSE, G. F. (1970). *Phytopathology* **60**, 1778. (*9*)

SEHGAL, O. P. and KRAUSE, G. F. (1968). *J. Virol.* **2**, 966. (*12*)

SELMAN, B. J. (1973). In GIBBS (1973a), p. 157. (*13*)

SEMAL, J. and HAMILTON, R. I. (1968). *Virology* **36**, 293. (*9, 11*)

SEMAL, J. and KUMMERT, J. (1970). *J. gen. Virol.* **10**, 79. (*9*)

SEMANCIK, J. S., MAGNUSON, D. S. and WEATHERS, L. G. (1973). *Virology* **52**, 292. (*2*)

SERGEANT, E. P. (1967). *Ann. appl. Biol.* **59**, 31. (*14*)

SEVERIN, H. H. P. (1921). *Phytopathology* **11**, 424. (*13*)

SEVERIN, H. H. P. (1950). *Hilgardia* **19**, 357. (*17*)

SHALLA, T. A. (1968). *Virology* **35**, 194. (*11*)

SHATKIN, A. J. (1968). In FRAENKEL-CONRAT (1968a), p. 351. (*5*)

SHATKIN, A. J., SIPE, J. D. and LOH, P. (1968). *J. Virol.* **2**, 986. (*9*)

SHAW, J. G. (1973). *Virology* **53**, 337. (*11*)

SHAW, J. G. and BEAVERS, J. B. (1970). *J. econ. Ent.* **63**, 850. (*16*)

SHEFFIELD, F. M. L. (1931). *Ann. appl. Biol.* **17**, 471. (*3*)

SHEPARD, J. F. (1970). *Phytopathology* **60**, 1669. (*8*)

SHEPARD, J. F., JUTILA, J. W., CATLIN, J. E., NEWMAN, F. S. and HAWKINS, W. H. (1971). *Phytopathology* **61**, 873. (*8*)

SHEPARD, J. F. and SECOR, G. A. (1969). *Phytopathology* **69**, 1838. (*8*)

SHEPHERD, R. J., BRUENING, G. E. and WAKEMAN, R. J. (1970). *Virology* **41**, 339. (*11*)

SHEPHERD, R. J. and WAKEMAN, R. J. (1971). *Phytopathology* **61**, 188. (*5, 6*)

SHEPHERD, R. J., WAKEMAN, R. J. and ROMANKO, R. R. (1968). *Virology* **36**, 150. (*1, 5*)

SHIH, D. S. and KAESBERG, P. (1973). *Proc. natn. Acad. Sci. U.S.A.* **70**, 1799. (*9, 11, 12*)

SHIH, D. S., LANE, L. C. and KAESBERG, P. (1972). *J. mol. Biol.* **64**, 353. (*11*)

SHIKATA, E. and MARAMOROSCH, K. (1965). *Virology* **27**, 461. (*13*)

SHIKATA, E. and MARAMOROSCH, K. (1967). *Virology* **32**, 363. (*11*)

SHOPE, R. E. (1948). *Am. J. Bot.* **35**, 803. (*16*)

SHOPE, R. E. (1953). *J. exp. Med.* **97**, 601. (*16*)

SIEGEL, A. (1965). *Adv. Virus Res.* **11**, 25. (*12*)

SIEGEL, A., HILLS, G. J. and MARKHAM, R. (1966). *J. mol. Biol.* **19**, 140. (*12*)

SIEGEL, A. and WILDMAN, S. G. (1954). *Phytopathology* **44**, 277. (*12*)

SIEGEL, A. and WILDMAN, S. G. (1956). *Virology* **2**, 69. (*11*)

SIEGEL, A., WILDMAN, S. G. and GINOZA, W. (1956). *Nature, Lond.* **178**, 1117. (*10*)

SIEGEL, A., ZAITLIN, M. and DUDA, C. T. (1973). *Virology* **53**, 75. (*11*)

SIEGEL, A., ZAITLIN, M. and SEHGAL, O. P. (1962). *Proc. natn. Acad. Sci. U.S.A.* **48**, 1845. (*11*)

SILVERSTEIN, S. C., ASTELL, C., LEVIN, D. H., SCHONBERG, M. and ACS, G. (1972). *Virology* **47**, 797. (*11*)

SIMMONDS, J. H. (1959). *Qd. J. agr. Sci.* **16**, 371. (*15*)

SIMONS, J. N. (1962). *J. econ. Entomol.* **55**, 358. (*13*)

SIMONS, T. J. and ROSS, A. F. (1971a). *Phytopathology* **61**, 293. (*11*)

SIMONS, T. J. and ROSS, A. F. (1971b). *Phytopathology* **61**, 1261. (*3, 11*)

SINDEN, J. W. and HAUSER, E. (1950). *Mushr. Sci.* **1**, 96. (*16*)

SINHA, R. C. (1960). *Virology* **10**, 344. (*13*)

SINHA, R. C. (1963). *Phytopathology* **53**, 1170. (*13*)

SINHA, R. C. (1965). *Virology* **26**, 673. (*13*)

SINHA, R. C. and CHIYKOWSKI, L. N. (1967). *Virology* **31**, 461. (*13*)

SINHA, R. C. and PALIWAL, Y. C. (1970). *Virology* **40**, 665. (*17*)

SINHA, R. C. and REDDY, D. V. R. (1964). *Virology* **24**, 626. (*8*)

SLACK, S. A. and SCOTT, H. A. (1971). *Phytopathology* **61**, 538. (*13*)

SLYKHUIS, J. T. (1953). *Can. J. agr. Sci.* **33**, 195. (*14*)

SLYKHUIS, J. T. (1955). *Phytopathology* **45**, 116. (*13, 14, 15*)

SMITH, F. F., JOHNSON, G. V., KAHN, R. P. and BING, A. (1964). *Phytopathology* **54**, 748. (*14*)

SMITH, F. L. and HEWITT, W. B. (1938). *Calif. Agr. Exp. Sta. Bull.* No. 621. (*13*)

SMITH, K. M. (1937). *A textbook of plant virus diseases*, 1st edition. London: Churchill. (*2, 12*)

SMITH, K. M. (1946). *Parasitology* **37**, 21. (*13*)

SMITH, K. M. (1972). *A textbook of plant virus diseases*. 3rd edition. London: Longman. (*12, 13*)

SMITH, S. H. and SCHLEGEL, D. E. (1965). *Virology* **26**, 180. (*11*)

SOKOL, F. and CLARK, H. F. (1973). *Virology* **52**, 246. (*9*)

SOUTHEY, J. F. (1965). *Tech. Bull. Min. Agr. Fish, Food No. 7.* London: HMSO (*4*)

SPIRE, D., FERAULT, A. C. and BERTRANDY, J. (1973). *2nd Int. Cong. Plant Path. Minneapolis.* Abstract no. 0912. (*16*)

SPIRIN, A. S. (1963). *Prog. Nucleic Acid Res.* **1**, 301. (*9*)

SREENIVASAYA, M. and PIRIE, N. W. (1938). *Biochem. J.* **32**, 1707. (*5*)

STAEHELIN, M. (1958). *Biochim. biophys. Acta* **29**, 410. (*10*)

STANLEY, W. M. (1934). *Phytopathology* **24**, 1055. (*10*)

STANLEY, W. M. (1935). *Science* **81**, 644. (*1*)

STANLEY, W. M. (1939). *J. biol. Chem.* **129**, 405. (*9*)

STANLEY, W. M. and LAUFFER, M. A. (1939). *Science* **89**, 345. (*5*)

STAPLES, R. and ALLINGTON, W. B. (1956). *Univ. Neb. Agr. Exp. Sta. Res. Bull.* No. 178. (*13*)

STAVIS, R. L. and AUGUST, J. T. (1970). *Ann. Rev. Biochem.* **39**, 527. (*11*)

STAYNOV, D. A., PINDER, J. C. and GRATZER, W. B. (1972). *Nature, New Biology* **235**, 108. (*9*)

STEERE, R. L. (1957). *J. biophys. biochem. Cytol.* **3**, 45. (*3, 9*)

STEERE, R. L. (1959). *Adv. Virus Res.* **6**, 1. (*6*)

STEHELIN, D., PETER, R. and DURANTON, H. (1973). *Biochim. biophys. Acta* **293**, 253. (*5*)

STENT, G. S. (ed.) (1960). *Papers on bacterial viruses.* London: Methuen. (*16*)

STEWART, P. R. and LETHAM, D. S. (eds.) (1973). *The ribonucleic acids.* Berlin: Springer-Verlag.

STODDARD, E. M. (1947). *Conn. Agr. Exp. Sta. Bull.* **506**. 1. (*17*)

STOLLAR, B. D. and DIENER, T. O. (1971). *Virology* **46**, 168. (*8*)

STONE, O. M. (1963). *Ann. appl. Biol.* **52**, 199. (*15*)

STOREY, H. H. (1932). *Proc. Roy. Soc. Lond. B.* **112**, 46. (*13*)

STOREY, H. H. (1938). *Proc. Roy. Soc. Lond. B.* **125**, 455. (*13*)

STUBBS, L. L. and GROGAN, R. G. (1963). *Aust. J. agric. Res.* **14**, 439. (*14*)

SUBAK-SHARPE, J. H. (1969). In MORGAN (1969), p. 242. (*16*)

SUBAK-SHARPE, J. H., BURK, R. R., CRAWFORD, L. V., MORRISON, J. M., HAY, J. and KEIR, H. M. (1966). *Cold Spring Harbor Symp. Quant. Biol.* **31**, 737. (*16*)

SUGIYAMA, T. (1966). *Virology* **28**, 488. (*5*)

SUGIYAMA, T., KORANT, B. D. and LONBERG-HOLM, K. K. (1972). *Ann. Rev. Microbiol.* **26**, 467. (*11*)

SUMNER, J. B. (1926). *J. biol. Chem.* **69**, 435. (*6*)

SVOBODOVA, J. (1966). In BEEMSTER and DIJKSTRA (1966), p. 48. (*15*)

SYLVESTER, E. S. (1962). In MARAMOROSCH (1962), p. 11. (*13*)

SYLVESTER, E. S. (1969a). *Virology* **38**, 440. (*13*)

SYLVESTER, E. S. (1969b). *Virology* **37**, 26. (*13*)

SYLVESTER, E. S. and RICHARDSON, J. (1969). *Virology* **37**, 26. (*13*)

SYLVESTER, E. S. and RICHARDSON, J. (1970). *Virology* **42**, 1023. (*13*)

SYMINGTON, J. (1969). *Virology* **38**, 317. (*5*)

SYMONS, R. H., REES, M. W., SHORT, M. N. and MARKHAM, R. (1963). *J. mol. Biol.* **6**, 1. (*5*)

TAHAMA, Y. (1968). *Bull. Kumamoto Sericultural Exp. Sta.* No. 40. (*17*)

TAKAHASHI, I. and MARMUR, J. (1963). *Nature, Lond.* **197**, 794. (*16*)

TAKAHASHI, W. N. and RAWLINS, T. E. (1932). *Proc. Soc. Exp. Biol. Med.* **30**, 155. (*1, 6, 9*)

TAKEBE, I. and OTSUKI, Y. (1969). *Proc. natn. Acad. Sci. U.S.A.* **64**, 843. (*1, 4, 7, 8, 11*)

TAO, M., SMALL, G. D. and GORDON, M. P. (1969). *Virology* **39**, 534. (*10*)

TAYLOR, C. E. (1968). *Rep. Scott. Hort. Res. Inst. for 1967*, p. 66. (*13*)

TAYLOR, C. E. (1972). *Pest Articles and News Summaries* **18**, 269. (*13*)

TAYLOR, C. E. and CADMAN, C. H. (1969). In MARAMOROSCH (1969), p. 55 *(16)*

TAYLOR, C. E. and MURANT, A. F. (1968). *Pl. Pathol.* **17**, 171. *(15)*

TAYLOR, C. E. and MURANT, A. F. (1969). *Ann. appl. Biol.* **64**, 43. *(14)*

TAYLOR, C. E. and ROBERTSON, W. M. (1969). *Ann. appl. Biol.* **64**, 233. *(13)*

TAYLOR, C. E. and ROBERTSON, W. M. (1970a). *Ann. appl. Biol.* **66**, 375. *(13)*

TAYLOR, C. E. and ROBERTSON, W. M. (1970b). *J. gen. Virol.* **6**, 179. *(13)*

TAYLOR, C. E. and THOMAS, P. R. (1968). *Ann. appl. Biol.* **62**, 147. *(14)*

TEAKLE, D. S. (1960). *Nature, Lond.* **188**, 431. *(13)*

TEAKLE, D. S. (1962). *Virology* **18**, 224. *(13)*

TEAKLE, D. S. (1972). In KADO and AGRAWAL (1972), p. 248. *(13)*

TEMMINK, J. H. M., CAMPBELL, R. N. and SMITH, P. R. (1970). *J. gen. Virol.* **9**, 201. *(13)*

TEZUKA, N., TANIGUCHI, T. and MATSUI, C. (1971). *Virology* **43**, 717. *(11)*

THIRUMALACHAR, M. J. (1954). *Phytopath. Z.* **22**, 429. *(15)*

THOMAS, C. E. (1969). *Phytopathology* **59**, 633. *(13)*

THOMAS, P. E. and FULTON, R. W. (1968). *Virology* **34**, 459. *(11)*

THOMSON, A. D. (1960). *Nature, Lond.* **187**, 761. *(11)*

THRESH, J. M. (1958). *West African Cocoa Res. Inst. Bull.* **5**. *(14, 15)*

THRESH, J. M. (1964). *Nature, Lond.* **202**, 1028. *(13)*

THRESH, J. M. (1967). *Ann. appl. Biol.* **60**, 455. *(14)*

THUNG, T. H. (1931). *Handel. Nederl. Ind. Natuurwetenshap. Congr. 6de, 1931*, 450. *(11)*

TODD, J. M. (1958). *Proc. 3rd Conf. Potato Virus Diseases Lisse-Wageningen, 1957*, 132. *(14)*

TODD, J. M. (1961). *Eur. Potato J.* **4**, 316. *(15)*

TOKUMITSU, T. and MARAMOROSCH, K. (1966). *Expl. Cell Res.* **44**, 652. *(4)*

TOMITA, K. and RICH, A. A. (1964). *Nature, Lond.* **201**, 1160. *(5)*

TOMLINSON, J. A. (1962). *Pl. Pathol.* **11**, 61. *(14)*

TOMLINSON, J. A. and CARTER, A. L. (1970). *Ann. appl. Biol.* **66**, 381. *(14)*

TOMLINSON, J. A. and WALKEY, D. G. A. (1967). *Virology* **32**, 267. *(10)*

TOSIC, M. and FORD, R. E. (1972). *Phytopathology* **62**, 1466. *(12)*

TOTH, R. and WILCE, R. T. (1972). *J. Phycol.* **8**, 126. *(16)*

TRAUTMAN, R. and HAMILTON, M. G. (1972). In KADO and AGRAWAL (1972), p. 491. *(9)*

TREMAINE, J. H. and ARGYLE, E. (1970). *Phytopathology* **60**, 654. *(12)*

TREMAINE, J. H. and GOLDSACK, D. E. (1968). *Virology* **35**, 227. *(5)*

TREMAINE, J. H. and WILLISON, R. S. (1961). *Can. J. Bot.* **39**, 1843. *(8, 9)*

TSUGITA, A. (1962). *J. mol. Biol.* **5**, 293. *(12)*

TURNER, W. F. (1952). *Pl. Dis. Reptr. Suppl.* **211**, 35. *(17)*

TWORT, F. W. (1915). *Lancet*, **1915** (II), 1241. *(16)*

USHIYAMA, R. and MATTHEWS, R. E. F. (1970). *Virology* **42**, 293. *(11)*

UYEMOTO, J. K., GROGAN, R. G. and WAKEMAN, J. R. (1968). *Virology* **34**, 410. *(11)*

VALENTA, V., MUSIL, M. and MISIGA, S. (1961). *Phytopath. Z.* **42**, 1. *(17)*

VAN DER PLANK, J. E. (1946). *Trans. Roy. Soc. S. Africa* **31**, 269. *(14)*

VAN DER PLANK, J. E. (1949a). *Empire J. expl. Agr.* **17**, 18. *(14)*

VAN DER PLANK, J. E. (1949b). *Empire J. expl. Agr.* **17**, 141. *(14)*

VAN DER PLANK, J. E. and ANDERSSEN, E. E. (1945). *Sci. Bull. Dep. Agric. S. Afr.* no. 240. *(15)*

VANDERVEKEN, J. (1968). *Virology* **34**, 807. *(15)*

VANDERVEKEN, J. and VILAIN, N. (1967). *Ann. Epiphyties* **18**, 125. *(15)*

VAN HOOF, H. A. (1958). Doctoral Thesis, Wageningen Agr. Univ. Van Putten & Oortmeijer, Alkmaar, The Netherlands. *(13)*

VAN KAMMEN, A. (1968). *Virology* **34**, 312. *(7)*

VAN RAVENSWAAY-CLAASEN, J. C., VAN LEEUWEN, A. B. J., DUIJTS, G. A. and BOSCH, L. (1967). *J. mol. Biol.* **23**, 535. *(11)*

VAN REGENMORTEL, M. H. V. (1964). *Virology* **23**, 495. *(6)*

VAN REGENMORTEL, M. H. V. (1967). *Virology* **31**, 467. *(8, 12)*

VAN REGENMORTEL, M. H. V. (1972). In KADO and AGRAWAL (1972), p. 390. *(6, 9)*

VAN REGENMORTEL, M. H. V. (1975). *Virology* **64**, 415. *(12)*

VAN REGENMORTEL, M. H. V. and LELARGE, N. (1973). *Virology* **52**, 89. *(6)*

VAN REGENMORTEL, M. H. V. and VON WECHMAR, M. B. (1970). *Virology* **41**, 330. *(8)*

VAN SLOGTEREN, D. H. M. (1955). *Proc. 2nd Conf. Potato Virus Diseases, Lisse-Wageningen*, p. 51. *(8)*

VAN SLOGTEREN, D. H. M. (1971). C.M.I./A.A.B. *Descriptions of plant viruses*, No. 71. *(3)*

VAN VLOTEN-DOTING, L., DINGJAN-VERSTEEGH, A. and JASPARS, E. M. J. (1970). *Virology* **40**, 419. *(7)*

VAN VLOTEN-DOTING, L., KRUSEMAN, J. and JASPARS, E. M. J. (1968). *Virology* **34**, 728. *(7)*

VARMA, A. (1967). Studies on red clover vein mosaic virus and some associated viruses. Ph.D. Thesis, University of London. *(14)*

VARMA, A., GIBBS, A. J. and WOODS, R. D. (1970). *J. gen. Virol.* **8**, 21. *(10)*

VARMA, A., GIBBS, A. J., WOODS, R. D. and FINCH, J. T. (1968). *J. gen. Virol.* **2**, 107. *(5, 9)*

VENEKAMP. J. H. (1972). In KADO and AGRAWAL (1972), p. 369. *(6)*

VODKIN, M., KATTERMAN, F. and FINK, G. R. (1974). *J. Bacteriol.* **117**, 681. *(16)*

VOLK, J. and KRCZAL, H. (1957). *NachrBl. dt. PflSchdienst, Braunschweig* **9**, 17. *(13)*

VOLKOFF, O. and WALTERS, T. (1970). *Can. J. Genet. Cytol.* **12**, 621. *(16)*

VON SENGBUSCH, P. (1965). *Z. Vererbungsl.* **96**, 364. *(8, 12)*

WALLACE, H. R. (1963). *The biology of plant parasitic nematodes.* London: Edward Arnold. *(14)*

WALLACE, J. M. (1944). *J. agr. Res.* **69**, 187. *(11)*

WANG, A. L. and KNIGHT, C. A. (1967). *Virology* **31**, 101. *(12)*

WARMKE, H. E. (1968). *Virology* **34**, 149. *(3)*

WARMKE, H. E. (1969). *Virology* **39**, 695. *(3)*

WARREN, R. C. and HICKS, R. M. (1971). *J. Ultrastruct. Res.* **36**, 861. *(9)*

WATSON. L. and GIBBS, A. J. (1974). *Ann. appl. Biol.* **77**, 23. *(12)*

WATSON, M. A. (1938). *Proc. Roy. Soc. Lond. B* **125**, 144. *(13)*

WATSON, M. A. (1940). *Proc. Roy. Soc. Lond. B* **128**, 535. *(13)*

WATSON, M. A. (1946). *Proc. Roy. Soc. Lond. B* **133**, 200. *(13)*

WATSON, M. A. (1959). *N.A.A.S. q. Rev.* **43**, 1. *(15)*

WATSON, M. A. (1966). *Pl. Pathol.* **15**, 145. *(14)*

WATSON, M. A. and HEATHCOTE, G. D. (1966). *Rept. Rothamsted exptl. Sta. for 1965*, p. 292. *(14)*

WATSON, M. A. and NIXON, H. L. (1953). *Ann. appl. Biol.* **40**, 537. *(13)*

WATSON, M. A., SERJEANT, E. P. and LENNON, E. A. (1964). *Ann. appl. Biol.* **54**, 153. *(13)*

WATSON, M. A. and WATSON, D. J. (1953). *Ann. appl. Biol.* **40**, 1. *(3)*

WATSON, M. A., WATSON, D. J. and HULL, R. (1946). *J. agr. Sci. Camb.* **36**, 151. *(15)*

WEATHERS, L. G. and CALAVAN, E. C. (1959). *Proc. 1st Conf. Int. Org. Citrus Virologists*, 197. *(13, 15)*

WEBER, K. and OSBORN, M. (1969). *J. biol. Chem.* **244**, 4406. *(9)*

WEBSTER, R. G. (1968a). *Immunology* **14**, 29. *(8)*

WEBSTER, R. G. (1968b). *Immunology* **14**, 39. *(8)*

WEBSTER, R. G., LAVER, W. G. and FAZEKAS DE ST GROTH, S. (1962). *Aust. J. exp. Biol.* **40**, 321. *(9)*

WEINSTEIN, D. B., MARSH, J. B., GLICK, M. C. and WARREN, L. (1969). *J. biol. Chem.* **244**, 4103. *(9)*

WEISSMANN, C., FEIX, G. and SLOR, H. (1968). *Cold Spring Harbor Symp. Quant. Biol.* **33**, 83. *(11)*

WELKIE, G. W. and POUND, G. S. (1958). *Virology* **5**, 362. *(11)*

WELTON, R. E., SWENSON, K. G. and SOHI, S. S. (1964). *Virology* **23**, 504. *(7)*

WESTERN, J. H. (ed.) (1971). *Diseases of crop plants.* London: Macmillan.

WETTER, C. (1961). *Proc. 4th Conf. Potato Virus Diseases*, Wageningen, p. 164. *(8)*

WHITCOMB, R. F. (1969). In MARAMOROSCH (1969), p. 449. *(7)*

WHITCOMB, R. F. (1973). *Proc. N. Cent. Branch Entomol. Soc. Amer.* **28**, 38. *(17)*

WHITCOMB, R. F. and BLACK, L. M. (1961). *Virology* **15**, 136. *(8, 13)*

WHITCOMB, R. F. and DAVIS, R. E. (1970a). *Infect. Immun.* **2**, 209. *(17)*

WHITCOMB, R. F. and DAVIS, R. E. (1970b). *Ann. Rev. Entomol.* **15**, 405. *(17)*

WHITCOMB, R. F., JENSEN, D. D. and RICHARDSON, J. (1966). *Virology* **28**, 454. *(17)*

WHITCOMB, R. F., JENSEN, D. D. and RICHARDSON, J. (1968). *J. invert. Pathol.* **12**, 202. *(17)*

WHITNEY, W. K. and GILMER, R. M. (1974). *Ann. appl. Biol.* **77**, 17. *(13)*

WILKINS, P. W. and CATHERALL, P. L. (1974). *Ann. appl. Biol.* **76**, 209. *(14)*

WILLIAMS, R. C. and BACKUS, R. C. (1949). *J. Am. Chem. Soc.* **71**, 4052. *(9)*

WILLIAMS, R. C. and SMITH, K. M. (1958). *Biochem. biophys. Acta* **28**, 464. *(16)*

WILLIAMS, R. C. and WYCKOFF, R. W. G. (1944). *J. appl. Phys.* **15**, 712. *(9)*

WILLS, J. B. (ed.) (1962). *Agriculture and land use in Ghana.* London: Oxford University Press.

WILSON, J. H. and ABELSON, J. N. (1972). *J. mol. Biol.* **69**, 57. *(11)*

WILSON, J. H. and KELLS, S. (1972). *J. mol. Biol.* **69**, 39. *(11)*

WINDSOR, I. M. and BLACK, L. M. (1972). *Phytopathology* **62**, 1112. *(17)*

WINDSOR, I. M. and BLACK, L. M. (1973). *Phytopathology* **63**, 1139. *(17)*

WINKLER, A. J. (1949). *Hilgardia* **19**, 207. *(17)*

WITTMANN, H. G. (1965). *Z. Vererbungsl.* **97**, 297. *(12)*

WITTMANN, H. G. (1970). In CHARLES and KNIGHT (1970), p. 55. *(11)*

WITTMANN, H. G. and WITTMANN-LIEBOLD, B. (1966). *Cold Spring Harbor Symp. Quant. Biol.* **31**, 163. *(12)*

WOOD, H. A. and BOZARTH, R. F. (1972). *Virology* **47**, 604. *(16)*

WOOD, H. A., BOZARTH, R. F. and MISLIVEC, P. B. (1971). *Virology* **44**, 598. *(16)*

WOODS, A. F. (1900). *Science* **11**, 17. *(3)*

WRIGLEY, N. G. (1968). *J. Ultrastruct. Res.* **24**, 454. *(9)*

WRIGLEY, N. G. (1970). *J. gen. Virol.* **6**, 169. *(16)*

WYATT, G. R. (1951). *Biochem. J.* **48**, 585. *(9)*

WYSS, U. (1971). *Nematologica* **17**, 505. *(13)*

YAMASHITA, S., DOI, Y. and YORA, K. (1973). *Virology* **55**, 445. *(16)*

YARWOOD, C. E. (1959). *Phytopathology* **49**, 220. *(16)*

YARWOOD, C. E. (1960). *Virology* **12**, 245. *(4)*

YARWOOD, C. E. (1971). *Pl. Dis. Reptr* **55**, 342. *(13, 16)*

YARWOOD, C. E. and FULTON, R. W. (1967). In MARAMOROSCH and KOPROWSKI (1967), Vol. 3, p. 238. *(4)*

YARWOOD, C. E., RESCONICH, E. C., ARK, P. A., SCHLEGEL, D. E. and SMITH, K. M. (1961). *Pl. Dis. Reptr* **45**, 85. *(3, 17)*

YOUNG, D. J. and WATSON, L. (1970). *Aust. J. Bot.* **18**, 387. *(12)*

YPHANTIS, D. A. (1964). *Biochemistry* **3**, 297. *(9)*

ZAITLIN, M., DUDA, C. T. and PETTI, M. A. (1973). *Virology* **53**, 300. *(11)*

ZAITLIN, M. and HARIHARASUBRAMANIAN, V. (1972). *Virology* **47**, 296. *(11)*

ZAITLIN, M., SPENCER, D. and WHITFELD, P. R. (1968). In *Biochemical regulation in diseased plants or injury.* Tokyo: Phytopathological Society of Japan, p. 91. *(9, 11)*

ZECH, H. (1963). Quoted by MUNDRY (1963). *(11)*

ZIMM, B. H. (1948). *J. Chem. Phys.* **16**, 1093. *(9)*

ZUCKERMAN, B. M., HIMMELHOCH, S. and KISIEL, M. (1973). *Nematologica* **19**, 117. *(16)*

Virus names index

This index contains the names of viruses, virus-like agents and virus groups, together with their cryptograms (see Table 2.1, page 10), and the page where each occurs in the book. Many of the viruses of organisms other than higher plants are listed under the virus groups to which they belong.

Abutilon mosaic agent.
194, 221.
Adeno-satellite viruses; [D/1:1.5/25:S/S:V/*].
243.
Adenoviruses; [D/2:22/13:S/S:V/0].
242, 243, 245, 247.
African swine fever virus; D/*:*/*:S/*:V/*.
243.
Alfalfa mosaic virus;
R/1:1.1/16 + 0.8/16 + 0.7/16:U/U:S/Ap.
9, 11, 12, 14, 16, 20, 33, 40, 50, 59, 60, 61, 64, 73, 74, 83, 84, 88, 91, 98, 108, 125, 138, 142, 145, 187, 213, 224, 261, 262.
Alga viruses.
235–7.
Alphaviruses; [R/1:Σ4/6:S/S:V,I/Di], including various equine encephalitides, Semliki Forest virus, Sindbis virus.
111, 244, 245.
Anthriscus yellows virus; */*:*/*:*/*:S/Ap.
203, 204, 207.
Apple chlorotic leaf spot virus; R/1:2.3/5:E/E:S/*.
13, 63.
Apple mosaic virus; (R)/*:*/(16):S/S:S/*.
15, 183, 217.
Apple stem grooving virus; */*:*/*:E/E:S/*.
14, 63.
Apricot mottle agent.
9.
Arabis mosaic virus; R/1:2.4/42 + 1.4/28:S/S:S/Ne:*Nepovirus*.
9, 14, 24, 109, 181, 199, 202, 229, 230, 241.
Arboviruses, including alphaviruses, bunyaviruses, flaviviruses, rhabdoviruses.
64, 241, 245.
Arenaviruses; [R/1:Σ3.2/*:S/*:V,I/0], including lymphocytic choriomeningitis virus.
246.

Baculoviruses; [D/2:100/10:U/U:I/0], including granulosis and nuclear polyhedrosis viruses.
241, 242.
Barley stripe mosaic virus; R/1:(0.9 to 1.4)/4:E/E:S/*.
63, 74, 83, 86, 89, 172, 183, 184.
Barley yellow dwarf virus; R/1:2/*:S/S:S/Ap.
7, 11, 15, 23, 73, 109, 151, 165, 172, 190, 191, 195, 203, 214, 218, 219, 261, 262.
Bean common mosaic virus; */*:*/*:E/E:S/Ap:*Potyvirus*.
13, 172, 184, 261.
Bean leaf roll virus; */*:*/*:*/*:S/Ap.
15, 19, 23, 207, 214, 226.
Bean pod mottle virus; R/1:2.5/37 + 1.5/30:S/S:S/Cl:*Comovirus*.
14, 18, 55.
Bean yellow mosaic virus; */*:*/*:E/E:S/Ap:*Potyvirus*.
13, 261.
Bee acute paralysis virus: R/1:2/30:S/S:I/0.
241.
Beet curly top virus; */*:*/*:S/S:S/Au.
23, 24, 38, 44, 150, 155, 186, 192, 194, 202, 208, 211, 215, 223, 225, 226.
Beet Krauselkrankheit agent.
196.
Beet mild yellowing virus; */*:*/*:*/*:S/Ap.
7, 15, 23, 29, 109, 206–8, 218, 223–5, 228, 231.
Beet mosaic virus; */*:*/*:E/E:S/Ap:*Potyvirus*.
192, 206.
Beet yellow net agent.
23.
Beet yellows virus; R/1:4.3/5:E/E:S/Ap.
7, 11–14, 24, 26, 28, 29, 63, 150, 186, 189, 192, 206, 208, 214, 223–5, 228, 231.
Belladonna mottle virus; R/1:2/36:S/S:S/*:*Tymovirus*.
25.
β phage; D/2:22/*:X/X:B/0.
233.
Black raspberry latent virus; R/1:(1.4 to 0.9)/*:S/S:S/*.
15.
Bottlegourd mosaic virus, closely related to cucumber green mottle mosaic virus.
Broad bean mottle virus;
R/1:1.1/22 + 1.0/21 + 0.9/22:S/S:S/*:*Bromovirus*.
15, 57, 71, 78, 101, 106, 144, 147, 153.

General Index

Note that only page numbers are given, and that we have not referenced the great variety of hosts named in the text of Chapter 16. All viruses and virus-like agents are listed in the Virus Names Index. Other pathogens are listed in this index.